Charles Seale-Hayne Library
University of Plymouth
(01752) 588 588
LibraryandITenquiries@plymouth.ac.uk

DEVELOPMENTS IN PETROLEUM SCIENCE
Advisory Editor: G.V. Chilingarian

Developments in Petroleum Science, 17A

enhanced oil recovery, I
fundamentals and analyses

Developments in Petroleum Science, 17A

enhanced oil recovery, I
fundamentals and analyses

Edited by

ERLE C. DONALDSON

*School of Petroleum and Geological Engineering, University
of Oklahoma, Norman, Oklahoma, U.S.A.*

GEORGE V. CHILINGARIAN

*Department of Petroleum Engineering, University of Southern California,
Los Angeles, California, U.S.A.*

and

TEH FU YEN

*Departments of Civil and Environmental Engineering,
University of Southern California, Los Angeles, California, U.S.A.*

ELSEVIER — Amsterdam — Oxford — New York — Tokyo 1985

ELSEVIER SCIENCE PUBLISHERS B.V.
1 Molenwerf
P.O. Box 211, 1000 AE Amsterdam, The Netherlands

Distributors for the United States and Canada:

ELSEVIER SCIENCE PUBLISHING COMPANY INC.
52, Vanderbilt Avenue
New York, NY 10017

Library of Congress Cataloging in Publication Data
Main entry under title:

Enhanced oil recovery.

    (Developments in petroleum science ; 17A)
    Bibliography: p.
    Includes indexes.
    Contents: 1. Fundamentals and analysis.
    1. Secondary recovery of oil. 2. Oil field flooding.
I. Donaldson, Erle C. II. Chilingarian, George V.,
1929-   . III. Yen, Teh Fu, 1927-   . IV. Series.
TN871.E527 1984   622'.3382   83-25288
ISBN 0-444-42206-4

ISBN 0-444-42206-4 (Vol. 17A)
ISBN 0-444-41625-0 (Series)

Printed in The Netherlands

Dedicated to

J. Robert Fluor, Robert T. Johansen and Paul Weisz

for their important contributions to the field of enhanced oil recovery

PREFACE

Estimates of ultimate World recoverable liquid petroleum are ranging from 10,440 to 21,170 Quads (one Quad is equal to 10^{15} Btu; one Quint is equal to 10^3 Quads). World oil production up to 1983 was approximately 2,800 Quads. In 1982 alone, the World production was about 110 Quads, which is equivalent to 19 billion barrels. The latter figure also can be interpreted as ca. 65 million barrels per day of oil consumption.

Based on the present oil consumption, in fifteen years our World recoverable petroleum will be exhausted. The conventional recovery methods, however, extract only about one-third of the original oil-in-place in a reservoir. Any effort spent on enhanced oil recovery, therefore, may aid in recovering a portion of the remaining approximately 50 Quints of locked-in oil on a worldwide basis.

Various enhanced oil recovery (EOR) techniques are targeting toward this oil-in-place in many producing reservoirs. This is an available near-term option for supplementing liquid petroleum from synthetic alternatives, discoveries of new fields, exploitation of heavy oil and asphalt, etc.

At this time, thermal, CO_2-miscible, and chemical methods have been employed by the oil industry. Based on current industrial trends, the projected production in U.S.A. during the next decade using thermal, CO_2-miscible, and chemical methods will be 0.2—0.7, 3.0—5.6 and 1.5—2.5 million barrels per day, respectively. The majority of this activity is centered in four states in the U.S.A.: Texas, California, Louisiana, and Oklahoma.

Enhanced oil recovery techniques are new and developing technologies, widespread success of which remains generally unproven on a large scale. Uncertainties in determining the oil content of reservoirs, a critical parameter for the application of enhanced oil recovery, and other uncertainties, such as oil quality, present major process limitations. Better methods for measuring residual oil saturation are needed. Environmental impacts of enhanced oil recovery methods are not thoroughly studied and need to be quantified and evaluated. Other research needs include development of novel methods, improvement of the process recovery efficiencies, and advancement of the technology of miscible and chemical methods to facilitate widespread process application.

We believe that more fundamental research by the worldwide community is needed to solve these problems. In view of this, the editors prepared this two-volume book, which reviews and summarizes some of the state-of-art progress. The first volume is devoted to the fundamentals and analyses, whereas the second volume is devoted to the processes and operations. We hope that this book will encourage the scientists and engineers to further contribute to this very difficult field. We also hope that in the years to come enhanced oil recovery will contribute significantly to the availability of petroleum as an energy source.

G.V. CHILINGARIAN
E.C. DONALDSON
T.F. YEN

LIST OF CONTRIBUTORS

C. BARKER — Department of Geosciences, University of Tulsa, Tulsa, Okla., U.S.A.

G.V. CHILINGARIAN — Department of Petroleum Engineering, University of Southern California, Los Angeles, Calif., U.S.A.

A.G. COLLINS — Department of Extraction Research, National Institute for Petroleum and Energy Research, Bartlesville, Okla., U.S.A.

E.C. DONALDSON — School of Petroleum and Geological Engineering, University of Oklahoma, Norman, Okla., U.S.A.

J.P. HELLER — Petroleum Recovery Research Center, New Mexico Institute of Mining and Technology, Socorro, N.M., U.S.A.

G.L. LANGNES — Mobil Oil Corporation, New York, N.Y., U.S.A.

F.G. McCAFFERY — Reservoir Properties Division, Chevron Oil Field Research Company, La Habra, Calif., U.S.A.

A. MEHDIZADEH — Aeronautical and Mechanical Engineering Department, California State Polytechnic College, San Luis Obispo, Calif., U.S.A.

N.R. MORROW — Petroleum Recovery Research Center, New Mexico Institute of Mining and Technology, Socorro, N.M., U.S.A.

A.S. ODEH — Mobil Oil Corporation, Dallas, Texas, U.S.A.

J.O. ROBERTSON, Jr. — Earth Engineering, Inc., Cudahy, Calif., U.S.A.

D.N. SARAF — Department of Chemical Engineering, Indian Institute of Technology, Kanpur, India

M.M. SHARMA — Petroleum Engineering Department, University of Southern California, Los Angeles, Calif., U.S.A.

J. TORABZADEH — Petroleum Engineering Department, University of Southern California, Los Angeles, Calif., U.S.A.

C.C. WRIGHT — Consultant, Tijuana, Baja California, Mexico

T.F. YEN — Departments of Civil and Environmental Engineering, University of Southern California, Los Angeles, Calif., U.S.A.

CONTENTS

Chapter 1

INTRODUCTION

ERLE C. DONALDSON, T.F. YEN and GEORGE V. CHILINGARIAN

Enhanced oil recovery is generally considered as the third, or last, phase of useful oil production, sometimes called tertiary production. The first, or *primary*, phase of oil production begins with the discovery of an oilfield using the natural stored energy to move the oil to the wells by expansion of volatile components and/or pumping of individual wells to assist the natural drive. When this energy is depleted, production declines and a *secondary* phase of oil production begins when supplemental energy is added to the reservoir by injection of water. As the water to oil production ratio of the field approaches an economic limit of operation, when the net profit diminishes because the difference between the value of the produced oil and the cost of water treatment and injection becomes too narrow, the *tertiary* period of production begins. Since this last period in the history of the field commences with the introduction of chemical and thermal energy to enhance the production of oil, it has been labeled as *enhanced oil recovery* (EOR). Actually, EOR may be initiated at any time during the history of an oil reservoir when it becomes obvious that some type of chemical or thermal energy must be used to stimulate production.

The combined total oil production by primary and secondary methods is generally less than 40% of the original oil in place. Thus, the potential target for EOR is greater than the reserves that can be produced by conventional methods.

Before initiating EOR, the operator must start from status quo and obtain as much information as possible about the reservoir and its oil saturation. This is done by using past analytical data and production history of the field and, within the limits of economic investment, new geophysical surveys, tracer analyses and core studies to define the reservoir as accurately as possible. This body of information furnishes the rational basis for prediction of recoverable oil reserves by various proven techniques for EOR. The EOR procedures involve the injection of chemical compounds dissolved in the injection water, miscible gas injection alternating with water injection, the injection of micellar solutions (microemulsions composed of surfactants, alcohols, and crude oils), the injection of steam, and in-situ combustion.

Perhaps the most critical datum for EOR is the existing oil saturation of the reservoir. The investor must weigh the estimated recoverable oil by EOR against the total cost of implementing these newer, or developing, technologies. The choice of the process also is dependent upon the amount of oil in place as well as other considerations such as depth, oil viscosity, etc. Consequently, numerous new logging methods have been developed recently as well as other methods, such as the single-well tracer, for the accurate determination of reservoir oil saturation.

The general procedure for chemical EOR is illustrated in Fig. 1-1, using the specific case of the alkaline—polymer technique. In general, the introduction of chemicals to a petroleum reservoir is preceded by a *preflush* (the injection of a low-salinity or controlled tapered salinity water) to place a compatible aqueous buffer of fluid between the highly saline reservoir brine and the chemical solutions, which may be adversely affected by the dissolved salts. Chemical additives are detergent-type compounds (frequently petroleum sulfonates), organic polymers (to increase sweep efficiency in a heterogeneous reservoir), and micellar solutions. The alkaline or other chemical solution is injected after the reservoir conditioning preflush, as illustrated in Fig. 1-1. Injection of the chemical solutions is followed by the injection of a polymer solution (usually a polyacrylamide or a polysaccharide) to increase fluid viscosity, to aid in displacement of the chemicals through the reservoir and to minimize loss due to dilution and channeling. Finally, the salinity of the injected water following injection of the polymer is gradually increased to the normal concentration of the oilfield fluids.

Another EOR technique utilizes the injection of gas for pressure maintenance and oil displacement by miscible or solution drive as illustrated in Fig. 1-2 for the injection of carbon dioxide, which has become a viable EOR procedure. The principal mechanisms for miscible gas mobilization of oil are: (1) reduction of the oil viscosity upon solution of the gas into the oil and (2) increase in the volume of the oleic phase. Solution of carbon dioxide, which is highly soluble in crude oil at high pressure, causes appreciable swelling of the oil. Three modes of carbon dioxide injection have been developed: (1) injection of the gas in slugs that are followed by water injection, as illustrated in Fig. 1-2; (2) injection of water saturated with carbon dioxide; and (3) high-pressure injection of the gas itself.

Several techniques have emerged for thermal methods of EOR and the selection of one over the other depends on the total evaluation of the oil reservoir and the economics. Thermal processes have thus far been confined to recovery of heavy oils, i.e., API° < 20. *Steam flooding*, as illustrated in Fig. 1-3, is a simple process in principle. Steam is generated at the surface and injected into a central well to drive the oil to peripheral production wells through a combination of (1) thermal release of the oil from the sand, (2) reduction of the oil viscosity, and (3) pressure drive of the oil to the production wells. The mechanism of oil displacement is a combination of interacting physical changes, such as viscosity reduction and steam distillation, which can be visualized as separate advancing fronts as shown in Fig. 1-3. Considerable effort is required to treat the boiler water and the stack gases result-

CHEMICAL FLOODING
(Alkaline)

The method shown requires a preflush to condition the reservoir and injection of an alkaline or alkaline/polymer solution that forms surfactants in situ for releasing oil. This is followed by a polymer solution for mobility control and a driving fluid (water) to move the chemicals and resulting oil bank to production wells.

Mobility ratio is improved, and the flow of liquids through more permeable channels is reduced by the polymer solution resulting in increased volumetric sweep.

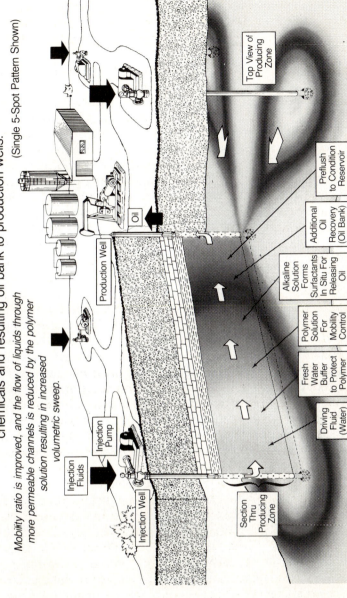

Fig. 1-1 Schematic diagram of chemical flooding (alkaline). (Courtesy of Mr. J. Lindley, U.S. Department of Energy, Bartlesville, Oklahoma.)

CARBON DIOXIDE FLOODING

This method is a miscible displacement process applicable to many reservoirs. A CO$_2$ slug followed by alternate water and CO$_2$ injections (WAG) is usually the most feasible method.

Viscosity of oil is reduced providing more efficient miscible displacement.

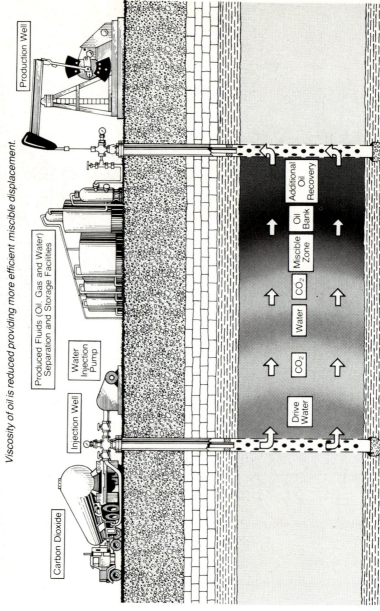

Fig. 1-2. Schematic diagram of carbon dioxide flooding. (Courtesy of Mr. J. Lindley, U.S. Department of Energy, Bartlesville, Oklahoma.)

STEAM FLOODING

Heat, from steam injected into a heavy-oil reservoir, thins the oil making it easier for the steam to push the oil through the formation toward production wells.

Heat reduces viscosity of oil and increases its mobility.

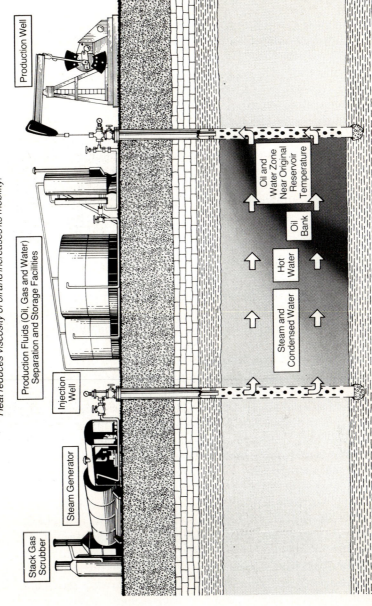

Fig. 1-3. Schematic diagram of steam flooding. (Courtesy of Mr. J. Lindley, U.S. Department of Energy, Bartlesville, Oklahoma.)

ing from burning of produced oil, which often contains sulfur and nitrogen compounds.

A second thermal EOR technique is illustrated in Fig. 1-4. This is a single-well injection/production scheme (referred to as steam soak), in which steam is injected and then allowed to transfer heat to the reservoir in the vicinity of the well before production of the oil is commenced by pumping. Pumping is continued until production declines below an acceptable level, and then the cycle is repeated.

The third thermal EOR method requires in-situ ignition of the reservoir oil and maintenance of a burning front by injection of air or oxygen. Several distinct zones develop within the reservoir as illustrated in Fig. 1-5. This mechanism is complex, but the burning front (and oil production) proceeds in an orderly fashion pushing ahead of it a mixture of combustion gases, steam, hot water, and mobilized oil. The process can be carried out until the burning front has extended to such a large radius from the injection well that continued injection of air is no longer technically or economically feasible.

Thus, *enhanced oil recovery* implies the use of one of many techniques that have been proven to be technically feasible. No method for enhancing oil recovery is so general, however, that it can be used in all situations; hence, this book is an effort to bring together a detailed analysis of the proven EOR techniques. It begins with an examination of the theories of the origin of petroleum deposits.

The evidence for a biogenic origin of petroleum rests strongly on the analyses of petroleums, which are complex mixtures of hydrocarbons containing nitrogen, sulfur and oxygen compounds in varying amounts depending on the source of the petroleum. The source rocks are generally aquatic sedimentary rocks containing marine fossils. Other supporting evidence for a biogenic origin comes from compounds found in all petroleums that have characteristic molecular structures closely related to those in living systems. The type of organic matter originally deposited in the source rock, however, exercises the initial control over the type of petroleum which is finally generated. As the source rocks were subjected to diagenetic and catagenetic processes and tectonic events (uplift, folding, burial, erosion, etc.) oil contained in them migrated into permeable traps. After initial accumulation in a trap, oil may be remobilized by geologic forces and relocated, frequently with considerable compositional changes. Major changes in the composition of petroleum can also be produced by intimate contact with flowing gases and water. The gases tend to cause precipitation of asphaltenes, whereas a major effect of contact with flowing water is removal of water-soluble components. If in the process of migration and redeposition the petroleum is exposed to the surface, evaporation of volatile components will occur together with bacterial degradation of low-molecular-weight paraffins. Thus, the chemical composition of today's oil deposits and the trace of oil in the source rocks of initial deposition lead to the theory of biogenic origins.

CYCLIC STEAM STIMULATION

Steam, injected into a well in a heavy-oil reservoir introduces heat that, coupled with alternate "soak" periods, thins the oil allowing it to be produced through the same well. This process may be repeated until production falls below a profitable level.

Schematic portrays one well during the 3 phases of this process. Flow pattern is stylized for clarity.

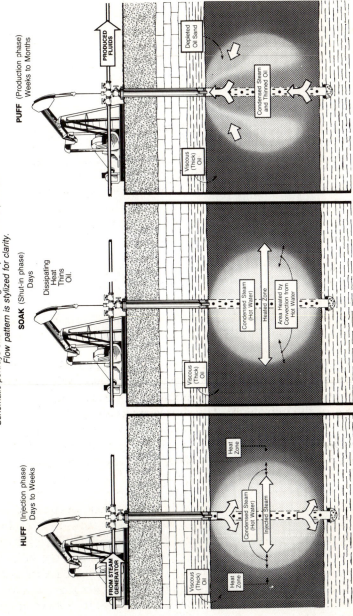

HUFF (Injection phase)
Days to Weeks

FROM STEAM GENERATOR

Viscous (Thick) Oil

Heat Zone

Condensed Steam (Hot Water)

Injected Steam

Heat Zone

SOAK (Shut-in phase)
Days

Dissipating Heat Thins Oil.

Viscous (Thick) Oil

Condensed Steam (Hot Water)

Heated Zone

Area Heated by Convection from Hot Water

PUFF (Production phase)
Weeks to Months

PRODUCED FLUIDS

Depleted Oil Sand

Condensed Steam and Thinned Oil

Viscous (Thick) Oil

Fig. 1-4. Schematic diagram of cyclic steam stimulation. (Courtesy of Mr. J. Lindley, U.S. Department of Energy, Bartlesville, Oklahoma.)

IN-SITU COMBUSTION

Heat is used to thin the oil and permit it to flow more easily toward production wells. In a fireflood, the formation is ignited, and by continued injection of air, a fire front is advanced through the reservoir.

Mobility of oil is increased by reduced viscosity caused by heat and solution of combustion gases.

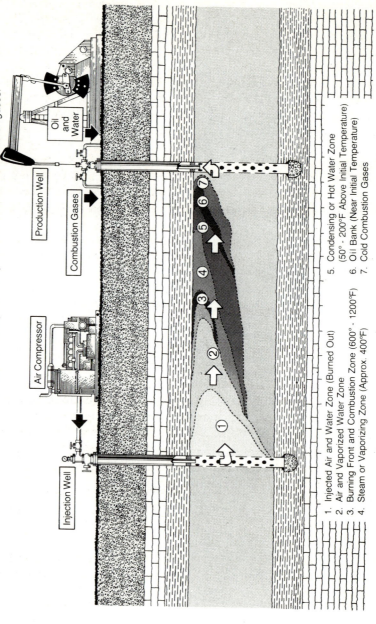

1. Injected Air and Water Zone (Burned Out)
2. Air and Vaporized Water Zone
3. Burning Front and Combustion Zone (600° - 1200°F)
4. Steam or Vaporizing Zone (Approx. 400°F)
5. Condensing or Hot Water Zone (50° - 200°F Above Initial Temperature)
6. Oil Bank (Near Initial Temperature)
7. Cold Combustion Gases

Injection Well

Air Compressor

Production Well

Combustion Gases

Oil and Water

Fig. 1-5. Schematic diagram of in-situ combustion. (Courtesy of Mr. J. Lindley, U.S. Department of Energy, Bartlesville, Oklahoma.)

The migration and accumulation of petroleum after its genesis is governed by natural phenomena of the flow of fluids in porous media. Three distinct phases are involved, frequently with simultaneous flow of either two or three fluids: aqueous, oleic and gaseous. Various aspects of multi-phase flow of fluids through porous media are discussed, therefore, with a detailed analysis of two- and three-phase relative permeability, which has been tested by numerous scientists under steady- and unsteady-state conditions. A fundamental microscopic description of multi-phase flow in porous media has not developed because of the complexity of the phenomena; however, empirical relations have been formulated based on Darcy's law relating fluid velocity to the pressure gradient and fluid viscosity and are used extensively in petroleum engineering. Coupled to this is the concept of relative permeabilities, which is basic to the analysis of the fluid flow problems; hence, this subject is discussed in considerable detail in Chapter 4 that includes a review of the literature.

The analysis of two- and three-phase flow in porous media is presented in a broad sense considering linear flow of fluids deep within a reservoir. An important consideration of production engineering, however, is the condition existing in the reservoir in the vicinity of the wellbore under static and dynamic conditions. A comprehensive analysis of the fundamental mathematical relationships and their interpretations is considered from the practical engineering perspective of well test analysis. A massive amount of data is obtained from pressure transient testing such as wellbore damage, wellbore volume, reservoir pressure, reservoir discontinuities, in-situ permeability, and from many other data required by the petroleum engineer for efficient field operation. The demonstrated utility of the equations, and their numerous solutions for widely different well and reservoir conditions, have made them the indispensable tools of production engineers. Their use for the evaluation of well and reservoir conditions for enhanced recovery application are very important; therefore, a concise presentation showing practical applications of these equations is given.

Flow of oil and gas within the reservoir and at the wells, as well as during original migration, is accompanied by the flow of aqueous solutions of salts (oilfield waters or brines). The flow is complicated by interstitial, immobile saturations of aqueous solutions. The chemical and physical properties of oilfield waters are important to primary and secondary production, because the produced waters require treatment with chemicals and filtration before they can be injected into disposal aquifers or returned to the producing formation for waterflooding. The chemical properties became even more important as soon as EOR became a viable production operation. The addition of chemical compounds to the injection waters was complicated by the electrolytes existing in all oilfield waters. Thus, a discussion of the problems of fluid compatibility, clay sensitivity, corrosion, and bacteria is included in Chapter 7. It covers an integral part of the technology of EOR.

The use of chemical additives and fuels in the oilfields introduced the new dimension of environmental concerns. EOR requires the processing of large quantities of chemical compounds in oilfields, which in many cases are near populated areas or on farm and ranch lands. The environmental concerns arise because of the large amounts of chemicals, such as detergents, caustics, organic polymers, alcohols and others, that must be stored and used within relatively small areas. New standards of air, surface water, and land contamination, and new policies and regulations, are more stringent than those used in the case of primary and secondary recovery techniques.

Air pollution caused by concentrated use of thermal processes for oil recovery near urban centers resulted in establishment of specific regulations restricting the allowable amounts of oxides of sulfur and nitrogen and hydrocarbons that could be released. This has had a severe impact on the economics of thermal recovery because treatment or recovery of pollutants is necessary in all cases.

Where liquid chemicals and gases are injected underground for oil recovery, controls are necessary to eliminate the emission of vapors from storage and pumping facilities. Even the injected chemicals must be considered as potential sources of pollution of fresh-water aquifers through communication with the deeper petroleum reservoirs by fractures, faults, abandoned wells, incompetent cement bonding to injection wells, etc. Hence, any EOR technique carries the added burden of surveys to protect against accidental ground water pollution, equipment to abate emissions of chemicals, and liability for unintentional, accidental pollution that may occur.

Institutional barriers, environmental concerns, and the technical and economic risks developing from these concerns play an important role in the commercial employment of enhanced oil recovery techniques and add an economic burden that cannot be ignored. Thus environmental concerns have become an important and integral part of enhanced oil recovery.

Predictions of the future trends of EOR are very difficult to make because it is still an incipient (developing) technology undergoing extensive fundamental and pilot-engineering research. A new approach that has the potential for development of an entirely new technology is the world-wide resurgence of interest for the applications of microorganisms for enhancement of oil production, in transportation of crude oils, and in refining processes.

The developments in EOR have thus far been encouraging, but a breakthrough of a new process with widespread economical application would make a tremendously favorable impact on the economics of oil recovery and its availability. It is this potential and exciting prospect that supports the continuing research effort by industry, universities, and governments.

Detailed treatment of surfactant, alkaline, polymer, carbon dioxide, and miscible flooding will appear in the second volume of this book. The application of microorganisms for enhancement of oil recovery and in-situ combustion are also discussed in Volume II.

Chapter 2

ORIGIN, COMPOSITION AND PROPERTIES OF PETROLEUM

COLIN BARKER

INTRODUCTION

Petroleum is a naturally occurring complex mixture made up predominantly of organic carbon and hydrogen compounds. It also frequently contains significant amounts of nitrogen, sulfur and oxygen together with smaller quantities of nickel, vanadium and other elements. It may occur in gaseous, liquid or solid form as natural gas, crude oil or asphaltic solids, respectively. The economic importance of crude oil and natural gas, and the problems associated with their production, have stimulated considerable interest in the origin of petroleum and the factors which control its physical properties and chemical compositions. Both the physical nature and chemical composition of petroleum are determined initially by the type of organic matter in the source rock where the petroleum is generated. There may be, however, considerable changes in composition during migration to the reservoir and as a result of maturation and alteration processes operating in the reservoir. In this chapter the following sequence of events is discussed: (1) generation in the source rock, (2) migration to the reservoir, (3) remobilization of petroleum to a new reservoir, and (4) possible subsequent alteration and migration which may affect the composition of petroleum in the reservoir.

GENERATION OF PETROLEUM

Biogenic origin

A biogenic origin for the carbonaceous material in petroleum is now widely (but not universally*) accepted. Inasmuch as inorganic processes generate racemic mixtures, the presence of optically active compounds in oils (especially the multiringed cycloalkanes (naphthenes)) provides strong support

---

* See Porfir'ev (1974) for some recent views on the inorganic origin of petroleum and Robinson (1963) for an outline of a dualist theory incorporating both biological and inorganic aspects. Recently Gold and Soter (1982) have proposed an upper mantle source for large quantities of methane.

12

for the biological hypothesis. Oils also contain so-called "chemical fossils" or "biological markers" which are compounds with characteristic structures that can be related to those in living systems. The compounds include isoprenoids, porphyrins, steranes, hopanes and many others. The relative abundances of members of homologous series are often similar to those in living systems with the strong odd preference in the long chain normal alkanes ($> C_{23}$) being particularly well documented (Bray and Evans, 1961). In addition, the lack of thermodynamic equilibrium among compounds (Bestougeff, 1967) and the close association of petroleum with sedimentary rocks formed in an aqueous environment suggests a low-temperature origin*. Finally, the elemental composition of petroleum (C, H, N, S, O), the isotopic composition of oils, and the presence of petroleum-like materials in recent sediments are consistent with a low-temperature origin even if taken alone they do not prove it. The evidence supporting a biological source for the material that generates petroleum (Table 2-1) appears to be overwhelming (Hunt, 1979) and is accepted in this chapter.

TABLE 2-1

Summary of evidence for a biogenic origin of petroleum

Optical activity Presence of porphyrins, isoprenoids, etc. Odd-even preference for *n*-alkanes	biogenic materials
Occurrence in sedimentary rocks Distribution of isomers Stability range of some compounds	low-temperature environment
Elemental composition Hydrocarbons in sediments Carbon isotope ratios	supporting

Thermal generation

Organisms produce a wide range of organic compounds, including major amounts of biopolymers like proteins, carbohydrates and lignins, together with a wide variety of lower-molecular-weight lipids (Hunt, 1979; Tissot and Welte, 1978). After the death of the organism, all or part of this organic material may accumulate in aquatic environments where the various com-

---

* In this context low temperature means less than a few hundred degrees as opposed to temperatures in the range 700—1200°C that characterize igneous processes involving silicate melts.

pounds have very different stabilities. Some organic material is metabolized in the water column by other organisms (including bacteria) and only the biochemically resistant material is incorporated into sediments. Survival here depends on many factors, particularly the oxidizing or reducing nature of the system, with preservation being strongly favored in anoxic sediments (Demaison and Moore, 1980). Whereas sediments deposited in oxygenated environments rarely have more than 4 wt.% organic matter, and usually very much less, anoxic sediments may exceed 20 wt.% organic matter. The latter accounts for more than 50% by volume of the rock. The formation of a petroleum accumulation requires more than just a concentration of bitumens that are present in recent sediments. This conclusion follows from the observation that many compounds which are quantitatively important in crude oils are not found in sediments. For example, organisms and recent sediments are almost devoid of hydrocarbons in the C_2–C_{10} range, yet these alkanes may account for up to 50% of some crudes. Other compounds that are not synthesized by organisms are also reported in crude oils. It is apparent that these compounds must have been formed from the available organic matter by reaction in the subsurface.* The organic matter originally incorporated into sediments is buried and subjected to increasing temperature and pressure as time goes by. Continued heating of organic matter in a reducing environment leads ultimately to final products of methane and graphite, i.e., it involves a redistribution of hydrogen to produce one hydrogen-rich and one hydrogen-poor product (Fig. 2-1). At the low temperatures typical of

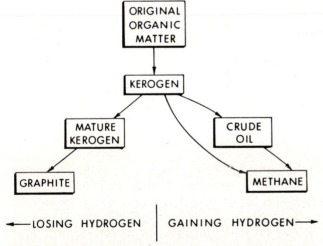

Fig. 2-1. Schematic representation of the redistribution of hydrogen between kerogen and bitumen during petroleum generation.

---

* Diagenetic and catagenetic processes (editorial comment).

14

sedimentary rocks, the organic matter progresses along this path and produces a high-molecular weight, solvent-insoluble material called "kerogen", that becomes more aromatic and carbon-rich as it evolves towards graphite. In contrast, the solvent-extractable, lower-molecular weight organics called "bitumens" (or simply "extractables") become more hydrogen-rich and eventually progress through compositions typical of crude oils to those of gas (Barker, 1979).

The type of organic matter (kerogen) in source rocks exercises initial control over the type of petroleum generated. In many cases organic matter type can be recognized by microscopic examination (after removing the mineral matrix with hydrofluoric and hydrochloric acids). Wood-derived materials that are lignin-rich appear to generate gas, whereas surface coatings of plants that are rich in cutin and esters of long-chain acids and alcohols produce waxy oils that frequently have high pour points. Organic matter derived from marine algae and other aquatic organisms seems to produce the normal

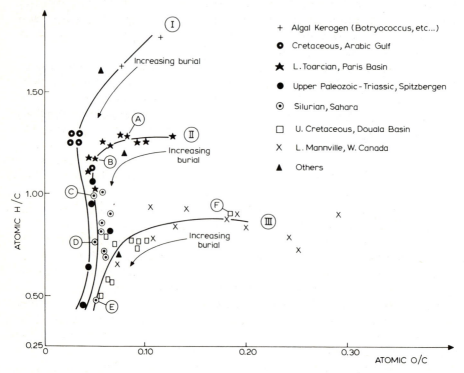

Fig. 2-2. A van Krevelen diagram showing examples of kerogen evolution paths. Path *I* includes algal kerogen and excellent source rocks from the Middle East; path *II* includes good source rocks from North Africa and other basins; path *III* corresponds to less oil-productive organic matter, but includes gas source rocks. Evolution of kerogen composition with depth is marked by arrows along each path (Tissot et al., 1974; courtesy of the *Bulletin, American Association of Petroleum Geologists*).

range of crudes (Barker, 1982). The latter organic matter frequently has no definite structure and is called amorphous.

The major organic matter types also have different elemental compositions. Algal debris and surface coatings are hydrogen-rich and relatively oxygen-poor, whereas in contrast woody organic matter is rich in oxygen and poor in hydrogen, because of its dominantly aromatic character. With rising temperature, generally produced by increasing depth of burial, all types of organic matter get steadily richer in carbon and follow distinct tracks across a diagram where the H/C ratio is plotted versus O/C ratio. This so-called "van Krevelen diagram" is shown in Fig. 2-2 along with the tracks followed by the major organic matter types upon burial. These three "diagenetic tracks" were orginally defined by Tissot et al. (1974) and the terminology is now widely used. The relative amounts of the various types of organic matter are controlled by the environment where the sediments were deposited. For example, rivers transport lignin and surface coatings derived from land plants to the marine environment, thus making deltas gas-prone environments (Halbouty et al., 1970). Frequently, high-pour-point crude oils, that are rich in long-chain alkanes from surface coatings, are also present in deltas (Evamy et al., 1978; Hedberg, 1968).

The generation of petroleum-like materials from kerogen is strongly dependent on temperature, but lower temperatures can be offset to some extent by longer heating times. The relationship closely approximates a doubling of reaction rate (i.e., halving of the reaction time) for every $10°C$ rise in temperature. Pressure, catalytic effects and radioactivity seem to be of no great importance.

The important role of temperature in petroleum generation was clearly shown by Philippi (1965). He found that in the Los Angeles and Ventura Basins of California the bitumen/kerogen ratio remained low and fairly constant in the shallower parts of the section, but then increased abruptly with increasing depth (Fig. 2-3). This increase occurred at a depth of about 10,500 ft in the Los Angeles Basin and about 16,000 ft in the Ventura Basin. The geothermal gradients are not the same in these two basins and Philippi noted that, although the depths are different, the marked increase in bitumen/kerogen ratio occurred at the same temperature in both basins. Inasmuch as the increased generation of bitumens is the process of petroleum generation, this significant study demonstrated clearly the important role of temperature. It also implied a minor role (if any) for pressure, because in the Los Angeles Basin the hydrostatic pressure at a depth of 10,500 ft is 4920 psi, whereas in the Ventura Basin it is about 50% higher with a value of 7520 psi at 16,000 ft. Subsequent analyses of samples from many wells worldwide have shown similar trends with relatively low bitumen/kerogen ratios for the shallower section and then rapidly increasing values deeper.

Single-well studies of vertical sections that penetrate many strata have inherent uncertainties, because the organic matter type may vary considerably from one layer to another. One way of circumventing this problem is to analyze a single rock unit which has been buried to different depths in different locations. This approach has been used by Tissot et al. (1971), who selected the Toarcian shale of the Paris Basin, since it is buried more deeply at the basin center than at its edges due to the basin configuration. By restricting the study to a single rock unit, the age, lithology, and amount and type of the organic matter remain relatively uniform. When uncertainties due to variations in these parameters are eliminated, the bitumen/kerogen ratio shows a depth trend similar to that observed in the Californian basins.

The essential features of the petroleum generation process can be duplicated in laboratory studies, but higher temperatures are needed to produce reactions in a few hours or days rather than the millions of years that are required in nature. Although dry pyrolysis in an inert atmosphere is widely used and provides useful information about source rock character, it does

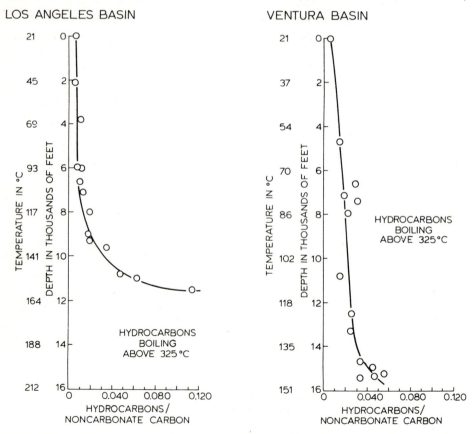

Fig. 2-3. Variation of the hydrocarbon/non-carbonate carbon ratios with depth in the Los Angeles and Ventura Basins, California (after Philippi, 1965; courtesy of *Geochimica Cosmochimica Acta*).

not simulate natural diagenesis and catagenesis and produces olefins which are rarely found in crude oils. In contrast, pyrolysis at elevated pressures in an aqueous medium appears to duplicate natural processes much more closely (Lewan et al., 1979).

The generation of petroleum is non-biological, is induced by temperature, and is influenced by available time. It follows essentially first-order kinetics (Connan, 1974). Attempts have been made to treat the generation process more quantitatively using various models based on the first-order kinetics. The most detailed one was presented by Tissot (1969) and Tissot and Espitalié (1975). The methodology developed by Lopatin (1971), however, is much easier to use and appears to give an adequate description, especially when properly calibrated (Waples, 1980).

Because the generation process is inorganic, the lower-molecular-weight compounds in the bitumen fraction that are generated from the kerogen show none of the biological characteristics typical of compounds in recent sediments. Inasmuch as the increasing amounts of thermally generated bitumens are formed with increasing depth (and hence increasing temperature), they overshadow the small amounts of bitumens preserved from recent sediments. Thus the bitumen fraction loses features such as odd-even predominan-

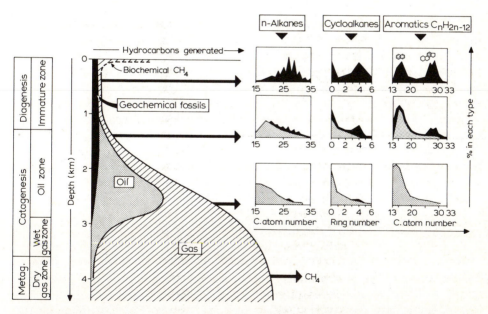

Fig. 2-4. General scheme of hydrocarbon formation as a function of burial of the source rock. The evolution of hydrocarbon composition is shown in insets for three different depths. The depth scale is only approximate and will depend on geothermal gradient. The values given correspond to an average for Mesozoic and Paleozoic source rocks. Actual depths will also depend on organic matter type and burial history (after Tissot et al., 1974, p. 185, fig. II.6.1; courtesy of the Springer-Verlag).

ce in the long-chain alkanes, optical activity, and the predominance of four- and five-ringed compounds in the naphthene fraction. There are also systematic trends in biomarker configurations. Many of these trends were documented initially in the Paris Basin, but have since been shown to be similar in many other areas. Both the quantitative and qualitative aspects of bitumen formation (i.e., petroleum generation) are well summarized in Fig. 2-4. The shaded area represents the biologically derived bitumens the content of which steadily decreases with depth, both in absolute terms and relative to the thermally generated bitumens. As they are diluted, the normal alkane distribution moves from a shallow distribution with marked odd-even predominance (especially in the longer chains) to a smoother distribution with the most abundant compounds being of progressively shorter chain lengths. The multi-ringed naphthenes initally show high abundances for compounds with four and five rings. This steadily diminishes, however, until in the deeper samples one- and two-ring compounds are most abundant.

Biogenic methane

An important exception to the thermal generation of petroleum is the bacterial formation of methane. The bacteria that generate this methane are anaerobic and are only effective in sulfate-free, anoxic conditions (Claypool and Kaplan, 1974). These bacteria have long been recognized for their role in forming "marsh gas". The methane they produce is extremely light isotopically, i.e., generally more negative than $-55^0/_{00}$ (relative to the PDB standard) for the carbon. This methane contains only trace amounts of ethane and lighter hydrocarbons, though it may have higher amounts of inorganic gases. Many of the large natural gas accumulations in Siberia (including the giant Urengoy gas field with reserves of 5.9×10^{12} m^3 are thought to be bacterially generated, because they are usually shallow, are isotopically lighter than $-59^0/_{00}$ (Yermakov et al., 1970), and have only $\leqslant 1\%$ of heavier hydrocarbons. Such deposits are also present in other parts of the world (Rice, 1975; Claypool et al., 1980; Rice and Shurr, 1980).

MIGRATION

Most petroleum is found in reservoir rocks that have high porosity and permeability. These properties are often developed by natural processes that result in the removal of fine-grained materials such as clays, micrite (in the case of carbonate reservoirs), and organic matter. In the case of carbonate rocks, dolomitization and solution are mainly responsible for increasing porosity and permeability. In general, reservoir rocks contain insufficient organic matter to generate commercially significant quantities of petroleum. It is now generally believed that petroleum generation occurs in organic matter-rich source rocks and that part of the bitumens then migrates to accumulate in reservoir rocks (Cordell, 1973). Thus migration has a critical role in linking the organic matter-rich source rocks to the reservoir. There is considerable discussion about the migration mechanism, with suggestions ranging from movement in true solution in subsurface waters to movement as a separate crude oil phase. Solubilities of hydrocarbons in water are low at room temperature (McAuliffe, 1978), but rise with increasing temperature, particularly above 100°C (Price, 1976). The effects of increasing pressure and of higher salinities are less well documented, but in general solubilities are around a few tens of parts per million and increase in the sequence alkanes < aromatics < "NSO's".* Within each group, solubilities decrease with increasing molecular size. There have been several suggestions that the migrating compounds move as some more soluble form such as micelles (Baker, 1960; Cordell, 1973), as "accommodated" molecules (Peake and Hodgson, 1967), or as hydrocarbon precursors like acids, alcohols, etc. If any of these mechanisms have an appreciable role in moving petroleum, the petroleum in the reservoir should be enriched in the more soluble components compared to the bitumens in the source rock. In fact, exactly the opposite trend is observed with the oils containing higher percentages of saturates and much lower concentrations of the NSO's (Fig. 2-5). This has been documented from basin-scale studies (Deroo et al., 1977; Combaz and

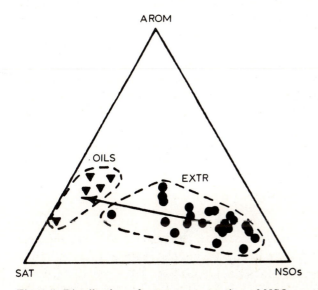

Fig. 2-5. Distribution of saturate, aromatic and NSO compounds in Jurassic crude oils and source rock extracts from the Parentis basin, France (Deroo, 1976).

---

* A convenient term used for compounds containing nitrogen, sulfur or oxygen (NSO) functional groups.

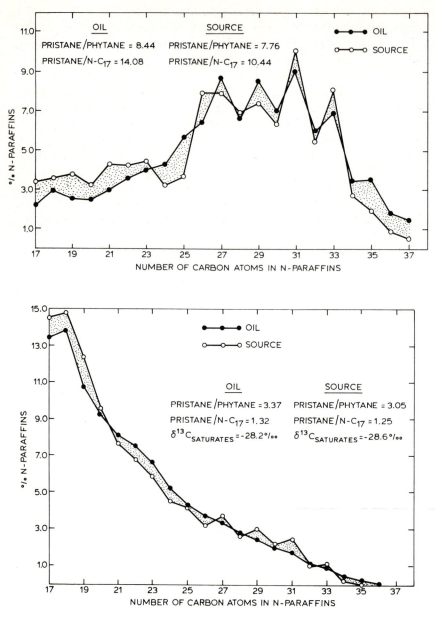

Fig. 2-6. Normal alkane distributions, pristane/phytane ratios, pristane/normal C_{17} ratios and carbon isotope ratios for crude oil—source rock pairs in the South Java Sea (from Sutton, 1977).

deMatharel, 1978) down to millimeter-scale (Barker, 1980). Within a compound type, such as the alkanes, the composition of oils over the range from C_4 to C_{35} is almost identical in source rocks (as bitumens) and in reservoirs (Williams, 1974), implying that oil migrates without large compositional fractionation, probably as a separate crude oil phase (Fig. 2-6).

Movement of a separate crude oil phase is not without its problems, because the forces produced by flowing water and by buoyancy are many orders of magnitude too small to deform oil droplets and force them past the constrictions in shale source rocks. It now appears that source rocks develop high internal pressures due to the thermal expansion of water ("aquathermal pressuring"; Barker, 1972) and to the volume increases during petroleum generation (Momper, 1978). The pressure gradient that develops squeezes oil out of source rocks into adjacent permeable "carrier beds". Since this gradient is much greater than the hydrostatic gradient, oils may be expelled downwards and out of the bottom of source rocks as well as out of the top. During migration through the source rocks, the alkanes are less strongly absorbed and so are preferentially expelled. At the other extreme, the NSO's

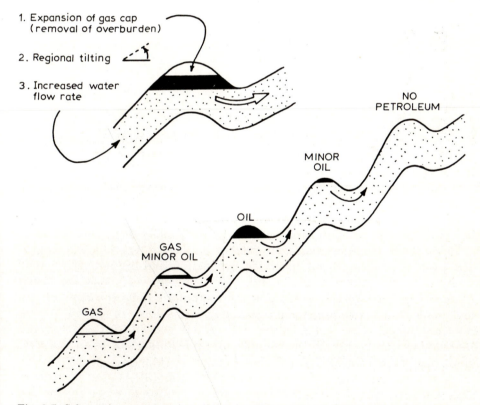

Fig. 2-7. Schematic representation of the possible results of differential entrapment.

are most strongly adsorbed and are depleted in the expelled oil (Tissot and Pelet, 1971). Buoyancy is the main driving force through the carrier beds and oils will continue to move up dip until stopped at a reversal of dip in a "structural trap" or where permeability decreases as in a "stratigraphic trap" (Cordell, 1976-77). Migration distances can be in excess of 100 km in structurally simple settings such as interior basins (Williams, 1974; Momper, 1978; Clayton and Swetland, 1980), but are much shorter when carrier beds are broken up by faulting as in deltas or block-faulted rifted continental margins (Barker, 1982). The overall process of migration seems to be very inefficient with somewhere around 5% of the available bitumens eventually being trapped in the reservoir (Hunt, 1977).

Oil may be remobilized after its initial accumulation in the reservoir. Although in the simplest case this may only involve a simple relocation, it can lead to major compositional changes if both gas and oil are involved. When an anticlinal reservoir is full to the spill point with a gas cap above oil, any spilling off the bottom will be oil and the next trap up dip will thus accumulate oil with no gas cap. This process was called "differential entrapment" by Gussow (1954) and leads eventually to oil in the up-dip (shallower) reservoirs and gas in the deeper ones (Fig. 2-7). Geological examples are given by Gill (1979).

COMPOSITIONAL CHANGES IN THE RESERVOIR

Thermal maturation

The composition of petroleum is not fixed and immutable. It changes and evolves in the reservoir in response to changing conditions. The increasing temperature that accompanies increasing depth of burial leads to thermal maturation of the crude oil. In this process large molecules are thermally cracked to smaller fragments and the trend is for the percentage of the lighter fractions to increase as the oil progresses to lower densities in the sequence from oil, to lighter oil, to wet gas, and finally dry gas (Connan et al., 1975), The increasing hydrogen content implied by this sequence is provided from parallel reactions involving cyclization and aromatization. The process is shown schematically in Fig. 2-8. The molecules that are losing hydrogen fuse into steadily larger and more aromatic units (possibly of colloidal dimensions) and become progressively less soluble in the oil. At the same time the oil is becoming lighter and a poorer solvent for the polycyclic aromatics, so that eventually these molecules precipitate out as an asphaltic residue in a process called "natural deasphaltening" (Milner et al., 1977).

This process is analogous to solvent deasphalting (SDA) used by refiners to improve oil quality by reducing the contents of sulfur, nitrogen and metals. Selected oil compositions before and after SDA are given in Table 2-2 and show that the deasphaltened oil is of higher API gravity and better quality. In nature, the quality of the produced oil improves and the asphaltic residue is left in the reservoir where it may have a serious adverse impact on porosity. For example, porosity has been completely plugged with asphaltic residue in parts of the Strachan Reef, Alberta (Hriskevich et al., 1980). This field is deeper than 14,000 ft and produces gas and condensate. Like many other fields that produce gas formed by thermally cracking oil, the gas here has high CO_2 and H_2S contents. In addition, gas reservoirs of this type are fre-

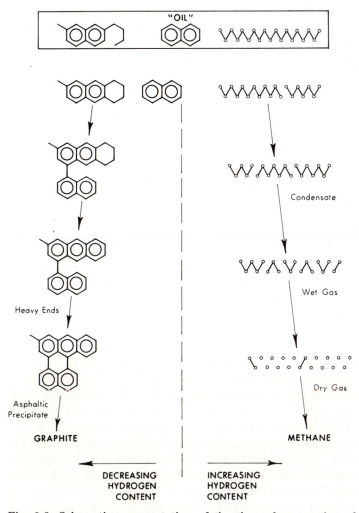

Fig. 2-8. Schematic representation of the thermal maturation of crude oil showing the parallel development of increasingly aromatic heavy molecules and progressively lighter alkanes (note the similarity to Fig. 2-1).

TABLE 2-2

Changes in chemical composition which accompany deasphaltening of selected oils (Billon et al., 1977)

	Murban	Arabian light	Buzurgan	Boscan
Original oil				
Specific gravity	0.982	1.003	1.051	1.037
S (%)	3.03	4.05	6.02	5.90
N (ppm)	3180	2875	4500	7880
Ni (ppm)	17	19	76	133
V (ppm)	26	61	233	1264
Asph. (wt.%)	1.2	4.2	18.4	15.3
Oil deasphaltened with propane				
Specific gravity	0.924	0.933	0.945	0.953
S (%)	1.80	2.55	3.60	4.70
N (ppm)	300	1200	930	1600
Ni (ppm)	1	1.0	2.0	6.0
V (ppm)	1	1.4	4.0	10.0
Asph. (wt.%)	0.05	0.05	0.05	0.05

quently abnormally pressured (e.g., the Thomasville field, Louisiana (Parker, 1973)).

Although the gas and light ends that precipitate asphaltenes are often generated in situ, gas introduced from some adjacent reservoir (for example during regional tipping) may also precipitate asphaltenes. Presumably for some oils natural gas injected into a reservoir for pressure maintenance could also lead to precipitation of heavy ends and produce dramatic changes in permeability.

Water washing

Major changes in petroleum composition can be produced by contact with flowing water (Milner et al., 1977). As the water moves past the oil in the reservoir it removes the most soluble components from the oil nearest to the oil—water contact. Since the most soluble compounds are the light ends, particularly the light aromatics, a heavier "tar" layer develops. The paraffins are among the least soluble components and as their concentration in the residual material of the tar layer increases, they may give rise to a considerably elevated pour point. In the case of the Sarir field in Libya, the tar layer is at a depth of 10,000 ft and is ultraparaffinic with a pour point of 160°F, whereas the bulk of the oil has a pour point of 70°F (Sanford, 1970).

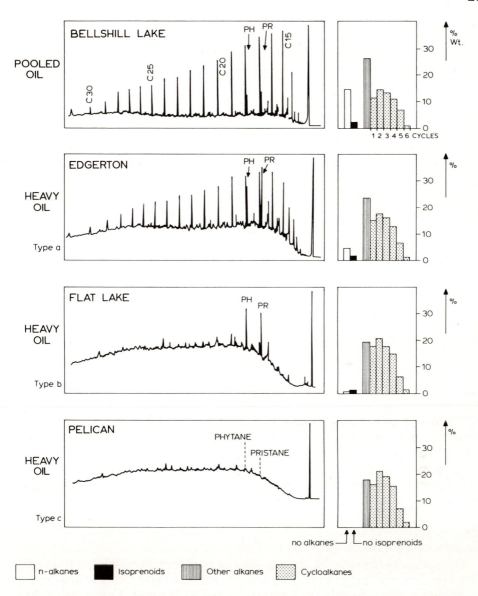

Fig. 2-9. Gas chromatograms and compositional data for Canadian oils showing the progressive changes produced by bacterial degradation. Unaltered oils are at the top, and the most severely degraded ones at the bottom (Deroo et al., 1977; courtesy of the *Geological Survey of Canada, Bulletin*).

Bacterial degradation

When the flowing water brings bacteria and oxygen into contact with oils at temperatures below about 75°C, bacterial degradation can occur producing major changes in crude oil composition. The bacteria that degrade oils in the reservoir are dominantly aerobic and preferentially consume normal alkanes so that the first indication of incipient biodegradation is a reduced paraffin content. With progressive degradation, all the paraffins are removed, whereas multi-branched compounds, such as the isoprenoids, remain. Ultimately even they are consumed. This sequence shows up very clearly in gas chromatographic traces. Fig. 2-9 presents data for a suite of oils from the area south of Athabasca, Canada, that show progressive degrees of biodegradation. This sequence of events can be duplicated exactly in laboratory studies (Jobson et al., 1972).

The bacteria that consume oil need free oxygen, which is very reactive in the subsurface and will not survive long unless water continues to flow in from the surface. Surface waters usually have lower salinities than deep subsurface ("connate") waters so that the influx of bacteria, oxygen, and nutrients is marked by flowing, fresher waters. The prediction of regional patterns of bacterial degradation, and hence of oil quality, can often be based on patterns of water salinities. This has been well documented by Bailey et al. (1973) for the Canadian part of the Williston Basin, and also by Bockmeulen et al. (1983) for the Bolivar coastal fields of Venezuela.

COMPOSITION OF PETROLEUM

Introduction

The composition of petroleum in a reservoir is initially controlled by the type of organic matter in the source rock; however, migration processes (and particularly adsorption in the source rock) tend to even out differences so that petroleums in reservoir rocks show a smaller range of compositions than do source rock extracts (Tissot and Welte, 1978).

The major changes in oil composition induced by alteration and maturation (but especially the latter) may cause marked differences in petroleum composition, even in different parts of a single reservoir. In the Bell Creek field, Wyoming, for example, API gravities range from 45° in the southwest to 32° in the center, because bacterial degradation has been effective in degrading only those parts of the field where water could penetrate. Vertical variations in oil composition have long been recognized and often attributed to "gravity segregation" or other processes. Many fields have layers of low-gravity, high-viscosity "tar" at the oil—water contact. The previous discussion suggests that tar layers of this sort may form in at least three different ways. They can be the result of water washing, which preferentially removes the more soluble components and the light ends. This process tends to leave the heavier paraffins of lower solubility, and their increased concentration may lead to elevated pour points. In contrast, bacterial degradation leads to a preferential loss of the normal paraffins. Since these components have the highest API gravities, their removal leaves material having low API gravities, a trend which is enhanced by removal of the more soluble light ends by the flowing water. Tar layers produced bacterially seem to be very common in reservoirs shallower than 7000 or 8000 ft, but bacterial degradation need not always lead to the development of tar layers. If the bacteria are active while oil is moving through the carrier bed system, the reservoirs will accumulate biodegraded crude and the composition will be uniform throughout the reservoir. Some of the large reservoirs in the Bolivar coastal fields of Venezuela appear to provide examples of this (Bockmeulen et al., 1983). In-situ deasphaltening can also lead to tar layer formation and in extreme cases this can decrease porosity to very low levels.

Thus, petroleum in the reservoir is frequently not uniform in composition, either in space or in time. This must be borne in mind when interpreting compositional data for crude oils and natural gases.

Sampling, storage and analysis

It is difficult to bring to the surface a representative sample of the petroleum in the subsurface reservoir, because normally only fluids that are pumped to the surface are available and in bringing oil up to the lower-pressure, lower-temperature near-surface environment, the volatile light ends can be lost in a variable and largely unpredictable way. At the other extreme, parts of the heavier fractions may be left behind in the reservoir. Although devices are available for obtaining samples under reservoir conditions, they are expensive, difficult to use, and are not employed routinely.

An additional sampling problem arises in fields that produce from multiple pay zones, because the oils from the different reservoirs are frequently commingled. In these cases only an average composition is obtained. If the oils in the producing zones are the same or similar, this may be acceptable, but when oil compositions vary widely, the composition of the mixture is of little geologic use. It is, however, still important to the refiner.

Contamination during production, sampling, or storage is always a potential source of uncertainty in determining petroleum composition. A wide variety of organic materials are used in drilling, testing, and production operations, such as mud additives, greases, lubricants, elastomer seals, and corrosion inhibitors, which are all possible contaminants in various degrees. Oils should not be shipped or stored in plastic bottles (other than teflon), because some of the light ends will be lost through the plastic. In addition, oils dissolve plasticizers that will contaminate the aromatic fraction in particular.

28

Until recently, analysis of the complex mixture of hydrocarbons in crude oils posed a major problem, but the rapid development of gas chromatography and mass spectrometry in the last few years has provided powerful analytical techniques for separating and characterizing many of the components in oils. Detailed analysis generally involves an initial separation into fractions. For some studies, distillation and collection of narrow boiling-point cuts is useful. For others, groups of related compounds are prepared using liquid chromatography. This is frequently carried out under gravity flow, or with low pressure differentials. Instruments for high-pressure liquid chromatography (HPLC), however, are now available and are being used more widely, because they offer advantages of smaller samples, quicker analysis, and better resolution. The fractions most often separated by liquid chromatography are (in order of elution from the column) the saturates, aromatics, nitrogen-, sulfur- and oxygen-containing compounds (generally called "NSO's"), and asphaltenes. For special purposes, further fractionation may be needed, such as the separation of the aromatics into mono-, di, tri-, and polynuclear aromatics. The fractions obtained in these various separation schemes can be characterized in detail by gas chromatography, mass spectrometry, or a combination of the two (GC-MS). Other analytical data, such as carbon isotope ratios and optical rotation, may also be acquired. A typical analytical scheme is outlined in Fig. 2-10.

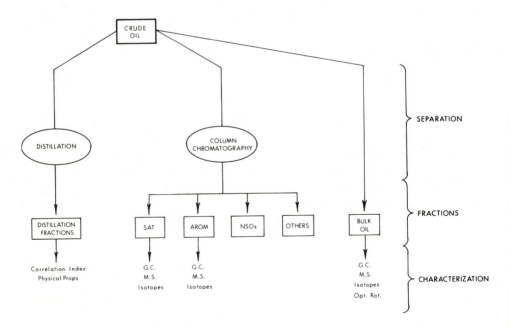

Fig. 2-10. Generalized analytical scheme.

COMPOUNDS IN CRUDE OIL

Introduction

Crude oils are complex mixtures, in some cases containing many hundreds of thousands of different compounds. Although these are predominantly hydrocarbons, they also include variable amounts of the compounds of sulfur, nitrogen, and oxygen. Oils vary widely in the absolute amounts of any given compound or compound type, but in spite of this the elemental composition shows rather a narrow range. In general, carbon accounts for 83—87% and hydrogen for 11—15%, whereas sulfur, oxygen, and nitrogen each lie in the range from 0 to 4 or 5%. Some inorganic materials ("ash") are also present in most oils but rarely exceed 0.05%. The elements are not uniformly distributed between the boiling-point ranges, and the nitrogen, sulfur, and oxygen compounds are always most abundant in the higher-molecular-weight fractions.

As methods of separating and analyzing oils have improved, an increasing amount of data has been accumulated for the compounds present in selected crude oils. A major step was taken in 1927 when the American Petroleum Institute (API) initiated its Research Project 6 for "The Separation, Identification, and Determination of the Chemical Constituents of Commercial Petroleum Fractions". A crude oil from the Ponca City field, Oklahoma, was selected and by 1953 about 130 hydrocarbons have been separated and identified. Since then, the number has increased rapidly, mainly due to the application of combined gas chromatography and mass spectrometry (GC-MS) analysis. Whitehead and Breger (1963) list over 200 hydrocarbons and about the same number of compounds containing oxygen, sulfur, and nitrogen, which have been identified in the Ponca City and other crude oils. Even though this list includes all the major components, it accounts for less than half of most crude oils. It does, however, help to delineate the most common type of compounds. For the hydrocarbons, these can be divided conveniently into the alkanes, naphthenes, aromatics and naphtheno-aromatics.

These groups show wide differences in relative abundances and also vary in their quantitative contribution to the various boiling-point fractions (Fig. 2-11). The straight-chain and branched alkanes are relatively more abundant in the lower boiling-point cuts, whereas naphtheno-aromatics, resins and asphaltenes become increasingly important as the boiling point of the fraction increases, and they dominate the highest-boiling-point fractions and the residuum. Naphthenic compounds are often one of the major groups of compounds and are significant over the whole boiling-point range, except for the very lightest and heaviest fractions. These comments are, of course, generalized and individual crude oils may vary widely from this pattern depending on the nature of the organic matter in the source rock and the past history of alteration and maturation in the reservoir.

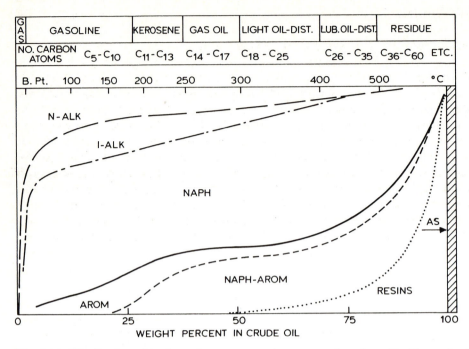

Fig. 2-11. Distribution of hydrocarbon compound types as a function of boiling point for a medium crude oil. *N-ALK* = normal alkanes; *I-ALK* = iso-alkanes; *NAPH* = naphthenics; *AROM* = aromatics; *AS* = asphaltenes.

TABLE 2-3

Distribution of n-alkanes between various carbon number ranges in various crude oils (data from Martin et al., 1963)

	John Creek, Kansas, U.S.A.	Kawkaw-lin, Mich., U.S.A.	Pine Unit, Mont., U.S.A.	Ponca City, Okla., U.S.A.	Uinta Basin, Utah, U.S.A.	Darius, Persian Gulf	North Smyer, Texas, U.S.A.	State Line, Wyo., U.S.A.
Age	Ord.	Dev.	Ord.	Ord.	Eo.	Cret.	Penn.	Paleo.
API gravity($^{\circ}$)	26.6	35.0	31.9	42.0	30.6	29.0	43.2	39.4
C_4-C_{10}	7.24	15.99	17.02	12.10	29.4	8.10	10.29	3.71
$C_{11}-C_{20}$	13.30	16.41	17.08	10.13	6.17	6.51	4.70	15.55
$C_{21}-C_{30}$	0.65	0.85	0.86	2.23	9.52	1.68	0.98	11.03
C_{30+}	0.04	0.07	0.09	0.25	2.11	0.20	0.07	0.93
Total	21.23	33.32	35.05	24.71	20.74	16.49	16.04	31.22

Hydrocarbons

Normal alkanes. Normal alkanes occur in most crude oils with an abundance generally in the range from 15 to 20%, but ranging as high as 35%. Relative abundances for individual compounds up to about C_{35} are reported routinely, because this is within the range of capability of normal gas chromatographic techniques. Also, there are reports of normal alkanes with 80 or more carbon atoms. The abundance of normal alkanes may be very low, or even zero, in oils which have been bacterially degraded. Table 2-3 shows the distribution of normal alkanes in various carbon number ranges for selected oils. The distribution is often irregular within the ranges and some oils show a slight predominance of *n*-alkanes with odd numbers of carbon atoms over those with even numbers (Fig. 2-6). The reverse (i.e., even greater than odd) is much less common and even then values are usually close to 1.0.

The Carbon Preference Index (CPI) provides a convenient number that summarizes the "odd-evenness" of the normal alkanes in a crude oil (or rock extract). It is defined as follows:

$$CPI = \frac{1}{2} \left[\frac{\text{sum of odd-chain length alkanes from } nC_{21} \text{ to } nC_{33}}{\text{sum of even-chain length alkanes from } nC_{20} \text{ to } nC_{32}} + \frac{\text{sum of odd-chain length alkanes from } nC_{21} \text{ to } nC_{33}}{\text{sum of even-chain length alkanes from } nC_{22} \text{ to } nC_{34}} \right]$$

It is important to note that other ranges of carbons numbers are also in use (Bray and Evans, 1961). Smooth distributions of normal alkanes give CPI values close to one, whereas odd predominance leads to CPI > 1.

Branched alkanes. Branched alkanes are present in most crude oils and can be the major component. The Anastasievskoye crude is reported to have nearly 85% branched alkanes (Smith, 1968); however, this is exceptional and is much higher than the average, which is around 25%. Like the normal alkanes, the branched compounds are not equally distributed among the various crude oil fractions but are more abundant in the light and middle boiling-point ranges (Fig. 2-11). In the Ponca City crude (studied in API Project 6) over half of the branched alkanes are in the C_6–C_{10} fraction (Table 2-3). For the lower-molecular-weight compounds, all the isomers are known. Higher fractions are less well characterized. It has been established that the monosubstituted branched alkanes predominate over polysubstituted ones and that the 2 or 3 positions are the preferred ones. Thus, Martin and Winters (1959) reported that, in general, for light oils the abundance of the C_6 isomers followed the sequence: 2-methylpentane > 3-methylpentane > 2,3-dimethylbutane > 2,2-dimethylbutane. For the C_7 isomers the sequence is: 3-methylhexane > 2-methylhexane > 2,3-dimethylpentane > ethylpentane > 2,2-dimethylpentane > 3,3-dimethylpentane > 2,2,3-trimethylbutane.

The presence of isoprenoids in crude oil is particularly interesting, because their characteristic branching (methyl groups on every fifth carbon atom, such as 2, 6, 10, ...) can be related to biochemically produced precursors. Pristane, phytane and farnesane were reported in the early sixties, but many others in the series are now known to be present (Illich et al., 1977). Pristane/phytane ratios range from less than one to more than ten with the higher values being characteristic of oils sourced from terrestrial organic matter or in more oxygenated environments (Powell and McKirdy, 1975).

Sufficient data have been accumulated to suggest that there might be relationships between the relative amounts of individual branched hydrocarbons. Smith (1968) has presented and discussed plots which show marked correlations between 2- and 3-methylpentane, and between 2- and 3-methylhexane. These compounds account for a major part of the isoheptanes and isohexanes, which therefore also show a correlation. The total amounts of the isohexanes and isoheptanes vary significantly and systematically with geologic age (Smith, 1968).

Cycloalkanes (naphthenes). The saturated cyclic hydrocarbons are the major component of crude oils, generally accounting for 30—60%. They are present in all fractions from C_5 and higher but show less variation between the various boiling-point ranges than the normal or branched alkanes (Fig. 2-11).

Cyclobutane has been reported in trace amounts in some crude oils, but the most abundant cyclic compounds are based on the five- or six-membered rings. Methylcyclohexane is especially abundant in naphthenic crude oils. It forms nearly 3% of Tertiary crudes from Texas and up to 20% of some Soviet crude oils. Although mono- and bicyclic compounds predominate, multi-ringed compounds become important in the higher-boiling fractions. Compounds containing up to nine fused rings have been identified by high-resolution mass spectrometry in these fractions. Cyclohexane, cyclopentane and their homologs have been studied in great detail. Some isomers of C_8 and C_9 compounds have been identified. The monocyclic compounds, which are mono-, bi-, or tri-substituted (generally in the 1, 3 or 1, 1, 3 positions), have also been identified. Smith (1968) has summarized many of the available data on the relative abundances of the compounds of cyclopentane and cyclohexane. Methylcyclohexane predominates over cyclohexane in almost all crudes, the methylcyclohexane/cyclohexane ratio ranging from 0.9 to 4.8. Methylcyclopentane also predominates over cyclopentane. Multi-ringed naphthenes, having ring structures that are directly traceable to biologically produced molecules, are receiving increasing attention. These "biomarkers" include the four-ring sterane and the five-ringed hopane series:

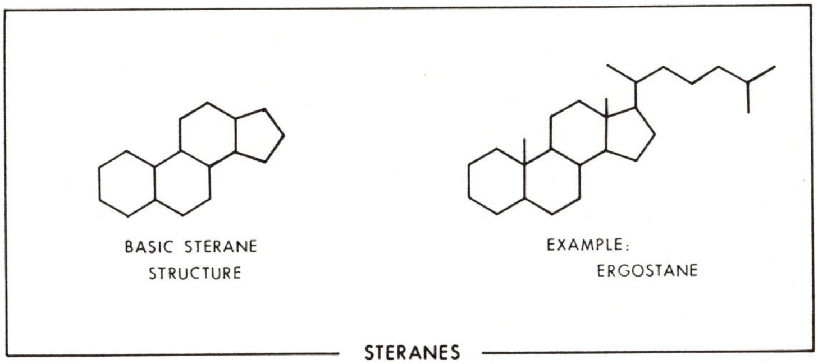

STERANES

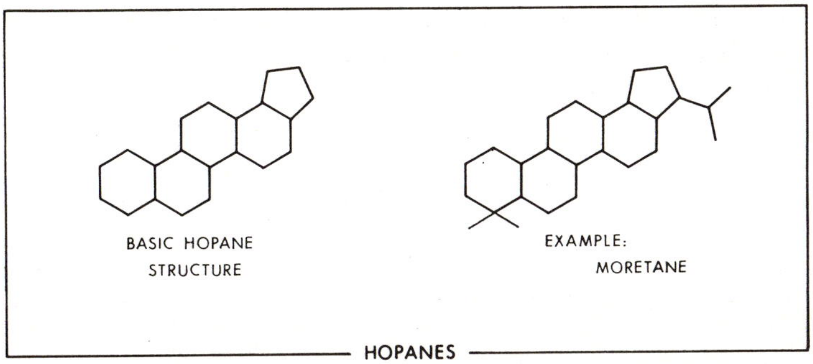

BASIC STERANE STRUCTURE

EXAMPLE: ERGOSTANE

HOPANES

BASIC HOPANE STRUCTURE

EXAMPLE: MORETANE

In each case, there is a whole series of compounds formed by different side chains and by varying spatial configurations. The relative abundances of members of the families are proving to be useful indicators of oil source, maturity, and migration history (Seifert and Moldowan, 1978).

Aromatics. Aromatic compounds are those containing only aromatic rings, which may have alkyl side chains or methylene bridges connecting them to other aromatic rings. In practice, aromatic fractions are frequently separated from other compound types by chromatography. The fractions obtained include compounds which, although predominantly aromatic in character, contain some saturated rings as well. Such mixed compounds, the naphtheno-aromatics, are discussed separately.

Lower aromatics are important constituents of most crude oils, particularly the light ones. Several detailed studies of crude oils have shown the presence of benzene and most of its isomers in the C_7 to C_9 range. The most important of these are toluene (which may reach nearly 2% content in some crude oils), m-xylene, and 1,2,4-trimethylbenzene. Benzene and xylene contents rarely exceed 1%. Aromatic compounds with more than one ring system have been reported in many crude oils. They range from naphthalene and diphenyl up to compounds with five or six fused rings. Methyl and ethyl side chains are common. These compounds are shown in Fig. 2-12 together with some

34

BASIC RING STRUCTURES

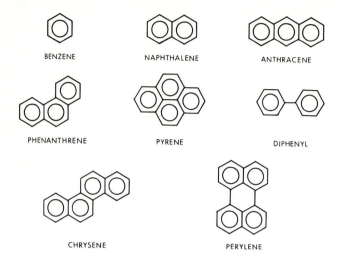

BENZENE   NAPHTHALENE   ANTHRACENE

PHENANTHRENE   PYRENE   DIPHENYL

CHRYSENE   PERYLENE

EXAMPLES OF COMPOUNDS WITH SIDE CHAINS

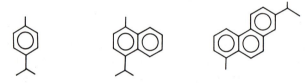

Fig. 2-12. Structures of selected aromatic hydrocarbons.

of the other compounds which have been isolated. In general, phenanthrene is present in larger quantities than anthracene.

Naphtheno-aromatics. Compounds containing both aromatic and saturated ring systems constitute a major component of crude oils, particularly in the higher-boiling-point fractions. It is often possible to relate their particular structures to biologic precursors, which has provided further impetus for their study. Some of the more important compounds identified so far in crude oils are shown in Fig. 2-13.

Bestougeff (1967) has pointed out that within a molecule the distribution of side chains on the aromatic and naphthenic rings follows the same pattern as in the single-ring compounds. He stated that "the short chains (methyl and ethyl) are characteristic constituents of the aromatic portion of the molecule, whereas a limited number (one or two) of rather long chains are commonly attached to the cycloparaffin rings". It also appears that the length and number of the chains increase with the molecular weight of the naphtheno-aromatic compound.

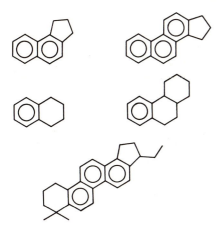

Fig. 2-13. Structures of some naphtheno-aromatic compounds present in crude oils.

Compounds with heteroatoms

Compounds containing nitrogen, sulfur and oxygen (in addition to carbon and hydrogen) are present in most crude oils and commonly account for as much as 10% of the total volume, though in extreme cases it can be much higher. The percentage of heteroatoms is often higher in shallow and degraded crudes, and in so-called "immature" crudes, than in deeper and more mature oils. The heterocompounds are distributed unevenly among the boiling-point ranges and predominate in the heavy residues. Heavy distillates contain intermediate amounts and the lighter fractions are poorest in heteroatoms.

Sulfur compounds. The sulfur content of crude oils is most commonly less than 5% and oils with more than 10% sulfur are rare. Hydrogen sulfide may be present in solution and some crudes contain free sulfur. In general, the most important sulfur compounds are the thiols (mercaptans), sulfides, disulfides, and sulfur-ring compounds (Mehmet, 1971; Rall et al., 1972). The relative abundances of groups of sulfur compounds in a selected suite of 78 oils is shown in Table 2-4. Ho et al. (1974) found non-thiophenic sulfur compounds to be less abundant in shallow "immature" crude oils and to increase with oil maturity. The benzothiophene/dibenzothiophene ratio also increased with maturity. In general, high-sulfur crude oils have lower API gravities (Nelson, 1972).

Nitrogen compounds. The nitrogen content of crude oils generally accounts for less than 1% of the total oil and is usually much lower than the sulfur content. Tissot and Welte (1978) gave an average value of 0.094 wt.%.

TABLE 2-4

Relative importance of the major sulfur compound types in 78 crude oils (data from Ho et al., 1974)

Compound type	Structure	Mean ± st. dev. (%)	Number of samples
Total sulfur		1.64 ± 1.97	78
Aliphatic sulfides	R—SH; R—S—R; —SH; R—S—; R—S—S—R	18.9 ± 9.2	25
Alkylaryl sulfides or thiaindans	—S—R	25.8 ± 9.7	24
Thiophene		3.3 ± 1.8	43
Benzothiophenes		5.8 ± 3.4	78
Dibenzothiophenes		9.0 ± 4.6	78
Benzonaphthothiophenes		5.9 ± 2.3	26
Sulfur not recovered		42.8 ± 13.6	78

Nitrogen compounds are frequently absent from the lower-boiling-point range fractions and, like the compounds of other heteroatoms, are concentrated in the heavier fractions. They are classified as "basic" if they can be extracted with dilute mineral acids, or "neutral" if they cannot be extracted (but see Costantinides and Arich, 1967). The ratio of basic compounds to neutral compounds in any given boiling-point range is approximately constant, even between different crude oils. This suggests that nitrogen compounds are not influenced by later alteration processes as is the case with sulfur compounds.

Oxygen compounds. It is more difficult to determine the total oxygen content of a crude oil than to determine nitrogen or sulfur contents. Until quite recently, oxygen was calculated by difference and, therefore, the older values should be used with caution. The development of direct methods for oxygen determination indicate that crude oils generally contain less than 1% oxygen. This is present mainly as naphthenic acids, phenols and fatty acids. The acids and their salts are surface-active compounds.

Metals

Most petroleums contain traces of metals in amounts ranging from 100 μg/g to 1 ng/g (Filby, 1975; Valkovic, 1978). The amount probably reflects the nature of the source materials, but the past history of migration, maturation, and alteration may have a modifying role. In general, the most degraded oils contain the highest metal contents. Although some metals could be present as inorganic particulate matter or are dissolved in emulsified formation waters, most appear to be in a true oil-soluble form as complexes, which are most abundant in the higher-boiling-point fractions. Nickel and vanadium are usually the most abundant metals and have been the most studied. Significant amounts of both occur coordinated to nitrogen in porphyrin complexes, but Yen et al. (1962) have suggested that the non-porphyrin nickel and vanadium (and possibly some of the other metals) are held in "holes" bordered by sulfur, nitrogen, or oxygen atoms in the asphaltene sheets. Other commonly reported metals include Fe, Co, Cr, Hg, Cu, and As.

Asphaltic components

The asphaltic fraction content varies widely among different crude oils ranging from a maximum of about 40% to a minimum of almost zero. Most chemical studies of this material have started with a preliminary separation on the basis of solubility. The fraction which dissolves in *n*-pentane is called the "petrolene fraction" (also the "maltene" or "malthene" fraction) and is usually a viscous liquid. The isoluble fraction, or "asphaltene", is generally a dark brown or black powdery solid. The elemental composition of asphaltenes separated from a variety of crude oils is given in Table 2-5 and shows that appreciable amounts of sulfur and oxygen are present. Other analyses have shown that nitrogen is also frequently an important consituent.

The asphaltic components in petroleum have molecular weights in the 900—4000 range, but appear to form aggregates with molecular weights over 100,000. These asphaltic components exist as colloidal particles in the crude oil. The chemical structure of such large entities is still incompletely known, but the available information is consistent with a structure in which a fairly condensed aromatic nucleus with peripheral naphtheno-aromatic groups is linked through methylene groups to other similar units. The heteroatoms are mostly in the ring structures of the asphaltene fraction (Yen, 1974).

TABLE 2-5

Elemental analyses of asphaltenes from various crude oils (Witherspoon and Winniford, 1967)

Crude oil and source	C (%)	H (%)	N (%)	S (%)	O (%)	C/H ratio
Baxterville, Miss., U.S.A.	84.5	7.4	0.80	5.60	1.7	0.95
Burgan, Kuwait	82.2	8.0	1.70	7.60	0.6	0.86
Lagunillas, Venezuela	84.2	7.9	2.00	4.50	1.6	0.89
Mara, Venezuela	83.5	8.3	0.98	2.68	1.5	0.81
Ragusa, Sicily	81.7	8.8	1.47	6.31	1.8	0.77
Salem, Ill., U.S.A.	88.2	8.1	1.71	0.62	1.3[a]	0.91
Wafra, Neutral Zone	81.8	8.1	1.03	7.80	1.5	0.84

[a] By difference.

PHYSICAL PROPERTIES

The physical properties of crude oils reflect the chemical composition in a complex and largely unpredictable way. Physical characteristics are particularly important in handling, transporting, and refining oils and are an important factor in establishing price. Density (as API gravity) is always determined and color is usually noted. Pour point, cloud point, viscosity, flash point, and other properties may be reported.

Density. Crude oil density is usually reported in degrees API and values vary from 7 or 8 for heavy tars up to 60° API or higher for very light condensates. Most oils are less dense than water; however, when API gravity falls below 10°, then oil densities exceed that of water and these oils will no longer rise by buoyancy in an aqueous system. A histogram of API gravities for 7386 oils is given in Fig. 2-14 and shows that oils commonly have gravities of around 35—40° API, but spread with decreasing frequency to lower and higher values.

Color. The color of crude oils in reflected light ranges from black to water white, with greenish-black, dark brown and reddish tinged oils being common. Although the highly paraffinic waxy crudes are frequently pale yellowish in color, they can be much darker and, in general, there is no simple relationship between chemical composition and color.

Viscosity. Crude oils vary widely in viscosity and at one extreme grade into highly viscous tars that are almost immobile and at the other into highly mobile condensates. In the reservoir, viscosity depends on dissolved gas and temperature as well as the nature of the oil, whereas values quoted are generally obtained under standard laboratory conditions, usually at elevated temperatures (e.g., 100°F). In general, oils of lower API gravities have higher viscosities (Levorsen, 1967).

Pour point. As crude oil temperature is lowered, a point is reached at which the oil will no longer flow. This is its pour point. Values vary from below −50°F up to temperatures in excess of 100°F. The high-pour-point oils are in general very rich in alkanes. Examples include oils from Brazil, Gabon, Libya, Sumatra and China. Many deltas, including the Niger, Mackenzie and Mahakam, have waxy, high-pour-point crudes.

CLASSIFICATION

Crude oils cover a wide range in physical properties and chemical compositions, and it is often important to be able to group them into broad categories of related oils. The classification scheme used depends on the application — whether for establishing price, refining characteristics, or nature of the organic source materials. Refiners are most concerned with the relative

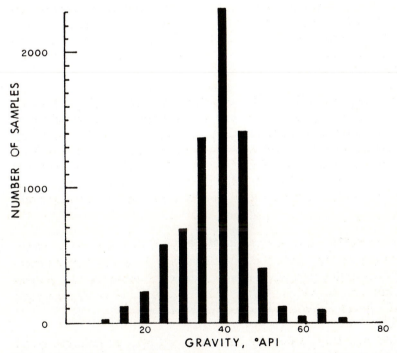

Fig. 2-14. Histogram showing frequency of occurrence of oils of various API gravities (data from U.S. Bureau of Mines: Smith, 1968).

abundances of various distillation fractions and their chemical character. Geochemists, on the other hand, are usually more interested in relating oils to their source rocks or in establishing their degree of thermal maturation or degradation. In this case, the characteristic biomarkers, that are only minor components, are often very useful.

For many years the U.S. Bureau of Mines has used an oil classification scheme based on the characteristics of two boiling-point fractions (Lane and Garton, 1935): Fraction I was collected between 250 and 275°C, and Fraction II was collected at a pressure of 40 mm of mercury and a temperature range of 275—300°C. The densities were used to characterize each of these fractions as "paraffin-base", "mixed base" or "naphthenic base". With improved analytical techniques alternative classification schemes became possible. Tissot and Welte (1978) have recently suggested a classification based on the relative amounts of various compound types in oil topped at 210°C. This scheme is based on their experience with approximately 600 oils

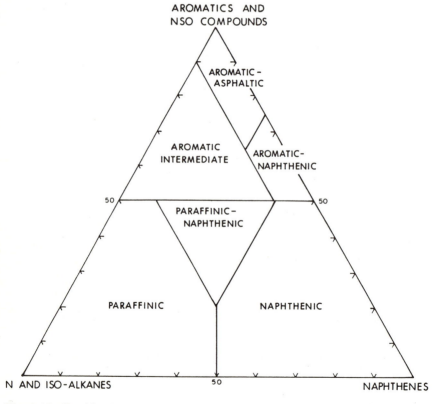

Fig. 2-15. Classification of crude oil based on chemical composition of the bulk in oils in terms of normal-plus-iso-alkanes, naphthenes and aromatics-plus-NSO compounds (modified after Tissot and Welte, 1978).

analyzed at the French Petroleum Institute. Fig. 2-15 shows how the classification relates to the ternary system with normal+iso-alkanes, naphthenes,
and aromatics+NSO compounds as the end members. ("Aromatics" include
all compounds with one or more aromatic rings and thus include the compound type called "naphtheno-aromatic" in the previous discussion.) Oils are
unevenly distributed among the classes and the most abundant types are the
paraffinic oils, the paraffinic-naphthenic oils, and the aromatic-intermediate
type oils. Earlier it was stressed that oils may change in composition in the
reservoir due to maturation or alteration. The trends produced by these
changes are shown in Fig. 2-16.

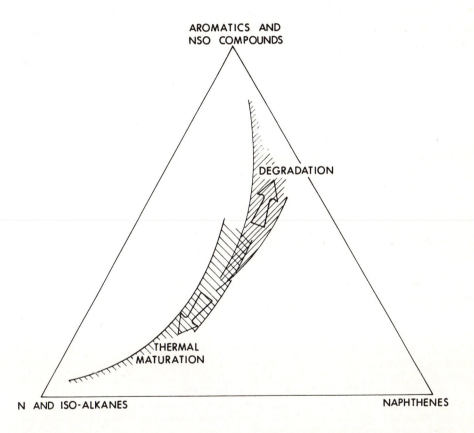

Fig. 2-16. Effects of thermal maturation and degradation on crude oil composition and
hence classification.

42

REFERENCES

Bailey, N.J.L., Krouse, H.R., Evans, C.R. and Rogers, M.A., 1973. Alteration of crude oil by waters and bacteria — evidence from geochemical and isotope studies. *Bull. Am. Assoc. Pet. Geol.*, 57: 1276—1290.

Baker, E.G., 1960. A hypothesis concerning the accumulation of sediment hydrocarbons to form crude oil. *Geochim. Cosmochim. Acta*, 19: 309—317.

Barker, C., 1972. Aquathermal pressuring — role of temperature in development of abnormal-pressure zones. *Bull. Am. Assoc. Pet. Geol.*, 56: 2068—2071.

Barker, C., 1974. Pyrolysis techniques for source-rock evaluation. *Bull. Am. Assoc. Pet. Geol.*, 58: 2349—2361.

Barker, C., 1979. Organic geochemistry in petroleum exploration. *Am. Assoc. Pet. Geol., Contin. Educ. Course Note Ser.*, 10: 159 pp.

Barker, C., 1980. Primary migration: the importance of water-mineral-organic matter interactions in source rocks. In: W.H. Roberts III and R.J. Cordell (Editors), *Problems of Petroleum Migration. Am. Assoc. Pet. Geol., Stud. Geol.*, 10: 19—31.

Barker, C., 1982. Oil and gas on passive continental margins. In: J.S. Watkins and C.L. Drake (Editors). *Am. Assoc. Pet. Geol., Mem.*, 34: 549—565.

Bestougeff, M.A., 1967. Petroleum hydrocarbons. In: B. Nagy and U. Colombo (Editors), *Fundamental Aspects of Petroleum Geochemistry*. Elsevier, Amsterdam, pp. 77—108.

Billon, A., Peries, J.P., Fehr, E. and Lorenz, E., 1977. SDA key to upgrading heavy crude. *Oil Gas J.*, 75 (4): 43—48.

Bockmeulen, H., Barker, C. and Dickey, P.A., 1983. Geology and geochemistry of crude oils in the Bolivar coastal fields, Venezuela. *Bull. Am. Assoc. Pet. Geol.*, 67: 242—270.

Bray, E.E. and Evans, E.D., 1961. Distribution of *n*-paraffins as a clue to recognition of source beds. *Geochim. Cosmochim. Acta*, 22: 2—15.

Claypool, G.E. and Kaplan, I.R., 1974. The origin and distribution of methane in marine sediments. In: I.R. Kaplan (Editor), *Natural Gases in Marine Sediments*. Plenum Press, New York, N.Y., pp. 99—139.

Claypool, G.E., Threlkeld, C.N. and Magoon, L.B., 1980. Biogenic and thermogenic origins of natural gas in the Cook Inlet basin. *Bull. Am. Assoc. Pet. Geol.*, 64: 1131—1139.

Clayton, J.L. and Swetland, P.J., 1980. Petroleum generation and migration in the Denver basin. *Bull. Am. Assoc. Pet. Geol.*, 64: 1613—1633.

Combaz, A. and deMatharel, M., 1978. Organic sedimentation and genesis of petroleum in Mahakam Delta, Borneo. *Bull. Am. Assoc. Pet. Geol.*, 62: 1684—1695.

Connan, J., 1974. Time-temperature relation in oil genesis. *Bull. Am. Assoc. Pet. Geol.*, 58: 2516—2521.

Connan, J., LeTran, K. and Van der Weide, B., 1975. Alteration of petroleum in reservoirs. *Proc. 9th World Pet. Congr.*, 2: 171—178.

Constantinides, G. and Arich, G., 1967. Non-hydrocarbon compounds in petroleum. In: B. Nagy and U. Colombo (Editors), *Fundamental Aspects of Petroleum Geochemistry*. Elsevier, Amsterdam, pp. 109—175.

Cordell, R.J., 1972. Depths of oil origin and primary migration: a review and critique. *Bull. Am. Assoc. Pet. Geol.*, 56: 2029—2067.

Cordell, R.J., 1973. Colloidal soap as proposed primary migration medium for hydrocarbons. *Bull. Am. Assoc. Pet. Geol.*, 57: 1618—1643.

Cordell, R.J., 1976-77. How oil migrates in clastic sediments. *World Oil*, Part 1, November 1976: 107—126; Part 2, December 1976: 75—80; Part 3, January 1977: 97—100; Part 4, February 1977: 36—38.

Demaison, G.J. and Moore, G.T., 1980. Anoxic environments and oil source bed genesis. *Bull. Am. Assoc. Pet. Geol.*, 64: 1179—1209.

Deroo, G., 1976. Correlations of crude oils and source rocks in some sedimentary basins. *Bull. Cent. Rech. Pau SNPA*, 10: 317—335 (in French).

Deroo, G., Powell, T.G., Tissot, B. and McCrossan, R.G., 1977. The origin and migration of petroleum in the Western Canadian sedimentary basin, Alberta — A geochemical and thermal maturation study. *Geol. Surv. Can., Bull.*, 262: 136 pp.

Espitalié, J., LaPorte, J.L., Madec, J., Marquis, F., Leplat, P., Paulet, J. and Boutefue, A., 1977. Méthode rapide de charactérisation des roches meres de leur potentiel et de leur degré d'évolution. *Rev. Inst. Fr. Pét.*, 32: 23—42.

Evamy, B.D., Haremboure, J., Kamerling, P., Knapp, W.A., Molloy, F.A. and Rowlands, P.H., 1978. Hydrocarbon habitat of Tertiary Niger delta. *Bull. Am. Assoc. Pet. Geol.*, 62: 1—39.

Filby, R.H., 1975. The nature of metals in petroleum. In: T.F. Yen (Editor), *The Role of Trace Metals in Petroleum*. Ann Arbor Science Publishers, Ann Arbor, Mich., pp. 31—58.

Gill, D., 1979. Differential entrapment of oil and gas in Niagaran pinnacle-reef belt of northern Michigan. *Bull. Am. Assoc. Pet. Geol.*, 63: 608—620.

Gold, T. and Soter, S., 1982. Abiogenic methane and the origin of petroleum. *Energy Explor. Exploit.*, 1: 89—104.

Gussow, W.C., 1954. Differential entrapment of oil and gas: a fundamental principle. *Bull. Am. Assoc. Pet. Geol.*, 38: 816—853.

Halbouty, M.T., Meyerhoff, A.A., King, R.E., Dott, R.H., Sr., Klemme, H.D. and Shabad, T., 1970. World's giant oil and gas fields, geologic factors affecting their formation and basin classification. *Am. Assoc. Pet. Geol., Mem.*, 14: 502—555.

Hedberg, H.D., 1968. Significance of high wax oils with respect to genesis of petroleum. *Bull. Am. Assoc. Pet. Geol.*, 52: 736—750.

Ho, T.Y., Rogers, M.A., Drushel, H.V. and Koons, C.B., 1974. Evolution of sulfur compounds in crude oils. *Bull. Am. Assoc. Pet. Geol.*, 58: 2338—2348.

Hriskevich, M.E., Faber, J.M. and Langton, J.R., 1980. Strachan and Ricinus West gas fields, Alberta, Canada. In: M.T. Halbouty (Editor), *Giant Oil and Gas Fields of the Decade: 1968—1978. Am. Assoc. Pet. Geol., Mem.*, 30: 315—327.

Hunt, J.M., 1977. Ratio of petroleum to water during primary migration in western Canada basin. *Bull. Am. Assoc. Pet. Geol.*, 61: 434—435.

Hunt, J.M., 1979. *Petroleum Geochemistry and Geology*. W.H. Freeman, San Francisco, Calif., 617 pp.

Illich, H.A., Haney, F.R. and Jackson, T.J., 1977. Hydrocarbon geochemistry of oils from Maranon Basin, Peru. *Bull. Am. Assoc. Pet. Geol.*, 61: 2103—2114.

Jobson, A., Cook, F.D. and Westlake, D.W.S., 1972. Microbial utilization of crude oil. *Appl. Microbiol.*, 23: 1082—1089.

Lane, E.C. and Garton, E.L., 1935. "Base" of a crude oil. *U.S. Bur. Mines, Rep. Invest.*, 3279.

Levorsen, A.I., 1967. *Geology of Petroleum*. W.H. Freeman, San Francisco, Calif., 2nd ed., 724 pp.

Lewan, M.D., Winters, J.C. and McDonald, J.H., 1979. Generation of oil-like pyrolyzates from organic-rich shales. *Science*, 203: 897—899.

Lopatin, N.V., 1971. Temperature and geologic time as factors of coalification. *Izv. Akad. Nauk SSSR, Ser. Geol.*, 3: 95—106 (in Russian).

Martin, R.L. and Winters, J.C., 1959. Composition of crude oils through seven carbons as determined by gas chromatography. *Anal. Chem.*, 31: 1954—1960.

Martin, R.L., Winters, J.C. and Williams, J.A., 1963. Composition of crude oils by gas chromatography: geological significance of hydrocarbon distribution. *6th World Pet. Congr., Sect. V*, pp. 231—260.

McAuliffe, C.D., 1978. Role of solubility in migration of petroleum from source. In: W.H. Roberts III and R.J. Cordell (Editors), *Physical and Chemical Constraints on Petroleum Migration. Am. Assoc. Pet. Geol., Contin. Educ. Course Note Ser.*, 8: C1—C39.

Mehmet, Y., 1971. The occurrence and origins of sulfur compounds in crude oils: a review. *Alberta Sulfur Research Ltd. Q. Bull.*, 8: 17 pp.

Milner, C.W.D., Rogers, M.A. and Evans, C.R., 1977. Petroleum transformations in reservoirs. *J. Geochem. Explor.*, 7: 101—153.

Momper, J.A., 1978. Oil migration limitations suggested by geological and geochemical considerations. In: W.H. Roberts III and R.J. Cordell (Editors), *Physical and Chemical Constraints on Petroleum Migration. Am. Assoc. Pet. Geol., Contin. Educ. Course Note Ser.*, 8: B1—B60.

Nelson, W.L., 1972. What's the average sulfur content vs. gravity? *Oil Gas J.*, 70: 59.

Parker, C.A., 1973. Geopressures in the deep Smackover of Mississippi. *J. Pet. Technol.*, 25: 971—979.

Peake, E. and Hodgson, G.W., 1967. Alkanes in aqueous systems. I. The accomodation of C_{12}—C_{36} n-alkanes in distilled water. *J. Am. Oil Chem. Soc.*, 44: 696—702.

Philippi, G.T., 1965. On the depth, time and mechanism of petroleum generation. *Geochim. Cosmochim. Acta*, 22: 1021—1040.

Porfir'ev, V.B., 1974. Inorganic origin of petroleum. *Bull. Am. Assoc. Pet. Geol.*, 58: 3—33.

Powell, T.G. and McKirby, D.M., 1975. Geologic factors controlling crude oil composition in Australia and Papua New Guinea. *Bull. Am. Assoc. Pet. Geol.*, 59: 1176—1197.

Price, L.C., 1976. Aqueous solubility of petroleum as applied to its origin and primary migration. *Bull. Am. Assoc. Pet. Geol.*, 60: 213—244.

Rall, H.T., Thompson, C.J., Coleman, H.J. and Hopkins, R.L., 1972. Sulfur compounds in crude oil. *U.S. Bur. Mines Bull.*, 659: 193 pp.

Rice, D.D., 1975. Origin of and conditions of shallow accumulations of natural gas. In: *Geology and Mineral Resources of the Bighorn Basin. Wyo. Geol. Assoc. 27th Annu. Field Conf. Guideb.*, pp. 267—271.

Rice, D.D. and Shurr, G.W., 1980. Shallow, low-permeability reservoirs of northern Great Plains — assessment of their natural gas resources. *Bull. Am. Assoc. Pet. Geol.*, 64: 969—987.

Rice, D.D. and Claypool, G.E., 1981. Generation, accumulation and resource potential of biogenic gas. *Bull. Am. Assoc. Pet. Geol.*, 65: 5—25.

Robertson, R., 1963. Duplex origin of petroleum. *Nature*, 199: 113—114.

Sanford, R.M., 1970. Sarir oil field, Lybia — desert surprise. In: M.T. Halbouty (Editor), *Geology of Giant Petroleum Fields. Am. Assoc. Pet. Geol., Mem.*, 14: 449—476.

Seifert, W.K. and Moldowan, J.M., 1978. Applications of steranes, terpanes and monoaromatics to the maturation, migration and source of crude oils. *Geochim. Cosmochim. Acta*, 42: 77—95.

Smith, H.M., 1968. Qualitative and quantitative aspects of crude oil composition. *U.S. Bur. Mines Bull.*, 642: 1—136.

Sutton, C., 1977. Depositional environments and their relation to chemical composition of Java Sea crude oils. *ASCOPE/CCOP Seminar on Generation and Maturation of Hydrocarbons in Sedimentary Basins, Manila*.

Tissot, B.P., 1969. Premières données sur les mécanismes et la cinétique de la formation du pétrole dans les sédiments. Simulation d'un schema réactionnel sur ordinateur. *Rev. Inst. Fr. Pét.*, 24 (4): 470—501.

Tissot, B. and Pelet, R., 1971. Nouvelles données sur les mécanismes de génèse et de migration du pétrole: simulation mathématique et application à la prospection. *Proc. 8th World Pet. Congr.*, 2: 35—46.

Tissot, B. and Espitalié, J., 1975. L'évolution thermique de la matière organique des sédiments: applications d'une simulation mathématique. *Rev. Inst. Fr. Pét.*, 30: 743—777.

Tissot, B.P. and Welte, D.H., 1978. *Petroleum Formation and Occurrence.* Springer, New York, N.Y., 538 pp.

Tissot, B., Califet-Debyser, Y. Deroo, G. and Oudin, J.L., 1971. Origin and evolution of hydrocarbons in early Toarcian shales, Paris Basin, France. *Bull. Am. Assoc. Pet. Geol.*, 58: 2177—2193.

Tissot, B., Durand, B., Espitalié, J. and Combaz, A., 1974. Influence of nature and diagenesis of organic matter in formation of petroleum. *Bull. Am. Assoc. Pet. Geol.*, 58: 499—506.

Valkovic, V., 1978. *Trace Elements in Petroleum.* Pennwell, Tulsa, Okla., 216 pp.

Waples, D.W., 1980. Time and temperature in petroleum formation: application of Lopatin's method to petroleum exploration. *Bull. Am. Assoc. Pet. Geol.*, 64: 916—926.

Whitehead, W.L. and Breger, I.A., 1963. Geochemistry of petroleum. In: I.A. Breger (Editor), *Organic Geochemistry.* pp. 248—332.

Williams, J., 1974. Application of oil-correlation and source rock data to exploration in the Williston basin. *Bull. Am. Assoc. Pet. Geol.*, 58: 1243—1252.

Witherspoon, P.A. and Winniford, R.S., 1967. The asphaltic components of petroleum. In: B. Nagy and U. Colombo (Editors), *Fundamental Aspects of Petroleum Geochemistry.* Elsevier, Amsterdam, pp. 261—297.

Yen, T.F., 1974. Structure of petroleum asphaltene and its significance. *Energy Sources,* 1: 447—463.

Yen, T.F., 1975. Chemical aspects of metals in native petroleum: In: T.F. Yen (Editor), *The Role of Trace Metals in Petroleum.* Ann Arbor Science Publishers, Ann Arbor, Mich., pp. 1—20.

Yen, T.F., Erdman, J.G. and Saraceno, A.J., 1962. Investigation of the nature of free radicals in petroleum asphaltenes and related substances by electron spin resonance. *Anal. Chem.*, 34: 694—700.

Yermakov, V.I. et al., 1970. Isotopic composition of carbon in natural gases in the northern part of the West Siberian Plain in relation to their origin. *Dokl. Akad. Nauk SSSR*, 190: 196—199.

Chapter 3

FUNDAMENTALS OF ENHANCED RECOVERY

N.R. MORROW and J.P. HELLER

INTRODUCTION

Enhanced oil recovery (EOR) generally refers to oil recovery over and above that obtained through the natural energy of the reservoir. Within this broad definition there are a variety of processes. These include various forms of waterflooding, caustic flooding, hydrocarbon injection, carbon dioxide flooding, micellar—polymer flooding, and several thermal methods. The widespread application of waterflooding (Craig, 1971) to boost production after initial decline in production has led to this process being referred to as secondary recovery and, in some respects, particularly for regulatory and pricing purposes, waterflooding has been set apart from other forms of enhanced recovery. In a typical waterflood, the "watercut" in the produced fluid continually increases, and the expenses of pumping, separation and disposal of the floodwater eventually exceed the income from the oil recovered. Then secondary recovery efforts are halted, even though oil remains in the reservoir.

Within the many enhanced recovery techniques, there is a class of processes, known as tertiary, designed to recover oil, commonly described as residual oil, left in the reservoir after both primary and secondary (water-flooding) recovery methods have been exploited to their respective economic limits. In addition to economic reasons for the existence of residual oil, there are also some technical ones. In the first place, an ordinary waterflood, operated at practical rates with ordinary water or brine, is physically incapable of displacing all the oil from reservoir rock. Capillary forces acting during the waterflood may cause part of the oil to be retained, in water-wet rock at least, as disconnected structures which do not flow under the pressure gradient that arises from flow of water. The detail of these structures is directly related to the microscopic mechanism of entrapment. Thus, even in those regions of the reservoir which are relatively well-swept, i.e., regions through which relatively large quantities of water have flowed, a residual oil saturation, S_{or}^*, which can range typically from 15 to 40% of pore space, will be retained. The residual oil saturation in these well-swept regions, of proven accessibility with respect to injected fluids, is an important target, albeit a difficult one, for tertiary oil recovery.

Secondly, oil is present in large areas of waterflooded reservoirs at saturations over and above those typical of the swept regions. These saturations are determined by macroscopic processes such as non-uniform flow and sweep in the reservoir. This oil, which is present in less accessible parts of the reservoir but at higher saturations (up to as high as the initial oil saturation, S_{oi}), is a second, challenging target for tertiary oil recovery.

In practice, of course, it is generally difficult to distinguish between these two categories of residual oil. Certainly in either case, for a successful EOR operation, the residual oil will have to be displaced from its present location and moved to the production wells. Enhanced recovery techniques have been applied to situations ranging from newly discovered reservoirs, the objective being to obtain increased oil recovery with respect to that given by conventional means, to recovery of tertiary oil from reservoirs which are close to the end of their normal, economically productive life. All share the common feature of fluid being displaced in a porous medium. The aim of this chapter is to provide a brief review of the fundamentals of enhanced recovery with emphasis on microscopic and macroscopic aspects of displacement.

PORE GEOMETRY

Many problems in enhanced recovery are related at the microscopic level to the pore geometry of reservoir systems. Fine detail of pore structure can be examined through electron micrography. Preparation of resin pore casts from rock samples provides a valuable supplementary approach to examination of pore structure. The casts are prepared by filling the pore space with liquid resin, and removing the mineral matter by acid leaching after solidifying the resin (Wardlaw, 1976; Gardner, 1980). Whereas micrographs of porous rocks and their pore casts are of increasing importance in investigation of pore structure, the detail they provide far exceeds that which can be used in developing a tractable mathematical model of pore space. They do, however, provide a useful indication as to geometries which are reasonably representative of pore shape.

Aside from the problem of defining pore shape, the closest approach yet attained to obtaining a realistic measure of pore size distribution is through computer assisted analysis of thin sections of core samples impregnated with resin or Wood's metal. The problem of analyzing two-dimensional images, presented by thin sections, to obtain three-dimensional properties is discussed in detail by Underwood (1970). Specific consideration of analysis of pore structure has been made by Dullien (1979) and Dullien and Dhawan (1974). This method, in principle at least, should provide fairly reliable estimates of pore size distribution, but has only been performed to date for a few samples.

In general, it has been much more common for pore structure to be inferred from fluid displacement measurements than for fluid displacement behavior to be predicted from detailed knowledge of pore geometry. The most common method of measuring pore size distribution is through mercury porosimetry (Allen, 1968). Pressure necessary to inject mercury into a given pore to which mercury has access is controlled by the size of the pore throat through which the mercury has to pass. Depending on the intended application of the results, there are some advantages to this. With respect to fluid flow and related properties, the effect of pore throat size will be dominant. This is reflected by a useful correlation (Chatzis, 1980) between injection pressure required for breakthrough, i.e., the pressure at which mercury establishes a continuous path through the sample (indicated by electrical conductivity), and permeability, for permeabilities ranging over 4 orders of magnitude (Chatzis, 1980):

$$P_c = 85.63 k^{-0.369} \qquad\qquad (3\text{-}1)$$

where P_c is the capillary pressure for breakthrough in psi, and k is the permeability in millidarcies. Breakthrough usually occurs when the saturation of the injected mercury reaches about 15% of the pore space.

A serious obstacle to deriving pore size distribution from mercury injection curves is that filling of larger pores may be delayed, because access through neighboring smaller pores is denied until their filling pressures are attained. When the larger pores do fill, they appear in the size distribution erroneously as an increased number of smaller pores. Comparisons of pore size distribution measurements obtained by mercury penetration and image analysis have been reported by Dullien and Dhawan (1974). It is unlikely that a satisfactory correction procedure can be used to obtain the true throat size distribution from mercury injection pressures (Chatzis and Dullien, 1977, 1981; Larson and Morrow, 1981).

Numerous mathematical models for transport in porous media have been proposed, the main objective being to explain and correlate observed behavior. In recent years, significant advances in porous media modeling have come from application of percolation theory (Levine et al., 1977; Larson et al., 1981) and from improved analyses of network models (Chatzis and Dullien, 1977). Common features of input to these models are pore size distribution, which may involve distributions for both pore throat and pore body sizes, and some form of topological distribution, which determines accessibility. Serial sectioning of test specimens is being developed as a method of obtaining a topological description of pore space (Pathak et al., 1980).

In practice, the two most commonly measured rock properties are porosity and single-phase permeability, with air or brine as the fluid. The complex mineralogy of rock samples often causes difficulty in obtaining reproducible permeability measurements. Aqueous fluids, in particular, can interact with the rock minerals to cause pronounced changes in measured permeabilities and even complete plugging (Jones, 1964). Movement of fine particles within the test specimen may contribute to difficulty in obtaining definitive measurements of permeability. Such behavior can distort or even completely obscure other phenomena which one may wish to investigate.

Packings of sand and beads provide the most widely used form of synthetic porous media for carrying out displacement studies. One advantage of such media is that the microscopic pore size can be scaled, at least in a statistical sense, according to particle size. In work on enhanced recovery it is now quite widely accepted (but for reasons that are not particularly well defined) that many forms of displacement tests should be carried out on consolidated porous media for more realistic simulation of reservoir conditions. In the United States, the most commonly used test material in enhanced recovery is broadly classified as Berea sandstone quarried from an outcrop area in Ohio. This material can vary by more than an order of magnitude in permeability, but generally slabs from a given part of the quarry have fairly consistent properties. A procedure of firing the rock to vitrify clays and burn out organic material is a common, but by no means universal, practice. Even procedures for cleaning and re-use of cores (this provides one approach to the ideal of carrying out a series of tests on a core of constant pore geometry) are not generally agreed upon. Much of the development work of oil recovery processes has been carried out using Berea sandstone cores because of its availability. Testing of processes designed for specific reservoirs, however, usually includes work with reservoir core samples taken from the test region.

MICROSCOPIC ASPECTS OF DISPLACEMENT

Displacement of oil from the reservoir entails the flow of both oil and displacement fluid(s) through the pores of the rock. The wide variation of sizes, shapes, and connectivity of these pores leads to a similarly wide distribution in the velocity of flow through the pore space. It is neither possible nor desirable to know the details of this extremely complex velocity distribution. Usually, a few rather general features are enough to yield a useful understanding of the microscopic aspects of the flow and recovery processes.

One such feature is that the flow velocities are low. This and the fact that the pores themselves are also small means that the Reynolds number is, in most areas of the reservoir, considerably less than unity. Thus, despite the fact that the streamlines of flow are not straight but quite tortuous in their paths through the pore space, the inertial forces, which might otherwise arise from these frequent changes of direction, are nevertheless generally negligible in comparison with the viscous ones. As a consequence, the resistance to flow can be viewed as entirely viscous and, therefore, proportional to the flow rate — a fact observed on a macroscopic scale and memorialized as Darcy's Law (Heller, 1972; Larson, 1981). This proportionality between the flow rate and pressure gradient is also generally observed when the pore space is occupied by two immiscible phases, such as oil and water.

In low-Reynolds-number or creeping flow, the energy of flow is dissipated by viscous forces, and the detailed velocity distribution arranges itself under the existing local constraints set by the pore geometry to minimize the rate of energy dissipation for a given flow rate. Generally, in a porous medium, the flow velocities in the distribution that satisfies this requirement will be largest along streamlines leading through the centers of the individual pores, and zero along the pore walls. This generalization can be expected to hold everywhere except near the displacement or frontal region itself, where the fluid content or saturation may be changing fairly rapidly.

In the frontal region, the microscopic aspects of the flow can be markedly influenced by the interaction of the fluids involved. Leaving out of consideration liquids between which there are violent chemical reactions, one can classify fluid pairs as immiscible, miscible, or conditionally miscible. In each of the three cases, there are characteristic differences in the microscopic displacement processes which produce different macroscopic manifestations.

IMMISCIBLE DISPLACEMENT AND THE FRONTAL REGION

Truly immiscible fluids are always separated by a well-defined interface, the thickness of which is of molecular dimensions. These bounding surfaces between the fluids play a crucial role in the microscopic motions of flow and displacement in porous media, and are particularly influential in the frontal region.

At the microscopic level, their existence and shape are determined by the combined effects of pore geometry and wetting behavior of the occupant fluids with respect to the surfaces of the pore spaces. Generally, the condition of strong water-wetting is taken as the base case against which the effects of other wetting conditions are compared. The nature or consequence of capillary effects important to oil recovery are first considered for systems which are strongly wetted by water.

The velocity and pressure distributions, which arise in the vicinity of interfaces that define the frontal region in a porous medium, provide interesting contrast to the velocity distribution associated with a moving interface in a tube. For the tube, the normal range of velocities from zero at the tube wall to a maximum at the tube center obviously cannot be continued through the interface region (Dussan, 1979). The presence of this moving boundary will give rise to substantial components of velocity perpendicular to the direction of interface motion. At low Reynolds numbers and normal interfacial tensions, the ratio of viscous to capillary forces is such that there is very little change in shape of the meniscus from that given by a stationary interface. Phenomena of change in interface shape and dynamic contact angle related to high capillary numbers have been reported (Hansen and Toong, 1971; Hoffman, 1975).

In a porous medium, the movements of interfaces which define the frontal region do not, in general, progress smoothly as in a tube. At a given instant, a large proportion of interfaces are essentially stationary with respect to their boundaries and apart from small changes in curvature, the interfaces present temporary barriers to flow. Thus, modification of streamlines in the vicinity of stationary menisci at the front is necessarily more drastic than that considered previously for the simple tube. As water is injected, there is a tendency for the curvatures of the barrier interfaces to increase. Associated with this trend is a deterministic sequence by which some of the menisci become unstable and penetrate neighboring pore bodies. These spontaneous local motions were first reported in connection with studies of soil moisture and are known as Haines jumps (Haines, 1930; Miller and Miller, 1956). Motion of the interface is tempered by the need to supply wetting phase to the pore body and at the same time remove the non-wetting phase. On the wetting phase side of the interface, a pressure deficiency is created because of local differences in capillary pressure. These local pressure gradients in the frontal region far exceed those associated with the low-Reynolds-number flow that was responsible for initiating the instability.

As a result of the local demand for water, neighboring menisci will retreat slightly or at least show transient reduction in curvature. Thus, the low-Reynolds-number flow patterns behind the front become modified by the combined effects of barrier interfaces and transient regional demands for fluid in the frontal region. Detailed description and modeling of these displacement processes was recently given by Mohanty et al. (1980).

RESIDUAL OIL

Included in the capillary instabilities, which can occur at the flood front, are those which involve trapping of residual oil. In a general way, it is thus evident why the microscopic processes of immiscible displacement are not completely efficient, and, even in well-swept regions, leave residual amounts of the displaced phase. In looking at these processes with the view of understanding EOR, two major examples of immiscible displacement are considered here. The first is the displacement of residual oil trapped as isolated blobs, and the second is attainment of increased efficiencies of immiscible displacement through reduced trapping.

A problem common to many displacement experiments in oil recovery is realistic simulation of the original microscopic fluid distribution in the reservoir. Oil reservoirs form by accumulation of oil, originating from source rocks, in geologic traps where the tendency for oil to rise under buoyancy forces is opposed by a capillary seal provided by the reservoir cap-rock. As oil accumulates in the trap, water is displaced. A substantial fraction of water, however, is found to remain along with the oil at heights well above the transition zone between the oil and its underlying aquifer. For wells completed above the transition zone, it is commonly observed that only oil is produced and that the connate water is, therefore, immobile.

In the laboratory, the original condition in the trap of 100% water saturation is simulated by saturating a core specimen with brine. Displacement of water by oil is usually carried out by oil flooding the test specimen until production of water appears to have ceased, using sufficient viscous pressure gradient and reversal of flow direction to largely overcome, or at least smear out, capillary end effects (Kyte and Rapoport, 1958). At this stage, a substantial amount of water, commonly about 20—30% of pore space, is retained by the core. This non-flowing water is assumed to represent both the saturation and microscopic distribution of connate water in the reservoir.

Waterflooding of the reservoir is simulated in the laboratory by waterflooding the core at a low rate until production of oil ceases. At this condition, a substantial fraction of oil remains trapped in the interstices of the rock as isolated capillary structures commonly referred to as blobs or ganglia. Two trapping mechanisms have been identified. One is snap-off, whereby an oil bridge in a pore throat becomes unstable and ruptures (Roof, 1970). A second mechanism is final separation of an isolated oil blob by filling of a pore body — a process described as bypassing. Pore geometries in which either bypassing or snap-off can predominate have been demonstrated with the aid of two-dimensional physical models of pore space known as micromodels (Chatzis et al., 1982). High aspect ratio, the ratio of pore-body to pore-throat size, has been shown to cause snap-off with the resulting blobs occupying single pore bodies. The shape of such blobs is usually close to spherical. High aspect ratio generally leads to high residual oil, the only recovery during waterflooding coming from the pore throats. Larger oil blobs are more commonly associated with heterogeneities in pore structure and subtleties in the mechanism by which oil becomes bypassed at low aspect ratios. Size distributions have been determined for residual oil ganglia formed by an oil which could be polymerized in place to form a solid and then separated from the rock (Chatzis et al., 1982).

Displacement of water by oil is generally described as drainage, whereas displacement of oil by water as imbibition. Displacements analogous to the laboratory waterflood can be carried out using a membrane, permeable to one phase only, to control the displacement (Amyx et al., 1960). This technique permits capillary pressures in the core sample to be balanced by an externally imposed pressure difference between oil and water. When relationships between capillary pressure and water content can be measured for both drainage and imbibition conditions, a hysteresis is revealed. At any given saturation, both the oil and water phases can be viewed, for porous media having smooth surfaces at least, as consisting of continuous and discontinuous parts which vary with saturation and with the path by which a given saturation condition is reached (Morrow and Harris, 1965). At intermediate saturations, simultaneous flow of oil and water suggests the existence of networks of oil and water, which have continuity through the porous media. Discontinuous oil blobs and hydraulically isolated water structures can also exist at intermediate saturations. The variation in trapped residual saturation with water content was investigated by Raimondi and Torcaso (1964). After establishing a given saturation, the flowing oil phase was changed to a solvent of density and viscosity closely similar to those of the oil phase. Observed displacement behavior led to identification of three kinds of oil: normal, lagging, and trapped. The normal oil is displaced in a manner resembling single-phase miscible displacement. Thereafter, a small amount of the original oil, described as lagging oil, is produced along with the injected oil over a much longer time period. A third component, the trapped oil, remains in the core.

Lagging oil is assumed to be held in dead-end or dendritic structures, which do not lie in the arteries of oil flow. This oil is assumed to be recovered by diffusional exchange with the solvent. More detailed investigation of the lagging non-wetting phase has been reported by El-Sayed and Dullien (1977) for conditions of primary drainage, for a mercury—vacuum system. The lagging fraction is referred to as pseudo dead-end pore volume, its existence being due to the presence of the other fluid. Estimates of non-wetting phase, pseudo dead-end pore volume from network calculations, have been reported by Chatzis (1980). The possible existence of trapped oil that may be shielded from injected chemical banks by the presence of a flowing (continuous) water phase could have serious effect on the efficiency of oil displacement.

RESIDUAL OIL — MAGNITUDE

The amount of trapped oil remaining after waterflood is considered the single most important factor in the economics of tertiary oil recovery (Interstate Oil Compact Commission, 1978). For a given rock sample, laboratory displacement tests have shown that the oil saturation prior to waterflood has direct influence on the fraction of trapped oil. In general, the fraction of trapped oil increases as the initial water content decreases. Braun and Blackwell (1981) ascribe an observed residual oil saturation of over 50% in Berea sandstone (this is unusually high) to a low initial water saturation. The relationship between initial and final saturation of the non-wetting phase has been investigated in detail by Wardlaw and Taylor (1976) using mercury injection and ejection curves. This relationship is interpreted as the relative efficiency with which a given initial oil saturation would be displaced by waterflooding.

In laboratory simulation of tertiary recovery processes, the sensitivity of residual oil to initial water saturation requires that careful consideration be given to how the connate water condition is established. It is often observed that water saturations obtained by oil flooding in the laboratory are higher than those measured for the reservoir by more direct methods. Lower water saturations can be achieved in the laboratory by some combination of prolonged flooding, increased injection rates, or use of high-viscosity oil. Use of pressure membranes and, for small core samples, centrifuging, offer alternative means of achieving low water saturation. Even if water saturations (and wetting conditions) are matched, a precise duplication of fluid distribution in the reservoir is not assured because a given saturation can correspond to many possible fluid configurations. There is no evidence to date, however, that feasible variations in water distribution between the laboratory and reservoir system would have serious effect on the magnitude or distribution of waterflood residual oil.

In summary, the residual oil saturation after waterflooding strongly water-wet rocks at normal rates depends on the initial water saturation and the pore geometry. The importance of knowing the residual oil saturation, in a reservoir being considered for enhanced recovery, is apparent from the effort that has gone into its determination (Interstate Oil Compact Commission, 1978). It is ironic (but typical of many necessary compromises in reservoir engineering) that laboratory waterflood procedures commonly used in estimating waterflood recoveries and in establishing residual oil saturations in reservoir core samples prior to testing tertiary recovery processes are regarded as unreliable for purposes of estimating residual oil in the reservoir (Interstate Oil Compact Commission, 1978).

RESIDUAL OIL — MOBILIZATION

Residual oil blobs, once isolated, remain trapped by capillary forces. It is necessary to examine the nature of these trapping forces and the conditions under which they can be overcome. When the water phase surrounding the blob is stationary, the interface of the blob, except where bounded by the solid, has constant curvature. If there is a pressure gradient in the water phase, the downstream curvature of the blob will increase and the upstream curvature will decrease. As long as the boundary conditions provided by the solid and contact angle permit curvatures of the oil blob to adjust locally to the imposed pressure gradient, the oil remains trapped. When the gradient is sufficient to mobilize the blob, a drainage process (oil displacing water) occurs at the leading end of the blob and imbibition (water displacing oil) takes place at the trailing end (Melrose and Brandner, 1974). Once a critical ratio of viscous to capillary forces is exceeded, mobilization occurs (Taber, 1969). Subsequent behavior depends on many additional effects, which include coalescence with other blobs, break-up, aggregation, and further trapping (Stegemeier, 1977).

An empirical approach to determining how these processes ultimately affect oil recovery in toto is made through study of capillary number relationships. Capillary number is a generic term for the ratio of viscous-to-capillary forces. The numerous forms of capillary number used in the past were recently reviewed by Taber (1981). Definition of two of the most well-known forms is given here.

From Darcy's equation, the superficial or Darcy velocity is:

$$v = \frac{k_{rw} K_w}{\mu} \frac{\Delta P}{L} \qquad (3-2)$$

where K_w is the absolute permeability to water, k_{rw} is the relative permeability to water at a given oil saturation (in this discussion at residual or reduced residual oil saturation), μ is the viscosity of water, and $\Delta P/L$ is the pressure drop per unit length.

A common form of capillary number, $v\mu/\sigma$, was mentioned previously. Consideration of Darcy's law and conditions for blob mobilization demonstrate that relationships between capillary number and reduced residual oil saturation should give a unique relationship for systems of different permeabilities but similar pore geometry. From Darcy's law, the capillary number $v\mu/\sigma$ can be expressed as:

$$\frac{v\mu}{\sigma} = \frac{k_{rw} K_w}{\mu} \frac{\Delta P}{L} \qquad (3-3)$$

A simple theory, presented by Melrose and Brandner (1974), predicts the range of capillary number over which oil recovery can be expected. The difference in drainage and imbibition pressures, P_{dr} and P_{imb}, respectively, required for mobilization is estimated from measurements of capillary pressure hysteresis. Measurements and calculations of conditions for mobilization, made for systems of simple pore geometry, demonstrated that this corresponds to the maximum pressure for mobilization. The volume constraint on blob shape permits mobilization pressure differences to be lower by up to a factor of two than those given by the simple theory, but confirm the concept that leading and trailing curvatures control mobilization (Morrow, 1979; Oh and Slattery, 1976). Thus, equation (3-3) can be presented as follows:

$$\frac{v\mu}{\sigma} = \frac{k_{rw} K_w}{\mu} \frac{(P_{dr} - P_{imb})}{L_B} \qquad (3-4)$$

where L_B is a characteristic blob length. For random packings of equal spheres having a typical porosity of 0.38, the permeability in absolute units is given by:

$$K_w = 0.00317 R^2 \qquad (3-5)$$

where R is the sphere radius of randomly packed equal-size spheres. The difference in drainage and imbibition capillary pressures is related to surface curvatures:

$$P_{dr} - P_{imb} = 2\sigma \left(\frac{1}{r_{dr}} - \frac{1}{r_{imb}} \right) \tag{3-6}$$

where r_{dr} is the radius of curvature for drainage and r_{imb} is the radius of curvature for imbibition.

For a given sphere size, r_{dr} and r_{imb} are proportional to R; hence:

$$P_{dr} - P_{imb} \propto \frac{1}{R} \tag{3-7}$$

If the microscopic mechanism is dominated by capillarity, then the blobs formed in geometrically similar packings will be similar and a typical blob length L_B can be expressed as:

$$L_B = nR \tag{3-8}$$

where n is a constant from one packing to another. Also, from similarity considerations, relationships between k_{rw} and oil saturation should be independent of bead size.

Thus, from the way in which R enters the various terms, it follows that the capillary number is independent of radius and it may be expected that relationship between $v\mu/\sigma$ and reduced residual oil content will give a universal curve for systems of different permeability but similar geometry. This has been demonstrated using random packings of equal spheres as the porous media (Morrow and Chatzis, 1982).

Extensive measurements for consolidated media showed that capillary number relationships for a wide variety of rocks were fairly closely correlated, but differed distinctly from those for random packings of spheres (Chatzis and Marrow, 1981). For bead packs, which were consolidated by sintering, capillary number relationships fell much closer to those for consolidated sandstones (Morrow and Chatzis, 1982).

For a given system, the product $v\mu$ is proportional to pressure drop, which is limited at the injection well by fracture pressures. Thus, the only practical method of raising the capillary number for the reservoir to give improved recovery is by reducing interfacial tension to very low values. Surfactant systems which give low interfacial tensions have been formulated, but there are many difficulties in propagating a chemical slug, which can maintain low interfacial tensions as it travels through a formation. The difficulties relate to adsorption, mobility control, dispersion, ion exchange, and other effects of salinity, temperature, and reservoir heterogeneity. There is presently widespread research activity aimed at overcoming these difficulties.

PREVENTION OF TRAPPING

Reduced residual oil saturations can also be achieved in principle by carrying out the initial waterflood at high capillary number. In general, lower reduced residual oil saturations are achieved at a given capillary number when the displaced oil is continuous (Stegemeier, 1977; Chatzis and Morrow, 1981). This is probably related to differences in mechanisms for displacement of discontinuous and continuous oil. Whereas blob mobilization requires that differences between drainage and imbibition must be overcome, prevention of trapping of a potential blob requires imbibition to occur at the trailing edge ahead of the time when the trailing oil is cut off from the main flowing body (Morrow and Songkran, 1981). Measurements for sphere packs showed capillary numbers for mobilization to be about twenty times greater than those for prevention of trapping (Morrow and Chatzis, 1982). For consolidated media, however, the separation was much less, and at 50% or greater reduction in residual oil, results for the two initial conditions (residual and continuous oil) fell close together. Furthermore, the large differences in capillary numbers for mobilization and entrapment in sphere packs were greatly reduced by consolidating the bead packs (Morrow and Chatzis, 1982). This change was assumed to be caused by the reduction in aspect ratio that accompanies sintering. As aspect ratio increases, the amount of oil trapped in single pore bodies increases and prevention of trapping becomes more difficult to achieve.

BUOYANCY FORCES AND PREVENTION OF TRAPPING

Prevention of trapping in sphere packs has been studied under conditions where no viscous or gravity forces were involved in reduction of the amount of trapped oil (Morrow and Songkran, 1982). Gravity forces are expressed by the Bond number, $\Delta \rho g R^2 / \sigma$, which is the ratio of a microscopic, hydrostatic head (proportional to $\Delta \rho g R$) to the capillary force (proportional to σ / R). In displacement experiments over the range where both viscous and buoyancy forces had influence on the trapping mechanism, reduction in residual non-wetting phase was correlated with a linear combination of capillary number and Bond number. Because the effect of buoyancy on trapping is rate independent, the observed linearity in the effect of viscous and gravity forces shows that the effect of viscous forces on the microscopic trapping mechanism at the necessary higher flooding velocities can be ascribed to the effect of flowing pressure gradient on interface curvature.

WETTABILITY

Wettability has a dominant effect on the microscopic distribution of phases and can cause drastic changes in the displacement mechanisms discussed previously for completely wetted systems (Raza et al., 1968). Contact angle measurements between crude oils and brines at selected mineral surfaces suggest that wetting conditions ranging from strongly water-wet to strongly oil-wet may be encountered from one reservoir to another (Treiber et al., 1972). Although contact angles are almost universally accepted as a basic measure of wettability, their application to reservoir systems is limited, because measurements are not made directly on reservoir rock surfaces. The mineralogical complexity of reservoir rocks could cause wettability to vary from place to place on the rock surface on a scale ranging from submicroscopic to macroscopic. Furthermore, at a given part of the surface, contact angles may well differ for advancing and receding interfaces and, particularly when adsorption is involved, can be sensitive to contact time and rate of interface movement across the solid surface.

In practice, the most generally accepted method of taking wettability effects into account in waterflooding is through making relative permeability measurements on reservoir core samples using reservoir fluids at reservoir conditions of temperature and pressure. Procedures used during the course of these measurements for core handling, preservation, cleaning, and restoration to reservoir conditions are subjects in themselves and no particular methods have received universal acceptance. Numerous possible reasons can be suggested as to why reservoir conditions may not be restored. An extensive study of relative permeability and associated contact angles has been reported (Treiber et al., 1972). Shifts in relative permeability caused by changing from well-defined wetting conditions to reservoir conditions were for the most part qualitatively consistent with observed contact angles at mineral surfaces believed to be representative of the pore linings of the rock. It was concluded that actual identification of reservoir wettability was not needed provided cores were available which were representative of reservoir conditions.

There is some danger in this approach because of the frequently capricious nature of wetting phenomena. Wetting properties are mainly determined by the outermost layer of molecules and their orientation on the rock surface. For example, contact angles for reservoir fluids have been reported to be highly sensitive to contamination by trace amounts of copper and nickel ions and to products formed by oxidation of crude oil (Treiber et al., 1972). The extent to which precautions must be taken in setting up reservoir condition displacement experiments has not been fully established. Current explanations of the phenomena which cause observed wettability effects are qualitative in nature and not particularly amenable to rigorous testing. Reservoir condition displacement experiments are so difficult and costly to carry out that, in general, careful checks on reproducibility cannot be made. As a bare minimum, it seems that having completed a reservoir conditions test, a further set of measurements should be made on the same core under strong wetting conditions. Even qualitative assessment of wettability from relative permeability is far more satisfactory if, in addition to the general form of these curves, relative shifts can be identified.

Fundamental studies of the effect of wetting on displacement have been hampered by inability to obtain satisfactory wettability control using well-defined systems (Morrow et al., 1973). Use of low-energy solid as porous media, and air and a pure liquid as fluids is the most reliable approach used to date. Provided effects of surface roughness on contact angle were taken into account, much of the observed effects of wettability on displacement behavior were consistent with the independent measure of wettability provided by contact angles (Morrow and McCaffery, 1978).

In studying systems with contact angles ranging from 0° to 180°, three main classes of wetting behavior were identified according to imbibition behavior, the porous media being initially 100% saturated with the phase through which the contact angle is measured. This phase is usually the water phase in most reservoir situations but, because designation of a wetting phase is not always obvious, has been described as the reference phase for general purposes (Morrow and McCaffery, 1978). The three main classes are:

(1) Wetted systems — If drained, these systems spontaneously imbibe the reference phase.

(2) Intermediate systems — Neither fluid spontaneously displaces the other.

(3) Non-wetted systems — The reference phase will drain spontaneously.

Detailed discussion of the interrelationships of contact angle and displacement behavior for uniformly wetted systems has been given by Morrow and McCaffery (1978) and is drawn on where necessary in the following brief discussion of problems in wettability relevant to enhanced recovery.

WETTABILITY AND RESIDUAL OIL

The most important feature of intermediate systems is the reversal in sequence of pore filling for drainage and imbibition. For drainage, from considerations of capillary pressures, larger pores tend to drain before the smaller pores. On the other hand, for imbibition, because the imbibing fluid must be forcibly injected, it tends to invade the larger pores first. The resulting gross differences in fluid distributions for drainage and imbibition at a given saturation cause large hysteresis in drainage and imbibition relative permeabilities for both the reference and non-reference phases (McCaffery and Bennion, 1974; Morrow and McCaffery, 1978).

This behavior has interesting implications with respect to distribution of liquids in a reservoir of intermediate wettability. For example, on considering a reservoir at connate water saturation, the water tends to be held in the fine pores, whereas oil is present in the remaining pores, which can be described in broad terms as middle-sized and large. When such a rock is waterflooded, because imbibition is forced, the water will tend to displace oil from the larger pores and remaining oil after waterflood will tend to be held in the intermediate-sized pores rather than the larger pores, as in the case of water-flooding a water-wet rock. Essentially the same distribution of fluids would be given by the mixed-wettability condition described by Salathiel (1973) in which rock surface underlying connate water is water-wet, whereas that in contact with oil is oil-wet and remains unchanged by waterflooding. For the oil-wet pores the same sequence of invasion can be expected as for intermediate-wettability systems.

The displacement behavior of systems, which retain connate water in the smaller pores and do not spontaneously imbibe water or, perhaps, imbibe only a small fraction of water, is characterized by early breakthrough and an extended period of oil production at increasingly high water/oil ratios. Sala-thiel (1973) demonstrated that oil production is an exponentially decreasing function of volumes of water injected, and even after passage of almost 5000 pore volumes of water, no well-defined residual oil had been attained. Thus, it appears that when water occupies the smaller and larger pores in the rock with the oil accommodated somewhere in between, the oil is able to maintain limited continuity through the porous medium. The possibility exists that, in at least some individual pores, there are stable capillary structures which permit flow of both phases. Thus, successful attempts at mathematical modeling such behavior will likely require duality of flow in single pores.

WETTABILITY AND OIL RECOVERY

Although wettability control or its alteration has been proposed as a method of enhanced recovery, the subject is not well developed, and there is no general agreement as to the optimum wetting condition for oil recovery. Apart from Salathiel's (1973) pioneering work, the detailed mechanics of displacement in oil—water—rock systems which are other than strongly water-wet has received little attention.

In early literature, it was often suggested that trapping by capillary forces would be nullified and oil recovery would be maximized if the contact angle were 90°. The rationale for this conclusion was that capillary pressure is proportional to the cosine of the contact angle and, therefore, capillary trapping forces would be eliminated. Blob mobilization, however, involves advancing and receding interfaces, and hysteresis in contact angle can cause an increase in the forces required for mobilization. Measurements on single blobs in systems of well-defined geometry support this prediction (Morrow, 1979). Melrose and Brandner (1974) predicted that, even without contact angle hysteresis, the effect of interaction of contact angle and pore geometry would cause forces for mobilization to increase.

Reduction in capillary pressures through reduction of contact angle could conceivably be of advantage with respect to prevention of trapping. In comparing the mechanism of mobilization with that for prevention of trapping for water-wet systems, differences in drainage and imbibition capillary pressures were compared with conditions for change in mechanism of imbibition. The supplemental force required to change the imbibition mechanism so that imbibition occurs at the trailing edge of a blob rather than causing the blob to disconnect from the main body of oil is given by the imbibition capillary pressures of these two extreme possibilities. Any reduction of imbibition pressures as a whole, these being approximately proportional to the cosines of advancing angles, will reduce the supplementary pressure difference and, therefore, help in prevention of entrapment (Morrow, 1976).

With the proposed mechanism, the optimum wetting condition for prevention of trapping corresponds to when imbibition capillary pressures approach zero. However, when the capillary imbibition pressure changes sign, the system assumes intermediate wetting properties and imbibition of the displacing fluid is no longer spontaneous. This change in sign is accompanied by drastic change in pore filling sequence with the larger pores now being filled before the smaller.

A variety of wetting phenomena could conceivably occur in such diverse processes as micellar and carbon dioxide flooding, both of which are capable of giving more than two fluid phases. Contact angle measurements in micellar fluids have been reported by Reed and Healy (1979). They observed little hysteresis on smooth, high-energy surfaces and noted that interfacial tension passed through a minimum when the contact angle was 68°. Because lowering of interfacial tension is a crucial factor in micellar flooding, it may be that the wettability condition for optimum recovery is mainly dictated by the optimum formulation for low interfacial tension. Thus, the wettability condition that gives the best recovery is not necessarily the most desired wetting condition if all other factors were equal (Reed and Healy, 1979). The 68° figure at high-energy surfaces should be compared with the value of 62° for low-energy surfaces, because both systems show no hysteresis on smooth surfaces. If the effects of surface roughness on contact angle hysteresis in reservoir rock are comparable with those measured at low-energy surfaces, the figure of 68° may be slightly too high for the system to benefit from mechanisms suggested for weakly wetted systems.

The increased difficulties of displacing oil immiscibly from strongly oil-wet rocks are indicated by the pressure gradients required to reduce connate water saturation from a water-wet or intermediate-wet rock by viscous pressure gradients or centrifuging (Jenks et al., 1969). Thus, in a reservoir that is oil-wet by the definition given previously, i.e., that oil will imbibe spontaneously, residual oil will be held in the smaller pores, and the requirements for mobilization are similar to those for connate water in water-wet systems.

A class of wetting behavior, which is quite distinct from the somewhat ideal uniformly wetted systems and which is often observed for reservoir systems, is described as mixed wettability. Systems of mixed wettability are capable of imbibing water spontaneously at high oil saturation, but will imbibe oil spontaneously at high water saturation. Such behavior, which would possibly be related to time-dependent adsorption, is completely outside the scope of behavior observed for uniformly wetted media for which either only one of the phases or neither can spontaneously imbibe. The phenomena of mixed wettability can be detected by a widely used measure of wettability called the Amott test, which is based on the fractions of fluid observed to imbibe relative to forced displacement for either phase (Amott, 1959). Another test, which can be applied to such systems, involves capillary pressure relationships measured by centrifuge (Donaldson et al., 1969).

Given the myriad of possible variations in displacement mechanisms that can be caused by wetting effects, extensive work is still needed in order to gain a reasonable working knowledge of the significance of wettability in various enhanced recovery processes.

MISCIBLE DISPLACEMENT

A quite different microscopic situation is involved in the displacement of fully miscible fluids. In this case, there is no discrete interfacial barrier between the fluids that prevents the interpenetration of their molecules. On the other hand, the difference between the adjectives "miscible" and "mixed" must be borne in mind. Whereas the former tells of a possibility, the latter represents the accomplished fact.

In the boundary zone between two regions occupied by distinguishable, but miscible fluids, two mechanisms may be active in blending them. At least one of these, molecular diffusion, is always present, resulting from the normal thermal motion of the molecules. The statistical effect of these continual, random motions is to drive the average molecular species densities (the fluid concentrations) towards uniformity in the space available to them. The relaxation process is exponential, with a characteristic half-time T_h (in seconds) that is proportional to the square of a linear distance (Crank, 1956):

$$T_h = x^2/SD \tag{3-9}$$

where x is the size of the space in cm, S is a numerical factor (of magnitude about ten) dependent on the space's dimensionality and shape, and D is the molecular diffusivity in cm^2/s. The implications of the power two in the equation are enormous. For instance, whereas the ratio of sizes of a typical reservoir (e.g., 1 km) to that of a typical pore (e.g., 10 μm) is only 10^8, the square of that number is involved in comparing the diffusion times. The half-time for decay of a concentration non-uniformity in the pore would be about 10 ms, whereas for the decay of a similar pattern in the reservoir it would be more than 3 million years. In these calculations the following fact was not taken into consideration: diffusion is slowed somewhat further by the fine structure of the reservoir rock. This might prolong the reservoir's non-uniform pattern by an additional factor of ten.

For widely spaced bodies of miscible fluids to become mixed within a shorter time, they must be brought into more intimate contact. Though the fluid contents of the reservoir cannot be literally stirred, significant variations in velocity naturally accompany flow through porous media, as was described previously. They might be expected to lead to rapid mixing of displacing and displaced fluids. As it turns out, however, the effect of the large inter-pore velocity variations is almost completely eliminated as a factor in extending the "mixing zone" between the two fluids. The reason for this apparent paradox is that molecular diffusion is very effective in homogenizing the contents of the individual pores during the time interval it takes the fluid to traverse the pore. A similar inverse influence of the molecular diffusivity in restraining the longitudinal mixing of miscible fluids during displacement in a capillary tube has been described by Taylor (1953).

Idealized miscible displacement experiments have been performed to study the growth of the mixing zone. In these experiments a macroscopically uniform rock sample is used, and the displacing and displaced fluids are chosen to be not only miscible, but hydrodynamically identical. Under these circumstances the thickness of the mixing zone (in the direction of flow) increases with the square root of time. If this distance Δx (in cm) is measured between the 10% and 90% iso-concentration surfaces, it is found that (Aronofsky and Heller, 1957):

$$\Delta x_{90,10} = 3.62 \sqrt{\mathfrak{D} t} \tag{3-10}$$

where t is the elapsed time in seconds since Δx was zero (it can also be expressed as the flow distance divided by flow velocity), and $\mathfrak{D}$ is a dispersion coefficient in cm^2/s (same units as those of the molecular diffusivity). This dispersion coefficient is a function of the ratio between the convective and diffusive effects, the so-called microscopic Peclet number Pe:

$$Pe = \lambda v/D \tag{3-11}$$

where D is the molecular diffusivity (or diffusion coefficient in cm^2/s), λ is a characteristic microscopic distance in cm, and v is the fluid velocity in cm/s. According to Heller (1963), the ratio of $\mathfrak{D}$ to D can be given as a quadratic function of Pe:

$$\mathfrak{D}/D = 1/F\phi + a_1 Pe + a_2 Pe^2 \tag{3-12}$$

where ϕ is the fractional porosity of the rock and F the independently measurable formation resistivity factor. The coefficients a_1 and a_2 depend on the intra-pore statistics of the rock, with a_2 being so small that in many rocks its influence is hard to detect.

Dispersion in a more heterogeneous rock — especially in one with correlated non-uniformities such as stratifications — often does not follow the simple square-root relations given in equation (3-10). For such a rock, the initial growth of the dispersed zone is more likely to be linearly related to the flow distance, with later expansion then settling down to a square root dependence. In this case, however, the dispersion coefficient $\mathcal{D}$ would be given by a larger value of the coefficient a_1 (or of the "microscopic distance" λ) in the equation for $\mathcal{D}$. More complex diffusion models of such rocks involving less well-connected porosity called dead-end pores have also been proposed (Coats and Smith, 1964).

In any case, however, the course of a miscible displacement is seen to be quite different than that of an immiscible one. In particular, the basis of the high, microscopic displacement efficiency is apparent — molecular diffusion is able to bring into the flow channels all or most of the fluid to be displaced. Thus, no residual oil will be left behind after miscible displacement by a good solvent, except that which is present in essentially inaccessible pore space.

CONDITIONALLY MISCIBLE DISPLACEMENTS

In the case of conditionally miscible displacements, there are two distinct situations which are often combined in actual cases to yield even greater complexity. The first and simplest situation, which is discussed in some detail below, involves a pair of pure liquids which are partially miscible, whereas the second situation occurs when at least one of the liquids is not pure and consists of a number of components.

In the first case, the two pure fluids can be thought of as being incompletely soluble in each other. At a particular temperature and pressure, a mixture of two partially miscible fluids may separate into two phases, or may not, depending on the composition of the mixture. If the mixture is mostly A with only a small fraction of fluid B, it will be a single-phase fluid. Similarly, a mixture of mostly B with just a small fraction of A will also consist of only a single phase. But a mixture of partially miscible fluids, which consists of nearly equal portions of A and B, will separate into two phases. These two end-point mixtures will be of limiting compositions. The situation can be diagrammed along a line, with zero content of A (and 100% B) at the left end, and 100% A (and zero B) at the right end. Then the two limiting concentrations are at x' and x''. A value of x in the range from 0.0 to 1.00 represents a mixture concentration. If x lies between 0.0 and x', a single phase is formed, which is predominately composed of component B. If x lies in the range x'' to 1.00, the result is also single phase — this time mostly A.

But if x is between x' and x'', then the mixture will separate into two phases and the compositions are x' and x''.

If a fluid of composition x' displaces another, the composition of which is x'', from a porous medium, then the fluids will act just like strictly immiscible fluids. A residual will be left behind, in accordance with the usual considerations of the capillary number as are applicable to other immiscible displacements.

Other possibilities are available as well. A single-phase liquid mixture in the range $0.0-x'$ is perfectly miscible with any other mixture with composition in the same range. The members of such a pair, of compositions x_1 and x_2, can then displace each other miscibly in a porous medium, with no residual. The same thing is true for a pair of fluids x_3 and x_4, such that $x'' \leqslant x_3 < x_4 \leqslant 1.0$.

But the last possibility is the one which presents new phenomena. It occurs when the two fluids are x_5 and x_6, where $0.0 \leqslant x_5 < x'$ and $x'' < x_6 \leqslant 1.0$. If these fluids were mixed and shaken in a test tube or beaker, they would separate into new fluids (immiscible phases) of composition x' and x''. But a stable displacement of x_5 by x_6 in a porous rock is a somewhat different matter. Mixing occurs only in a narrow zone in the rock. The fluids are immiscible and form interfaces between themselves, which partake of the fluid motion as noted in the section "Immiscible displacement and the frontal region". But a new situation exists as well. Substantial mass transfer of the molecules of components A and B takes place across these interfaces, until the compositions of the fluids on opposite sides of each of the interfaces are x' and x''. Upstream from the interface, the composition of the flowing displacing fluid changes gradually from x' to x_5 (in the direction opposite to fluid flow), the concentration gradients being in accord with the dispersion coefficient discussed in the section "Miscible displacement". On the downstream side, away from the two-phase region, one could also expect a gradual concentration change from x'' to x_6.

One further characteristic of partially miscible displacements can also be pointed out here. As the displacement proceeds and fluid of composition x_5 continually enters the input end of the rock, it is apparent that the residual fluid (which is left after the immiscible displacement) having composition x' is soluble in the input fluid x_5. If x' and x_5 are close in composition, then the solubility may be quite small. Nevertheless, the residual phase will be dissolved away by the input fluid, and the displacement will eventually be complete.

The presence of such a dissolution front, behind which complete displacement occurs, is also observed in more complex, partially miscible floods involving ternary mixtures. Such displacements in which alcohols were used as mutual solvents of oil and water were extensively studied in the early years of EOR from both experimental and theoretical points of view (Gatlin, 1959; Taber et al., 1961). The relationship between displacement behavior and the variations in ternary phase diagram geometry, which are brought about by the substitution of one alcohol for another, was described by Taber and Meyer (1964).

In this manner a partially miscible displacement can be considered to be the general or intermediate case, of which the earlier discussed displacements (miscible and immiscible) are the limiting cases.

Further complications are introduced into conditionally miscible displacements, as noted at the beginning of this section, if a wide range of components is present in one of the fluids. Crude oil is of course such a mixture, and additional dimensions would be needed to describe all details of the fluid composition changes during a displacement. In addition to the possibility of slow dissolution of the residual phase, variations of the composition can occur in the flow direction. This is indeed what happens during displacement of crude oil by CO_2: the lower-molecular-weight hydrocarbon components are quite soluble in dense-phase CO_2 and are extracted from the crude to move with the displacing fluid (Gardner et al., 1981). As a result of this process, a miscible front is developed between the CO_2 and the crude oil (Orr et al., 1982). The pressure dependence of this miscibility development has been extensively studied by Yellig and Metcalfe (1980) and Holm and Josendal (1982). The process by which such a developed miscibility front forms, and the phase behavior requirements for it, were first set down in 1961 by Hutchinson and Braun (1961). These earlier considerations were in connection with miscible displacement projects in which mixtures of light hydrocarbon gases were used as displacement fluids. Even in the absence of first contact miscibility, such developed or multiple-contact miscibility can lead to a displacement process which can reach very high flooding efficiency at least on a microscopic scale, or one-dimensional basis, as in a slim tube displacement.

Although familiarity with the microscopic pore level processes, as discussed in the preceding paragraphs, is vital to an understanding of oil recovery processes, these same topics should also be viewed in a larger perspective. The production of crude oil has been economically attractive in the market place, because it is a large-scale activity that can justify the large investments required. Thus, in considering scientifically the basic technical problems of displacement, it is not sufficient to consider only their microscopic or small-scale aspects. In the next section, some of the same subjects are discussed from a larger-scale point of view. This is not only useful for the description of both reservoir-wide phenomena and of many laboratory experiments, but has also some intrinsic scientific interest.

MACROSCOPIC ASPECTS OF DISPLACEMENT

The central goal of a useful macroscopic description of displacement is to make possible the numerical prediction of the gross, measurable features of flood processes. Furthermore, it must be possible to perform these calculations without detailed considerations of the microscopic situation. The quantities of interest in the macroscopic view are, therefore, averages over rock volumes which contain large numbers of pores. These quantities are rock parameters like porosity, permeability, relative permeabilities, wettability, and residual oil and connate water saturations, and variables like fractional flow rates and saturation, which are treated as if they were point functions of space and time. In the classical macroscopic treatments, it is assumed, without catastrophic results, that these quantities are related by algebraic or differential equations. Darcy's equation and equations of mass balance and of fractional flow are used rather successfully to describe the macroscopic phenomena, without reference to the microscopic variations that lie behind them. Inasmuch as equations and their uses are the subject of elementary reservoir engineering courses, it is not necessary to repeat them here.

There are also macroscopic variations in the flow velocities and saturation distributions, which do need to be considered here. These variations have their effect, of interest in EOR, in reservoir-scale and in many laboratory-scale displacement situations. Such larger-scale variations occur in both space and time, and for several reasons:

(1) One such variation, which takes different forms in immiscible, miscible, and partially miscible displacements, is manifested in the stretching out of the saturation gradients in the direction of flow during the course of the displacement.

(2) Distinct from, but interacting with, these longitudinal changes are spatial variations in the flow variables as observed in a direction perpendicular to flow. Both of these affect the time variation of the ratio of displacing to displaced fluid production rates. The latter variations, however, have much more serious implications, because they cause early breakthrough of the displacing fluid and decrease the overall efficiency of displacement.

The variations in fluid velocity, which have been referred to rather generally above without consideration of their origin, have in fact two distinct, but interacting causative mechanisms. The first of these is geometric, including the effects of permeability inhomogeneities in the rock, stratification, and variations of reservoir thickness. These geometric factors also include the effect of well layout, which results in extreme variations in the magnitude and direction of fluid velocities between one part of the reservoir and another. All of these geometric causes of velocity variations are well described in the literature, as are their resultant effects on the production, or "breakthrough" curves (Muskat, 1949; Collins, 1976). They cause a stretching out of the post-breakthrough production curve, and increase the amount of injected fluid which must be produced along with the oil. In this sense, all of the geometric causes of velocity variations are unfavorable to the economic production of petroleum.

Distinct from, but interacting with the above, is a second class of causes of velocity variation. They are the result of hydrodynamic differences between the displaced and displacing fluids. In general, whenever there is a difference in flow properties across the displacement front, there is the possibility of coupling between the equation that describes the compositional distribution of the fluids and Darcy's equation which gives the velocities themselves. Because of this coupling, the pattern of fluid velocities is influenced by the distribution of fluid compositions. Simultaneously, though, the continual changes of the composition distribution result from the velocities of the fluids. This feedback can be positive or negative, resulting in either instability or "super-stability" of the frontal regions (Heller, 1963).

Chief among the fluid properties that can lead to such coupling are the viscosity and the density. Under unfavorable conditions, the resulting frontal instability is manifested in large-scale non-uniformities of the flow velocity. Instability results when a less viscous fluid displaces a more viscous miscible one, or when a denser fluid overlies a less dense one. In either of these cases, the non-uniform velocities at the front result in the formation and growth of "fingers" or channels along which the displacing phase moves at a higher flow rate than does the bypassed, displaced fluid in the spaces between them. The above picture is drawn for miscible fluids, where because of the lack of residual oil the situation can be described analytically in terms of linear differential equations (Perrine, 1961; Heller, 1966).

A similar situation occurs for immiscible displacements as well, with some modifications. Chief among these is the need to consider the effect of relative permeability. This need is resolved by the use of the mobility ratio instead of the inverse viscosity ratio as the indicator of the force driving frontal instability. Mobility of a phase is defined as the ratio of the effective permeability to the viscosity. The sum of the mobilities of the flowing phases is just the ratio of the total darcy velocities to the pressure gradient. Thus, for phase a:

$$\lambda_a = k_{ea}/\mu_a \tag{3-13}$$

and for the total mobility, λ_{tot} is equal to:

$$\lambda_{tot} = \lambda_a + \lambda_b = \frac{(Q_a + Q_b)/A}{\Delta P/L} \tag{3-14}$$

for a flow system of cross-sectional area A and length L, in which both oil and water are mobile.

The effective mobility ratio, M, for waterfloods is equal to:

$$M = \frac{\lambda_{displacing}}{\lambda_{displaced}} = \frac{k_{rw}}{k_{ro}} \frac{\mu_o}{\mu_w} \tag{3-15}$$

the frontal region in a horizontal flow system is unstable and will break up into fingers if $M > 1$. The initial growth of the fingered zone increases exponentially in time (Chuoke et al., 1959) and depends both on M and on other factors, such as the saturation change across the front and the average spacing of the fingers. After the initial transient period, however, the flow velocity of the displacing fluid within the fingers will exceed that of the displaced fluid between them by a factor equal to the effective mobility ratio M. The length of the fingered zone will then increase in direct proportion to the injected volume.

This generalization is useful for both miscible and immiscible displacements, and can be used to estimate the extent to which an adverse mobility ratio will aggravate sweep efficiency. Conversely, it can also be used to estimate the gains to be made in EOR by the use of mobility control techniques. Like the methods which can be used for mobility control, this question is one which needs further study.

SUMMARY AND CONCLUSIONS

The fundamentals of enhanced oil recovery involve many wide-ranging phenomena, some of which have been discussed only briefly in this chapter. These phenomena are all based on the slow flow of fluids in porous media in close proximity to solid or fluid boundaries, which dictate the details of the motion. Because all oil reservoirs contain water as well as oil, the study of the distribution of these two phases is important and has formed a principal part of this chapter. The two fluids and their interactions with the rock strongly influence the amount of the residual saturations of fluids, as well as the course of their changing saturations during displacement. The various enhanced oil recovery processes entail microscopic displacements which may be classified as immiscible, conditionally miscible, or miscible. In the former case, capillarity is usually dominant in the neighborhood of fluid—fluid interfaces and is responsible for distinctive microscopic and macroscopic patterns of flow and displacement. On the other hand, there is no such force in miscible displacement. In the latter case, viscosity plays a major role in determining the flow patterns even in the frontal region. Mass transport of different components by both convection and molecular diffusion are most important in miscible displacement, but may play a significant role in the other two situations as well.

Finally, some attention has been given in this chapter to questions of macroscopic variations in velocity. Such variations can be due both to geometric inhomogeneities and dimensional effects, and to the growth of flow instabilities caused by viscosity or density differences between the displaced and displacing fluids. From the point of view of EOR project design, it is most important to recognize these macroscopic effects and to minimize the effect of frontal instabilities during displacements in order to achieve recovery.

REFERENCES

Allen, T., 1968. Determination of pore size distribution. In: *Particle Size Measurement*. Chapman and Hall, London, 248 pp.

Amott, E., 1959. Observations relating to the wettability of porous rock. *Trans. AIME*, 216: 156—162.

Amyx, J.W., Bass, D.M., Jr. and Whiting, R.L., 1960. *Petroleum Reservoir Engineering*. McGraw-Hill, New York, N.Y., 610 pp.

Aronofsky, J.S. and Heller, J.P., 1957. A diffusion model to explain mixing of flowing miscible fluids in porous media. *Trans. AIME*, 210: 345—349.

Bond, D.C. (Editor), 1978. *Determination of Residual Oil*. Interstate Oil Compact Commission, Oklahoma City, Okla., 302 pp.

Braun, E.M. and Blackwell, R.J., 1981. A steady-state technique for measuring oil-water relative permeability curves at reservoir conditions. *56th Annu. Fall Technol. Conf., SPE-AIME, San Antonio, Texas, October 5—7*, SPE 10155: 10 pp.

Chatzis, I., 1980. *A network approach to analyze and model capillary and transport phenomena in porous media*. Ph.D. Thesis, University of Waterloo, Waterloo, Ont.

Chatzis, I. and Dullien, F.A.L., 1977. Modelling pore structure by 2D and 3D networks with application to sandstones. *J. Can. Pet. Technol.*, 16: 97—108.

Chatzis, I. and Dullien, F.A.L., 1981. Mercury porosimetry curves of sandstones. Mechanisms of mercury penetration and withdrawal. *Powder Technol.*, 29: 117—125.

Chatzis, I. and Morrow, N.R., 1981. Correlation of capillary number relationships for sandstones. *56th Annu. Fall Technol. Conf., SPE-AIME, San Antonio, Texas, October 5—7*, SPE 10114: 15 pp.

Chatzis, I., Morrow, N.R. and Lim, H.T., 1982. Magnitude and detailed structure of residual oil saturation. *3rd Joint SPE/DOE Symp. on Enhanced Oil Recovery, Tulsa, Okla., April 4—7*, SPE 10681: 20 pp. (*Soc. Pet. Eng. J.*, 23 (2): 311—326, 1983).

Chuoke, R.L., Van Meurs, P. and Van der Poel, C., 1959. The instability of slow, immiscible, viscous liquid-liquid displacements in permeable media. *Trans. AIME*, 216: 188—194.

Coats, K.H. and Smith, B.D., 1964. Dead-end pore volume and dispersion in porous media. *Soc. Pet. Eng. J.*, 4 (1): 73—84.

Collins, R.E., 1976. *Flow of Fluids Through Porous Materials*. Petroleum Publishing Company, Tulsa, Okla., 270 pp.

Craig, F.F., Jr., 1971. *The Reservoir Engineering Aspects of Waterflooding. SPE-AIME Monogr. Ser.*, 3: 134 pp.

Crank, J., 1956. *The Mathematics of Diffusion*. Clarendon, Oxford, 347 pp.

Donaldson, E.C., Thomas, R.D. and Lorenz, P.B., 1969. Wettability determination and its effect on recovery efficiency. *Soc. Pet. Eng. J.*, 9 (1): 13—20.

Dullien, F.A.L., 1979. *Porous Media: Fluid Transport and Pore Structure*. Academic Press, New York, N.Y., 306 pp.

Dullien, F.A.L. and Dhawan, G.K., 1974. Characterization of pore structure by a combination of quantitative photomicrography and mercury porosimetry. *J. Colloid Interface Sci.*, 47: 337—349.

Dussan V., E.B., 1979. On the spreading of liquids on solid surfaces: static and dynamic contact lines. *Annu. Rev. Fluid Mech.*, 11: 371—400.

El-Sayed, M.S. and Dullien, F.A.L., 1977. Investigation of transport phenomena and pore structure of sandstone samples. *28th Annu. Meet., Pet. Soc. CIM, Edmonton, Alta., May 30—June 3.*

Gardner, K.L., 1980. Impregnation technique using colored epoxy to define porosity in petrographic thin sections. *Can. J. Earth Sci.*, 17: 1104—1107.

Gardner, J.W., Orr, F.M., Jr. and Patel, P.D., 1981. The effect of phase behavior on CO_2-flood displacement efficiency. *J. Pet. Technol.*, 33 (11): 2067—2081.

Gatlin, C., 1959. *The miscible displacement of oil and water from porous media by various alcohols.* Ph.D. Thesis, Pennsylvania State University, University Park, Pa., 173 pp.

Haines, W.B., 1930. Studies in the physical properties of soil, Part V. *J. Agric. Sci.*, 20: 97—116.

Hansen, R.J. and Toong, T.Y., 1971. Dynamic contact angle and its relationship to forces of hydrodynamic origin. *J. Colloid Interface Sci.*, 37: 196—207.

Heller, J.P., 1963. The interpretation of model experiments for the displacement of fluids through porous media. *AIChE J.*, 9 (1): 452—459.

Heller, J.P., 1966. Onset of instability patterns between miscible fluids in porous media. *J. Appl. Phys.*, 37: 1566—1579.

Heller, J.P., 1972. Observations of mixing and diffusion in porous media. In: *Proceedings, Second Symposium on Fundamentals of Transport Phenomena in Porous Media.* IAHR-ISSS, University of Guelph, Guelph, Ont., pp. 1—26.

Hoffman, R.L., 1975. A study of advancing interface, 1. Interface shape in liquid-gas systems. *J. Colloid Interface Sci.*, 50: 228—240.

Holm, L.W. and Josendal, V.A., 1982. Effect of oil composition on miscible-type displacement by carbon dioxide. *Soc. Pet. Eng. J.*, 22 (1): 87—98.

Hutchinson, C.A. and Braun, P.H., 1961. Phase relations of miscible displacement in porous media. *AIChE J.*, 7 (1): 64—74.

Jenks, L.H., Huppler, J.D., Morrow, N.R. and Salathiel, R.A., 1969. Coring for reservoir connate water saturations. *J. Pet. Technol.*, 21 (8): 932.

Jones, F.O., 1964. Influence of chemical composition of water on clay blocking of permeability. *J. Pet. Technol.*, 16 (4): 441—446.

Kyte, J.R. and Rapoport, L.A., 1958. Linear waterflood behavior and end effects in water-wet porous media. *J. Pet. Technol.*, 10 (10): 47—50.

Larson, R.G., 1981. Derivation of generalized Darcy equations for creeping flow in porous media. *I&EC Fundamentals*, 20: 132—137.

Larson, R.G. and Morrow, N.R., 1981. Effects of sample size on capillary pressures in porous media. *Powder Technol.*, 30: 123—138.

Larson, R.G., Davis, H.T. and Scriven, L.E., 1981. Displacement of residual nonwetting phase from porous media. *Chem. Eng. Sci.*, 36: 57—73.

Levine, S., Reed, P., Shutts, G. and Neale, G., 1977. Some aspects of wetting/dewetting of a porous medium. *Powder Technol.*, 17: 163—181.

McCaffery, F.G. and Bennion, D.W., 1974. The effect of wettability on two-phase relative permeabilities. *J. Can. Pet. Technol.*, 13: 42.

Melrose, J.C. and Brandner, C.F., 1974. Role of capillary forces in determining microscopic displacement efficiency for oil recovery by waterflooding. *J. Can. Pet. Technol.*, 13 (4): 54—62.

Miller, E.E. and Miller, R.D., 1956. Physical theory for capillary flow phenomena. *J. Appl. Phys.*, 27: 324—332.

Mohanty, K.K., Davis, H.T. and Scriven, L.E., 1980. Physics of oil entrapment in water-wet rock. *55th Annu. Fall Technol. Conf., SPE-AIME, Dallas, Texas, September 21—24,* SPE 9406: 16 pp.

Morrow, N.R., 1976. Capillary pressure correlations for uniformly wetted porous media. *J. Can. Pet. Technol.*, 15 (4): 49—69.

Morrow, N.R., 1979. Interplay of capillary, viscous and buoyancy forces in the mobilization of residual oil. *J. Can. Pet. Technol.*, 18 (3): 35—46.

Morrow, N.R. and Harris, C.C., 1965. Capillary equilibrium in porous materials. *Soc. Pet. Eng. J.*, 5 (1): 15—24.

Morrow, N.R. and McCaffery, F.G., 1978. Displacement studies in uniformly wetted porous media. In: J.F. Padday (Editor), *Wetting, Spreading and Adhesion*. Academic Press, New York, N.Y., pp. 289—319.

Morrow, N.R. and Songkran, B., 1981. Effect of viscous and buoyancy forces on nonwetting phase trapping in porous media. In: D.O. Shah (editor), *Surface Phenomena in Enhanced Oil Recovery*. Plenum, New York, N.Y., pp. 387—411.

Morrow, N.R. and Chatzis, I., 1982. Measurement and correlation of conditions for entrapment and mobilization of residual oil. *Annu. Rep., U.S. Dep. Energy*, Rep. No. DOE/BC/10310-20, 59 pp.

Morrow, N.R., Cram, P. J. and McCaffery, F.G., 1973. Displacement studies in dolomite with wettability control by octanoic acid. *Soc. Pet. Eng. J.*, 13 (4): 221—232.

Muskat, M., 1949. *Physical Principles of Oil Production*. McGraw-Hill, New York, N.Y., 922 pp.

Oh, S.G. and Slattery, J.C., 1976. Interfacial tension required for significant displacement of residual oil. In: *Proceedings, ERDA Symposium on Enhanced Oil and Gas Recovery, 1. Oil.* pp. D2/1—D2/29.

Orr, F.M., Jr., Silva, M.K., Lien, C.L. and Pelletier, M.T., 1982. Laboratory experiments to evaluate field prospects for CO_2 flooding. *J. Pet. Technol.*, 34: 888—898.

Pathak, P., Winterfeld, P.H., Davis, H.T. and Scriven, L.E., 1980. Rock structure and transport therein: unifying with Voronoi models and percolation concepts. *SPE/DOE Symp. on Improved Oil Recovery, Tulsa, Okla., April 20—23*, SPE 8846: 17 pp.

Perrine, R.L., 1961. The development of stability theory for miscible liquid-liquid displacement. *Soc. Pet. Eng. J.*, 1 (1): 17—25.

Raimondi, P. and Torcaso, M.A., 1964. Distribution of the oil phase obtained upon imbibition of water. *Soc. Pet. Eng. J.*, 4 (1): 49—55.

Raza, S.H., Treiber, L.E. and Archer, D.L., 1968. Wettability of reservoir rocks and its evaluation. *Prod. Mon.*, 33 (4): 2—7.

Reed, R.L. and Healy, R.N., 1979. Contact angles for equilibrated microemulsion systems. *59th Annu. Fall Meet., SPE-AIME, Las Vegas, Nev., September 23—26*, SPE 8262: 10 pp.

Roof, J.G., 1970. Snap-off of droplets in water-wet cores. *Soc. Pet. Eng. J.*, 10 (1): 85—90.

Salathiel, R.A., 1973. Oil recovery by surface film drainage in mixed-wettability rocks. *Trans. AIME*, 225: 1216—1224.

Stegemeier, G.L., 1977. Mechanisms of entrapment and mobilization of oil in porous media. In: *Improved Oil Recovery by Surfactant and Polymer Flooding*. Academic Press, New York, N.Y., pp. 55—91.

Taber, J.J., 1969. Static and dynamic forces required to remove a discontinuous oil phase from porous media containing both oil and water. *Soc. Pet. Eng. J.*, 9 (1): 3—12.

Taber, J.J., 1981. Research on enhanced oil recovery, past, present and future, In: D.O. Shah (Editor), *Surface Phenomena in Enhanced Oil Recovery*. Plenum, New York, N.Y., pp. 13—52.

Taber, J.J. and Meyer, W.K., 1964. Investigations of miscible displacements of aqueous and oleic phases from porous media. *Soc. Pet. Eng. J.*, 4 (1): 37—48.

Taber, J.J., Kamath, I.S.K. and Reed, R.L., 1961. Mechanism of alcohol displacement of oil from porous media. *Soc. Pet. Eng. J.*, 1 (1): 195—212.

Taylor, G.I., 1953. Dispersion of soluble matter in solvent flowing slowly through a tube. *Proc. R. Soc. London, Ser. A*, 219: 186—203.

Treiber, L.E., Archer, D.L. and Owens, W.W. 1972. Laboratory evaluation of the wettability of fifty-five oil producing reservoirs. *Soc. Pet. Eng. J.*, 12 (6): 531—540.

Underwood, E.E., 1970. *Quantitative Stereology*. Addison-Wesley, Reading, Mass., 274 pp.

Wardlaw, N.C., 1976. Pore geometry of carbonate rocks as revealed by pore casts and capillary pressure. *Bull. Am. Assoc. Pet. Geol.*, 60: 245—257.

Wardlaw, N.C. and Taylor, R.P., 1976. Mercury capillary pressure curves and the interpretation of pore structure and capillary behavior in reservoir rocks. *Bull. Can. Pet. Geol.*, 24 (2): 225—262.

Yellig, W.F. and Metcalfe, R.S., 1980. Determination and prediction of CO_2 minimum miscibility pressures. *J. Pet. Technol.*, 32 (1): 160—168.

Chapter 4

RELATIVE PERMEABILITIES

D.N. SARAF and F.G. McCAFFERY

INTRODUCTION

   Production from petroleum reservoirs under primary, secondary, or tertiary processes usually involves the simultaneous flow of two or more fluids. Multi-phase flow, particularly three-phase flow, is not well understood and has not been adequately described analytically, even for pipeline flow. With natural porous media that generally have complex geometry, a microscopic description of the multi-phase fluid flow process is not possible. Empirical macroscopic descriptions based on Darcy's (1856) pioneering work, which relates fluid velocity to pressure gradient and viscosity of the fluid through a constant called permeability, have, however, found good utility with petroleum engineers for making the needed fluid-flow calculations. Multiphase flow of fluids through porous media can be related to a relative permeability of each phase, fluid viscosities, pressure drop, capillary pressure, and permeability. Of these, relative permeabilities are the least understood and the most difficult quantities to measure.
   Assuming that accurate methods of measuring relative permeabilities are available, it is often considered worthwhile to use the laboratory data obtained on "representative" reservoir rock samples with reservoir fluids and to stimulate reservoir conditions of temperature and pressure as closely as possible. As discussed later, however, laboratory techniques are not straightforward and are subject to a variety of uncertainties. Relative permeability determinations involve specialized equipment and procedures, particularly for three-phase flow situations. It is often difficult to maintain the core samples in their reservoir state because of contact with drilling muds during coring, and changes of surface characteristics during cooling, depressuring, and handling. Alternatively, petroleum engineers have used information obtained from past reservoir performance to make future predictions by extrapolation. However, because conditions of saturation history and fluid flow behavior generally differ for primary, secondary, and tertiary recovery regimes, this past experience is often of limited value for the projection of performance with a new recovery method. Another way of achieving the information desired for design purposes is to use data published on similar porous media, but the assumption of similarity is again open to question. Finally, if mathematical models were available to reliably predict the relative

76

NOMENCLATURE

Symbols

A	area of cross-section
c	geometric constant in Kozeny equation
f	fractional flow
g	acceleration due to gravity
I_r	relative injectivity = $(q_{w1}/\Delta P)/$ $(q_{w1}/\Delta P)$ at start of injection
k	permeability (absolute)
k_g, k_o, k_w	effective permeability to gas, oil, and water, respectively
k_{ro}	three-phase oil relative permeability
k_{rg}	three-phase gas relative permeability
k_{rw}	three-phase water relative permeability
$k_{rw,nw}$	relative permeability to wetting phase (two-phase)
$k_{rnw,w}$	relative permeability to non-wetting phase (two-phase)
$k_{ro,w}$	relative permeability to oil (two-phase)
$k_{rw,o}$	relative permeability to water (two-phase)
L	length of core
N_B	Bond number
N_c	capillary number
ΔP	pressure drop
P_c	capillary pressure
P_b	a constant
q	volumetric flow rate
Q_w	cumulative water injection in pore volumes
r	radius of capillary
R	particle radius
s	specific surface area per unit of pore volume
s_o	specific surface exposed to the fluid per unit of grain volume
S	fluid saturation
S_{om}	minimum residual oil saturation
S_{wi}	irreducible wetting phase saturation
S_{hr}	residual hydrocarbon saturation
S_{oe}	effective oil saturation = $(S_o - S_{or})/(1 - S_{or})$

S_{or}	residual oil saturation
S_{of}	flowing oil saturation = $S_o - S_{ot}$
S_{ot}	trapped oil saturation
S_{Lr}	residual liquid saturation
S_{gt}	trapped gas saturation
S_{ot}^*	reduced trapped oil saturation = $S_{ot}/(1 - S_{wi})$
S_{gt}^*	reduced trapped gas saturation = $S_{gt}/(1 - S_{wi})$
S_{of}^*	reduced flowing oil saturation = $(S_o - S_{ot})/(1 - S_{wi})$
S_{wf}^*	reduced flowing water saturation = $(S_w - S_{wi})/(1 - S_{wi})$
S_{gf}^*	reduced flowing gas saturation = $(S_g - S_{gt})/(1 - S_{wi})$
S_o^*	reduced oil saturation = $(S_o - S_{om})/(1 - S_{wi} - S_{om})$
S_w^*	reduced water saturation = $(S_w - S_{wi})/(1 - S_{wi} - S_{om})$
S_g^*	reduced gas saturation = $(S_g - S_{gi})/(1 - S_{wi} - S_{om})$
t	time
V	superficial velocity
θ	contact angle
ϕ	porosity
ρ	density
λ	a constant
σ	interfacial tension
μ	viscosity
τ	tortuosity factor

Subscripts

1	inlet of core
2	outlet of core
av	average
g	gas
o	oil
w	water, wetting phase
nw	non-wetting phase
L	liquid
i	ith capillary
ow	oil/water
imb	imbibition
r	relative, residual
t	total

permeability characteristics of a given reservoir rock by making use of easily measured properties of the rock in question, the purpose would be well served. A discussion of this approach, involving an examination of mathematical models and their capabilities, follows.

MATHEMATICAL MODELS FOR PREDICTING TWO- AND THREE-PHASE RELATIVE PERMEABILITIES

Navier-Stokes equations, along with the equation of continuity and appropriate initial and boundary conditions, give a complete description of flow of Newtonian fluids. However, the equations are coupled, non-linear partial differential equations and cannot be solved in general. Moreover, porous media have a very complex geometry requiring boundary conditions to be specified at all grain surfaces and at interfaces between fluids (multiphase flow), which are unknown, making the task impossible. Different workers have, therefore, proceeded to idealize the porous medium in some way to enable simplification of the Navier-Stokes (N-S) equations to find a solution. Kozeny (1927) considered a porous medium to be an ensemble of flow channels of various cross-sections but of the same length. Solving N-S equations for all flow channels and using the concept of hydraulic radius (related to specific surface area*, s, through ϕ/s where ϕ is the porosity of the medium), he obtained:

$$k = \frac{c\phi^3}{\tau s^2} \tag{4-1}$$

where k = permeability, and c = geometric constant which varies between 0.50 and 0.67. The tortuosity factor, τ, in fact was introduced later in Kozeny equation and was taken to be $(L_e/L)^2$, where L_e is the effective path length a fluid element has to traverse in a porous medium of physical length L. Although fundamental in nature, the Kozeny equation in its final form contains three parameters, c, τ, and s, which are extremely hard to determine independently to provide an experimental verification. Carman (1937) modified the Kozeny equation to:

$$k = \frac{\phi^3}{5\,s_o^2\,(1-\phi)^2} \tag{4-2}$$

where s_o is the specific surface area exposed to the fluid per unit of grain volume. This is the so-called Kozeny-Carman equation with essentially

---

* Specific surface area = area of rock surfaces facing the pore space per unit of pore volume.

similar limitations except that the unknowns have been reduced to one. Wyllie and Gardner (1958) developed a generalized Kozeny-Carman equation by dropping the concept of hydraulic radius, ϕ/s, as a fundamental property of porous media. An excellent review of work on the Kozeny theory by various researchers, and its criticism, has been given by Scheidegger (1974).

Another approach is related to drag theory where use is made of N-S equations to estimate the drag of the fluid on the walls of the media, and this in turn is equated to the resistance to flow (μ/k in Darcy's law). The geometrical bodies causing drag have been looked upon as either fibers or spheres. Both of these approaches lead to final equations which resemble Darcy's law but have additional terms indicating that Darcy's law is only a limiting equation, valid for low permeabilities (Happel and Brenner, 1965). Although Darcy's law was found empirically, one can derive a similar expression from N-S equations by dropping inertia and time-dependent terms (for derivation, see King Hubbert, 1956). But, when inertial forces are comparable to viscous forces, the validity of Darcy's law begins to break down. Fortunately, most fluid flows of practical interest to petroleum engineers fall in the category where Darcy's law is applicable.

The simplest picture of a porous medium is a bundle of uniform capillary tubes parallel to the direction of flow. The N-S equations, in one-dimension form for steady incompressible flow, is reduced to the Hagen-Poiseuille law applicable to a single capillary tube:

$$q_i = \frac{\pi r_i^4 \, \Delta P}{8 \mu L} \tag{4-3}$$

where q_i = volumetric flow rate; r_i = radius of the capillary; ΔP = pressure drop across the capillary; μ = viscosity of the fluid; and L = length of the capillary. In a porous medium, however, all pores are not of equal size and a pore size distribution exists. To overcome this constraint, Purcell (1949) introduced the concept of capillary pressure which is a measure of pore size distribution. Capillary pressure for a single capillary is given by:

$$P_{ci} = \frac{2\sigma \cos \theta}{r_i} \tag{4-4}$$

where P_{ci} = capillary pressure; σ = interfacial tension; θ = contact angle; and r_i = radius of the capillary. Substitution for r_i from equation (4-4) in (4-3) gives:

$$q_i = \frac{V_i \Delta P}{2 \mu L^2} \left(\frac{\sigma \cos \theta}{P_{ci}} \right)^2 \tag{4-5}$$

Here, use has been made of the fact that capillary volume = $V_i = \pi r_i^2 L$. Assuming there are dN capillaries of radius between r_i and $r_i + dr_i$, and integrating over the entire range of capillaries, the total flow is:

$$q = \frac{P(\sigma \cos \theta)^2}{2\mu L^2} \int_0^N \frac{V_i}{P_{ci}^2} \, dN \qquad (4\text{-}6)$$

Saturation of a liquid, S, in a bundle of capillaries can be defined as a ratio of volume of filled capillaries to total void volume, i.e.:

$$dS = \frac{V_i \, dN}{V} = \frac{V_i \, dN}{\phi AL}$$

where the volume of all pores is ϕAL, with A being the area of the cross-section. Equation (4-6) can now be written as:

$$q = \frac{\Delta P \phi A (\sigma \cos \theta)^2}{2\mu L} \int_0^1 \frac{dS}{P_{ci}^2} \qquad (4\text{-}7)$$

Comparison of this with Darcy's law yields:

$$k = \frac{(\sigma \cos \theta)^2 \phi}{2} \int_0^1 \frac{dS}{P_c^2} \qquad (4\text{-}8)$$

P_c has been used instead of P_{ci}, which was the capillary pressure in the capillary of radius r_i. One can visualize that, in an actual porous medium, the overall capillary pressure will be determined by the smallest capillary because it will have the highest pressure. Equation (4-4), therefore, represents capillary pressure in a porous medium due to all capillaries of radii greater than or equal to r_i. Consequently, this pressure is a measurable quantity.

Purcell (1949) introduced an adjustable parameter, the lithology factor F, in the above expression to account for the non-circular nature of capillaries and other non-idealities of a real porous medium. This factor F was adjusted to obtain a match with the experimental results. It is analogous to the tortuosity factor in the sense that both are adjustable parameters and account for deviations of real porous media from assumed models. The integral term in Purcell's model could be viewed as related to the square of the hydraulic radius in terms of the Kozeny theory. Burdine et al. (1950) obtained an expression similar to Purcell's, but assumed a saturation-dependent tortuosity factor which remained inside the integral (equation (4-8)).

Several other capillary models have been proposed in addition to the above discussed parallel type model and these include a series type model

(Scheidegger, 1953) and network models (Fatt, 1956; Schopper, 1966). The reader is referred to Scheidegger (1974), Corey (1977), and Dullien (1979) for details.

Statistical theory has been used to describe flow of fluids through porous media using both random walk and random media models (see Scheidegger, 1974), but with limited success. These models primarily considered single-phase flow and were never extended to cover multi-phase regimes because of inherent limitations of assumed analogies.

Although, as mentioned above, several models and methods of attack have been attempted to describe flow of a single fluid, multiple-phase flow has been studied to a lesser extent. The only attempts in this direction are extensions of Purcell's parallel type capillaric model. Darcy's law for each phase results in:

$$q_w = \left(\frac{k_w A}{\mu_w} \right) \left(\frac{\Delta P_w}{L} \right) , \quad \text{and} \quad q_{nw} = \left(\frac{k_{nw} A}{\mu_{nw}} \right) \left(\frac{\Delta P_{nw}}{L} \right) \tag{4-9}$$

where the subscript "w" refers to wetting phase and "nw" refers to non-wetting phase. It is assumed here that, of the two phases flowing, one wets the rock surface preferentially over the other phase. Defining:

$$k_{rw,nw} = \frac{k_w}{k}, \quad \text{and} \quad k_{rnw,w} = \frac{k_{nw}}{k} \tag{4-10}$$

as the two-phase relative permeabilities to the wetting and non-wetting phases, respectively, and making use of Purcell's model (Gates and Lietz, 1950):

$$k_{rw,nw} = \frac{\int_0^{S_w} dS_w / P_c^2}{\int_0^1 dS_w / P_c^2} \tag{4-11}$$

Fatt and Dykstra (1951) suggested that the tortuosity factor, τ, would be dependent on the saturation and hence radius of the capillary and, as such, should be inside the integral of Purcell's equation. They assumed τ to be inversely proportional to radius of the capillary and obtained:

$$k_{rw,nw} = \frac{\int_0^{S_w} dS_w / P_c^3}{\int_0^1 dS_w / P_c^3} \tag{4-12}$$

which, they claimed, fitted their own as well as Gates and Lietz data better. Burdine (1953), while extending the earlier work (Burdine et al., 1950) to two-phase flow, observed that the tortuosity factor depends on the extent of wetting phase saturation and could be approximated as:

$$\tau_w = \left(\frac{S_w - S_{wi}}{1 - S_{wi}} \right)^2 \tag{4-13}$$

where S_{wi} = irreducible wetting phase saturation. The wetting phase relative permeability, then, will be given by:

$$k_{rw,nw} = \left(\frac{S_w - S_{wi}}{1 - S_{wi}} \right)^2 \frac{\int_0^{S_w} dS_w/P_c^2}{\int_0^1 dS_w/P_c^2} \tag{4-14}$$

This expression seemed to fit their experimental wetting phase data reasonably well. However, defining a similar tortuosity factor for the non-wetting phase as:

$$\tau_{nw} = \left(1 - \frac{S_w - S_{wi}}{S_m - S_{wi}} \right)^2 \tag{4-15}$$

where S_m is the lowest wetting phase saturation for which the non-wetting tortuosity is infinite (Corey, 1954), the non-wetting phase relative permeability is given as:

$$k_{rnw,w} = \left(1 - \frac{S_w - S_{wi}}{S_m - S_{wi}} \right)^2 \frac{\int_0^{S_w} dS_w/P_c^2}{\int_0^1 dS_w/P_c^2} \tag{4-16}$$

This equation did not fit the observed data as well as that for the wetting phase (equation 4-14). Rose (1949) and Rose and Bruce (1949), while working with Kozeny theory, showed that capillary pressure was a measure of fundamental characteristics of a rock and could also be used to predict the relative permeabilities. Rose and Bruce also developed a multi-core displacement cell for measuring capillary pressure curves.

Wyllie and co-workers (Wyllie and Spangler, 1952; Wyllie and Gardner,

1958) claimed to have developed a modified capillaric model, the so-called generalized Kozeny-Carman model, where the bundle of capillaries of Purcell was cut in a large number of thin slices with the pieces of tubes in each slice randomized and then the slices reassembled. However, it is not clear how this modification of the model leads to an improved expression for relative permeability since their final expression for k_{rw} resembles that of Burdine (1953) except that the lower limit of integration has been changed from 0 to S_{wi} by defining an effective porosity $\phi_e = \phi(1 - S_{wi})$ and visualizing that the irreducible wetting phase saturation is part of the rock matrix. The three-phase relative permeabilities to oil and gas are given by equations (4-17) and (4-18), respectively:

$$k_{ro} = \left(\frac{S_L - S_w}{1 - S_{wi}} \right)^2 \frac{\int_{S_w}^{S_L} dS_L / P_c^2}{\int_{S_{wi}}^{1.0} dS_L / P_c^2} \tag{4-17}$$

$$k_{rg} = \left(\frac{1 - S_L}{1 - S_{wi}} \right)^2 \frac{\int_{S_L}^{1.0} dS_L / P_c^2}{\int_{S_{wi}}^{1.0} dS_L / P_c^2} \tag{4-18}$$

Wyllie and Spangler (1952) developed an electrical resistivity method to estimate the tortuosity factor on the assumption that the resistance to flow of a fluid is similar to resistance to flow of electricity.

Corey (1954) observed, based on experiments on capillary pressure—oil desaturation measurements, that the expression:

$$\frac{1}{P_c^2} = \begin{cases} \dfrac{c(S_o - S_{or})}{1 - S_{or}} & \text{for } S_o > S_{or} \\[4mm] 0 & \text{for } S_o < S_{or} \end{cases} \tag{4-19}$$

was generally a good approximation. Based on this relationship, equation (4-16) could be integrated for gas—oil two-phase flow to give the drainage relative permeability as:

$$k_{rg,o} = \left(1 - \frac{S_o - S_{or}}{1 - S_{or}}\right)^2 \left[1 - \left(\frac{S_o - S_{or}}{1 - S_{or}}\right)^2\right] \qquad (4\text{-}20)$$

$$= (1 - S_{oe})^2 \ (1 - S_{oe}^2)$$

and:

$$k_{ro,g} = \left(\frac{S_o - S_{or}}{1 - S_{or}}\right)^4 = S_{oe}^4 \qquad (4\text{-}21)$$

where S_{oe} (the effective oil saturation) $= (S_o - S_{or})/(1 - S_{or})$ and S_m has been taken as equal to one. This provides a relationship between relative permeability to gas and to oil. Since gas relative permeability may be determined comparatively easily using a stationary liquid method (to be discussed later), the above analysis permits calculation of the relative permeability to oil through the use of equations (4-20) and (4-21). Corey et al. (1956) extended the above analysis to three-phase relative permeabilities by assuming that equation (4-19), for three-phase flow, could be approximated as:

$$\frac{1}{P_c^2} = \begin{cases} c \, (S_L - S_{Lr}) & \text{for} \ \ S_L > S_{Lr} \\ 0 & \text{for} \ \ S_L \leqslant S_{Lr} \end{cases} \qquad (4\text{-}22)$$

The three-phase oil relative permeability is then given by:

$$k_{ro} = \frac{(S_L - S_w)^3}{(1 - S_{Lr})^4} \ (S_w + S_L - 2S_{Lr}) \qquad (4\text{-}23)$$

It is well-known that direction of change of saturation is a very important parameter in multi-phase flow of fluids through porous media. While the above expressions can reasonably describe the drainage situation (displacement of wetting phase by non-wetting phase), the imbibition (wetting phase being spontaneously imbibed, displacing a non-wetting phase) relative permeability to the non-wetting phase is generally found to be lower than that predicted by the above equation. Naar and coworkers (Naar and Henderson, 1961; Naar and Wygal, 1961) developed a mathematical model for consolidated porous media based on the concept that the advancing wetting phase (imbibition) gradually traps some of the non-wetting phase (oil and/or gas) making it immobile. They incorporated a trapping modification in Wyllie-Gardner theory (1958) by assuming that one-half of the initial non-wetting phase saturation will finally be trapped. They defined a free, or flowing, oil saturation ($S_{of} = S_o - S_{ot}$) and a reduced saturation ($S_{of}^* = (S_o - S_{ot})/(1 - S_{wi})$, where S_{ot} is the trapped oil saturation). Thus, the imbibition k_{ro} can be determined as follows:

$$k_{ro(imb)} = S_{of}^{*3} \left[S_{of}^* + 3(S_{wf}^* + S_{ot}^*) \right] \tag{4-24}$$

where $S_{wf}^* = (S_w - S_{wi})/(1 - S_{wi})$ and $S_{ot}^* = (S_{ot})/(1 - S_{wi})$.

In deriving the above equation, $1/P_c^2 = cS^*$ has been assumed to hold. The relative permeabilities to water and gas are given by:

$$k_{rw(imb)} = S_{wf}^{*4} \tag{4-25}$$

and:

$$k_{rg(imb)} = S_{gt}^{*3} (3 - 2S_{gf}^*) \tag{4-26}$$

where $S_{gf}^* = (S_g - S_{gt})/(1 - S_{wi})$.

While Naar and coworkers have provided a model which accounts, qualitatively, for trapping of the non-wetting phase in an imbibition process, the quantification of trapped phase saturation seems rather arbitrary and is not likely to be satisfactory in different systems. Naar et al. (1962) observed that, for unconsolidated sands, the imbibition non-wetting phase relative permeability is higher than that for drainage. This observation is opposite to the hysteresis observed with consolidated sands. This means that the trapping model would completely fail in the case of unconsolidated systems. This difference with unconsolidated or poorly consolidated sands may possibly be attributed to differences in pore size distribution and cementation. Morrow and Chatiudompunth (1980) calculated the hydraulic radius of each individual phase from drainage and imbibition capillary pressure and surface area measurements on packs of spheres and found that the hydraulic radius of the non-wetting phase had hysteresis behavior similar to that shown by non-wetting phase relative permeabilities in unconsolidated media.

Ashford (1969) modified the Naar-Henderson (1961) theory by observing that trapping provided by their theory gave the upper limit but was often less. He introduced a calculable correction factor so as to make the theory applicable to a particular matrix. Further, he introduced the concept of a pseudo-irreducible water saturation to be used in place of S_{wi} in all calculations on the grounds that effective drainage permeability of the wetting phase is very nearly zero at a value of S_w greater than S_{wi} (Burdine, 1953). The pseudo-irreducible wetting phase saturation is, therefore, the minimum wetting phase saturation at which the wetting phase drainage relative permeability approaches (and is essentially) zero.

Brooks and Corey (1964, 1966) modified Corey's original capillary pressure versus saturation relationship (equation (4-19)) into a two-parameter expression:

$$S_e = \left(\frac{P_b}{P_c} \right)^{\lambda} \tag{4-27}$$

where λ and P_b are constants, characteristic of the medium, λ being a measure of pore size distribution and P_b a measure of maximum pore size. Using this relationship, two-phase relative permeabilities are given by:

$$k_{rw,nw} = (S_e)^{(2+3\lambda)/\lambda} \tag{4-28}$$

and:

$$k_{mw,w} = (1 - S_e)^2 \, (1 - S_e)^{(2+\lambda)/\lambda} \tag{4-29}$$

These equations reduce to equations (4-20) and (4-21) for $\lambda = 2$. The commonly encountered range for λ is between 2 and 4 for various standstones (Brooks and Corey, 1966). Talash (1976) obtained similar equations with somewhat different exponents.

Following the work of Corey and coworkers, Land (1968) developed expressions for imbibition relative permeabilities by providing for trapping of the non-wetting phase. He assumed that the maximum residual hydrocarbon saturation, $S_{hr} = (S_{or} + S_{gr})$, is independent of whether the initial hydrocarbon saturation is gas, oil, or both. The validity of this assumption cannot be ascertained at this stage, but there is some reason to believe that it may not always be true. However, following this reasoning, Land developed three-phase imbibition relative permeabilities as:

$$k_{ro(imb)} = S_{of}^{*2} \; \frac{\displaystyle\int_{S_w^* + S_{ot}^*}^{S_w^* + S_o^*} dS_L^* / P_c^2}{\displaystyle\int_0^1 dS^* / P_c^2} \tag{4-30}$$

$$k_{rw(imb)} = S_{wf}^{*2} \; \frac{\displaystyle\int_0^{S_{wf}^*} dS_{wf}^* / P_c^2}{\displaystyle\int_0^1 dS^* / P_c^2} \tag{4-31}$$

and:

$$k_{rg(imb)} = S_{gf}^{*2} \frac{\int_{1-S_{gf}^*}^{1} dS^*/P_c^2}{\int_{0}^{1} dS^*/P_c^2} \qquad (4\text{-}32)$$

where $S^* = 1 - S_{gf}^*$, i.e., the reduced flowing liquid saturation. Since these imbibition equations are extensions of Corey's model, one would expect that the corresponding drainage equations would be those presented by Corey et al. (1956).

Stone (1970, 1973) considered two two-phase flow conditions as limits of three-phase flow. For example, water—oil—gas flow will be bounded by water—oil flow at one end and oil—gas flow at the other. In a water-wet system, gas behaves as a completely non-wetting phase but oil has an intermediate wettability. The relative permeability to oil in a water—oil—gas system will, therefore, be bound by relative permeability to oil in a water—oil system at one end and in a gas—oil system at the other. He attempted to combine these two terminal relative permeabilities to obtain a three-phase result by using simple probability models. Water and gas three-phase relative permeabilities, according to Stone, are the same as their corresponding two-phase relative permeabilities. In his first model, Stone (1970) developed the expression:

$$k_{ro} = S_o^* \beta_w \beta_g \qquad (4\text{-}33)$$

where

$$S_o^* = \frac{S_o - S_{om}}{1 - S_{wi} - S_{om}}, \quad S_{om} = \text{minimum residual oil saturation,}$$

$$\beta_w = \frac{k_{ro,w}}{1 - S_w^*} \ (2\text{-phase}), \ S_w^* = \frac{S_w - S_{wi}}{1 - S_{wi} - S_{om}}, \quad \text{and}$$

$$\beta_g = \frac{k_{ro,g}}{1 - S_g^*} \ (2\text{-phase}), \ S_g^* = \frac{S_g}{1 - S_{wi} - S_{om}}.$$

Stone's second model (Stone, 1973) gave:

$$k_{ro} = (k_{ro,w} + k_{rw})(k_{ro,g} + k_{rg}) - (k_{rw} + k_{rg}) \qquad (4\text{-}34)$$

Although it seems reasonable that one should be able to combine the two two-phase oil relative permeabilities to arrive at three-phase data, at least for water-wet systems, the manner in which they have been combined in these models may not account for the total physics of the process. These probability models strongly depend on the assumption that there is at most only one mobile fluid in any channel. Dietrich and Bondor (1976) applied these models to published three-phase data and found them to be only partially successful. They found it necessary to modify Stone's model for the case where gas/oil relative permeability was measured in the presence of connate water. The modified model (see also Aziz and Settari, 1979) was given as:

$$k_{ro} = \frac{1}{k_{ro,cw}}(k_{ro,w} + k_{rw})(k_{ro,g} + k_{rg}) - (k_{rw} + k_{rg}) \qquad (4\text{-}35)$$

where $k_{ro,cw}$ is the oil relative permeability at connate water and zero gas saturation.

In general, two-phase relative permeabilities are measured in the laboratory in preference to the use of prediction procedures, and the experimental values obtained are subsequently used in reservoir simulations. Measurement of three-phase permeabilities is seldom attempted, however, primarily because of the enormous experimental difficulties, and estimation of the data is made using one of the following models:

(1) Corey or Brooks-Corey equations for drainage.
(2) Naar-Wygal equations for imbibition.
(3) Land's equations for both drainage and imbibition.
(4) Stone's equations for both drainage and imbibition.

Because extensive experimental three-phase relative permeability data are not available, each has not been tested for more than one or two sets, so that presently it is difficult to recommend any one of them. However, Land's model, although computationally more demanding due to the integral nature of the equations, seems to hold greater promise because of a more sound physical basis (Schneider and Owens, 1970).

In the absence of a widely tested and accepted mathematical model, it is still considered advisable to measure relative permeabilities in the laboratory and use these (with a higher degree of confidence) in reservoir engineering calculations. A review of various experimental methods used to date for obtaining relative permeabilities in the laboratory follows in the next section.

EXPERIMENTAL METHODS FOR RELATIVE PERMEABILITY MEASUREMENT

Both steady- and unsteady-state methods have been used extensively for the measurement of two- and three-phase relative permeabilities. Steady-state methods provide flexibility in controlling changes in saturations and give reliable results which are comparable to those obtained by displacement methods (Terwilliger et al., 1951; Richardson, 1957; Johnson et al., 1959; Schneider and Owens, 1970; and Declaud, 1972). Calculation of relative permeability from experimental data is straightforward and does not involve questionable assumptions except that the test specimen is homogeneous. However, the experiments are difficult to set up, requiring elaborate equipment due to the presence of end effects (Caudle et al., 1951; Geffen et al., 1951) and the difficulties associated with obtaining saturation measurements. In-situ saturation measurement techniques are at times of questionable accuracy and sometimes depend, in addition to saturation, on the distribution of fluids. Unsteady or displacement techniques are considered to be more representative of reservoirs, because saturation changes may occur fairly rapidly and not allow equilibrium to be attained (Levine, 1954). Levine found that the displacing-phase dynamic relative permeability became zero only at zero saturation of the displacing phase, which is unlike that obtained in equilibrium experiments. Experimental set-ups and measurements are simpler in the case of displacement experiments compared to the steady-state case. The Buckley-Leverett type of frontal advance theory, however, is assumed to hold in unsteady-state tests and this may be questionable under certain circumstances because of inhomogeneity of porous media (Huppler, 1970) or flow instability (Chouke et al., 1959; Rachford, 1964; Peters and Flock, 1981). Loomis and Crowell (1962) found that in certain reservoir rocks the unsteady- and steady-state methods gave comparable results, whereas in others the two results deviated significantly. They observed that calculation of relative permeability by the unsteady-state method of Welge (1952) was valid only for homogeneous cores, and the cores where the results deviated from those obtained by the steady-state method may have been heterogeneous. Amaefule and Handy (1981) have recently compared two-phase relative permeabilities obtained by both steady- and unsteady-state methods for a water-wet sandstone. Whereas the residual oil saturations were found to be the same, oil relative permeabilities were somewhat lower at the same oil saturations for unsteady-state systems than for steady-state. The water relative permeabilities were higher indicating unsteady-state floods to be less water-wet than steady-state ones. This effect may be a manifestation of the different ways in which the wetting and non-wetting phases get distributed during the two types of floods (Handy and Datta, 1966). However, one might visualize unsteady flow to be made up of a series of steady states and, to that extent, the use of steady-state measurements can be justified at least for homogeneous samples (Leverett and Lewis, 1941).

Steady-state methods

These methods can be broadly classified under four different categories:

Stationary fluid (static) methods. Semi-permeable barriers are placed at both ends of the core and only one fluid is allowed to flow. Osoba et al. (1951) used barriers which held the liquid phase stationary inside the core and only gas was allowed to flow, facilitating the measurement of gas relative permeability only. Rapoport and Leas (1951), on the other hand, allowed liquid to flow holding the gas phase stationary. A combination of both would enable one to measure relative permeability to both phases (Loomis and Crowell, 1962). Corey et al. (1956) extended this method to the three-phase case and measured relative permeabilities on Berea plugs by holding the water phase stationary between membranes permeable only to oil and gas. The fact that all mobile fluids are not permitted to flow simultaneously, however, makes the technique unrealistic.

Capillary pressure (Hassler) method. This method was developed by Hassler (1944). Semi-permeable membranes are provided at each end which keep the fluids separated, except inside the core where the fluids flow simultaneously. Pressures are measured in each phase through semi-permeable barriers and the pressure differences between the phases are maintained constant throughout the medium so as to eliminate capillary end effects and to ensure a uniform saturation along the core. Saturation can be altered by applying capillary pressures across the non-wetting phase ports and wetting phase semi-permeable membranes. Osoba et al. (1951) reported that this method gave relative permeability results that were lower than those of any other method they used. However, Josendal et al. (1952) found that this method gave results comparable to the dynamic method. Gates and Leitz (1950), Fatt and Dykstra (1951), and Corey (1954) used this technique for measuring two-phase relative permeabilities. Later, Corey et al. (1956) used the same technique for three-phase work also. Although the method has been employed periodically in the past, its use is declining, because the method is slow and time consuming. While this method works well for strongly wetting porous media, capillary control becomes difficult under intermediate wetting conditions. Rose (1980) has recently discussed problems in applying this method of measurement. Further, the mechanism of flow is somewhat different than that of the dynamic method and may be best suited to understanding problems at the time of oil accumulation in the reservoirs (Brownscombe et al., 1950).

Dynamic (Penn-State) method. Morse et al. (1947) developed the method wherein all the fluids flow simultaneously and no barriers are used. They used three pieces of core; the front piece serves as a fluid distributor and the last piece absorbs undesirable capillary end effects. Saturation as well as pressure drop was measured in the middle test section. Caudle et al. (1951), Osoba et al. (1951), and Geffen et al. (1951) used this method, or somewhat modified versions, for two-phase relative permeability studies and obtained consistent and reliable results. Although the method was reasonably fast and was more representative of reservoir flow conditions than others, it had limitations related to the saturation measurement which was made by weighing the test section. This meant that a test had to be interrupted every time a saturation measurement was made. Also, part of the fluid was lost due to expansion of gas when pressure in the core became atmospheric during dismantling, thus causing error. Obviously, measurements could not be made under reservoir conditions using this method. Evaporation of fluids during weighing also introduced errors. Moreover, capillary contact, which was assumed to exist between the different pieces of core during flow, may not always have occurred, particularly in loosely consolidated media which, in turn, causes capillary end effect errors. Osoba et al. (1951) also used a single core instead of three cores and showed that end effects could be eliminated or at least minimized by using higher flow rates (Leverett and Lewis, 1941). Alternatively, if in-situ saturation and pressure measurements could be made at points away from the ends, this single core method would hold greater promise, because capillary end effects are known to be most severe only near the ends (Geffen et al., 1951; Josendal et al., 1952). Saraf and Fatt (1967), Schneider and Owens (1970, 1980), and others have used this modified Penn-State method successfully in two- as well as three-phase relative permeability studies. Under these conditions, this is by far the best method available for steady-state measurements.

Quasi-steady state methods. (1) Gas drive method (Hassler et al., 1936) — In this method, liquid and gas both flow but only gas is introduced as a flowing fluid for short periods of time. Liquid flows because it is displaced by gas. Thus, it is really gas drive over short bursts of time and could be considered to be a quasi-steady state method. A high-viscosity liquid allows slow flow and more accurate measurement. Since equilibrium is not allowed to occur, it is only an approximate method, but it has the advantage of being quite fast.

(2) Solution-gas drive method — This is also not a steady-state method because liquid is displaced from the core as gas comes out of solution inside the pore space as a result of lowering the pressure of the liquid, which contains gas below its bubble point (Brownscombe et al., 1950). The mechanism of flow is different from dynamic displacement, but is similar to reservoirs producing under solution-gas drive, and hence the results would be most appropriately used for those conditions only.

One major problem encountered in these steady-state methods is due to capillary end effects. The capillary end effect causes an accumulation of the preferentially wetting phase in the region of the outflow boundary during both steady- and unsteady-state multiphase immiscible flow. This boundary effect arises because the pressure of both fluids in the porous medium must be continuous with the pressure outside the medium. The importance of the outlet end effect, which is localized, decreases as the length, flow rate, and fluid viscosities are increased (Kyte and Rapoport, 1958). Significant end effects can cause erroneously high pressure drops to be measured across the core during steady-state experiments. In addition, the assumption of a uniform saturation based on an average saturation measurement made by weighing the core is likely to lead to further errors in the calculated relative permeability relations. Some of the ways in which end effects are eliminated or minimized are summarized as follows:

(1) Control of capillary pressure at both ends by use of semi-permeable end plates (Hassler method).

(2) Use of high flow rates that cause the capillary pressure gradient to be small compared to the applied pressure gradient.

(3) Penn-State, three-core method enables the end piece to absorb end effects.

(4) Allowing only one phase to flow while keeping other phases stationary.

(5) Making the pressure drop and saturation measurements within the core, far removed from the ends, and at a point where end effects do not persist.

Scheidegger (1974) has compared the various methods of measuring steady-state relative permeability and recommends Penn-State or Hassler methods because of the excellent reliability of the results obtained. The use of a single core with an in-situ saturation measurement, however, has the advantage of simplicity and reliability, and it is recommended for two-phase as well as three-phase relative permeability studies.

Saturation measurement has been a major problem in steady-state methods. Earlier workers (Geffen et al., 1951; Caudle et al., 1951; Osoba et al., 1951; Morse et al., 1947) used weighing for two-phase experiments and weighing in conjunction with distillation for three-phase experiments (Caudle et al., 1951). Brownscombe et al. (1950) and Rapoport and Leas (1951) used movement of liquid in a flow meter tube to estimate, by material balance, the saturations inside the core. Problems with weighing methods have already been discussed in the three-core method. Distillation is destructive, and the material balance method has the inherent assumption that end effects are not important and saturations are uniform throughout the core.

A variety of in-situ saturation determination methods have been developed and used in the past. Some of these work simply by the measurement of resistance or capacitance of the fluid-filled cores, whereas others use the more sophisticated neutron scattering, NMR, or microwave techniques. A detailed discussion can be found in a recent Petroleum Recovery Institute (PRI) report on methods of in-situ saturation determination during core tests involving multi-phase flow (Saraf, 1981). As will be seen later, such measurements are not a prerequisite to the unsteady-state method of obtaining relative permeabilities.

In addition to obtaining saturations of each fluid phase, one needs to measure the flow rates of each fluid and the pressure drop while attempting to get at the steady-state relative permeability measurements. Positive displacement, constant-rate pumps have been used by most workers for displacing oil, water and sometimes, gas. High-pressure gas has been throttled through a needle valve (McCaffery, 1973) to get a reasonably constant gas flow. A constant-flow controller which allows a constant pressure drop across a needle valve has also been successfully used (Saraf, 1966). The gas flow rate can be measured (a) by collecting gas by displacement of water for a known length of time, (b) with a soap bubble meter, or (c) using a wet test meter. For pressure drop measurements, it is desirable that the displacement of fluid into any device during measurement be as small as possible so as to minimize error. Hence, a regular manometer is not desirable. Differential pressure transducers have been successfully used to measure differential pressure to the desired accuracy. Whenever pressure drop is measured in each phase separately, the connections can be made through semipermeable membrane ports. In the single-core dynamic method, the capillary tube connected to the transducer is inserted about one inch from each end and cemented to minimize backflow. Some of the above requirements, such as pressure drop measurement and constant rate fluid injection, are necessary for the unsteady-state methods which are discussed in the following section.

Unsteady-state (displacement) methods

Darcy's law, coupled with a statement of capillary pressure in differential form, results in Leverett's fractional flow equation (Leverett, 1941):

$$f_{w2} = \frac{1 + \dfrac{k_o}{q_t \mu_o}\left[\dfrac{\partial P_c}{\partial x} - g\,\Delta\rho\,\sin\theta\right]}{1 + \dfrac{k_o}{k_w}\cdot\dfrac{\mu_w}{\mu_o}}, \tag{4-36}$$

where f_{w2} is the fraction of water in the output stream; superficial flow rate of total fluid, $q_t = q_o + q_w$; θ is the angle between direction x and horizontal; $\Delta\rho = \rho_w - \rho_o$ = the density difference between displacing and displaced fluids. Assuming θ to be zero and neglecting capillary pressure in equation (4-36), Welge (1952) obtained:

$$S_{w,av} - S_{w2} = f_{o,2}\,Q_w, \tag{4-37}$$

where the subscript "2" refers to the exit end, and "av" refers to the average saturation. The Q_w is the cumulative water injection in pore volumes. Experimental measurement of Q_w versus $S_{w,av}$ in a water displacing oil test enables calculation of $f_{o,2}$ and S_{w2} as slope and intercept. f_{o2} is related to the ratio of relative permeabilities $k_{ro,w}/k_{rw,o}$ through:

$$f_{o,2} = \frac{k_{ro,w}/k_{rw,o}}{\dfrac{k_{ro,w}}{k_{rw,o}} + \dfrac{\mu_o}{\mu_w}} = \frac{1}{1 + \dfrac{\mu_o/k_{ro,w}}{\mu_w/k_{rw,o}}} \tag{4-38}$$

A similar expression can, of course, be written for the case of gas displacing oil. Buckley and Leverett (1942) developed the well-known frontal advance equation:

$$\left(\frac{\partial L}{\partial t}\right)_{S_w} = \frac{q_t}{A\phi}\left(\frac{\partial f_w}{\partial S_w}\right)_t \tag{4-39}$$

which states that any water saturation, S_w, during displacement with water, moves along the flow path at a velocity given by the right-hand side of equation (4-39). Q_w in equation (4-37) is related to the slope of f_w versus S_w curve as:

$$Q_w = \frac{1}{\left(\dfrac{df_w}{dS_w}\right)_{S_{w2}}} \tag{4-40}$$

Derivation of the Buckley-Leverett frontal advance equation from Leverett's fractional flow equation assumes capillary and gravitational forces to be negligible compared to viscous forces. The former requirement means that as $\partial S_w/\partial x \to 0$, $\partial P_c/\partial S_w \cdot \partial S_w/\partial x$ also approaches zero (Terwilliger et al., 1951). A $\partial S_w/\partial x$ (or $\partial S_g/\partial x$, with gas as the displacing fluid) of zero means that all points of saturation in this region move down the core at the same rate, and a stabilized zone is established. Capillary pressure effects can be neglected either when the flow system is large such as in reservoirs or when high rates of displacement are used (Owens et al., 1956). McEwen (1959) retained the capillary pressure term in Leverett's fractional flow equations and solved them using numerical techniques. He found that his solution approached a non-capillary solution only at high flow rates. Hovanessian and Fayers (1961) retained both capillary and gravity terms and have discussed the effect of both terms on pressure distribution curves. Bentsen (1977) has discussed conditions under which the capillary term may be neglected. He has shown the non-capillary, Buckley-Leverett solution to be a steady-state solution of a second-order, non-linear, parabolic, partial differential equation representing fraction flow. According to this analysis, the transient solution approaches the Buckley-Leverett solution at small capillary number and high mobility ratio. Yortsos and Fokas (1980) have presented an analytical solution for a linear waterflood including capillary pressure effects for particular functional forms of relative permeability ratio and capillary pressure curves. The porous medium is assumed to be semi-infinite. The solution approaches the Buckley-Leverett expression at very high injection rates, as expected. Richardson (1957) has shown that relative permeabilities obtained by steady-state methods compared favorably with those calculated from displacement experiments using Welge's approach.

If the capillary pressure gradient is comparable to the applied gradient in a laboratory displacement test, the resulting flow is unstabilized because of capillary end effects. In a water-wet system, water is at a lower pressure when it arrives at the outflow face, and its pressure, which depends on saturation, has to build up to breakthrough causing a delay between arrival and breakthrough. Rapoport and Leas (1953) found that waterflooding was a linearly scalable process, i.e., the fluid saturation, at any time and at a given position, is a function of only the number of pore volumes of that fluid injected with respect to that position (see also Parsons and Jones, 1976; Corey, 1977). Based on this concept, they defined a scaling coefficient, $LV\mu_w$, and found experimentally a critical value beyond which a flood is stabilized. This is done in practice by plotting oil recovery at breakthrough versus $LV\mu_w$. The value at which recovery becomes independent of the scaling coefficient gave the critical value. Kyte and Rapoport (1958) reported that $LV\mu_w \geqslant 1$ (in units of cm · cm/min · cp) resulted in fairly small end effects which means that, for a given displacement system, use of high flow rates can help stabilize the flow. Owens et al. (1956) observed in their experiments on gas displacing oil that, as a rule of thumb, a pressure drop high enough to make 0.5 pore volumes of gas flow at downstream pressure conditions in less than one minute resulted in consistent results on two-phase relative permeabilities.

It is, therefore, important when designing displacement experiments for measurement of relative permeability that: (a) the pressure gradient is large (scaling coefficient is greater than critical) to minimize capillary pressure effects, (b) the pressure drop is small compared to total operating pressure so that the incompressible fluid assumption is valid, (c) the core is homogeneous, and (d) the driving force and fluid properties are held constant.

A somewhat different theory for displacement was developed by Dietz (1953), who visualized the horizontal displacement of oil by water as a tongue formed along the bottom of a core. Using this theory, one could get the ratio of oil/water relative permeabilities from a knowledge of cumulative total production as a function of cumulative oil production and residual oil saturation. Croes and Schwarz (1955) experimented with laboratory waterfloods and showed that, while the Dietz theory was applicable to only some of the experiments under certain conditions, the Buckley-Leverett theory was able to yield relative permeabilities with greater precision if dimensionally scaled experiments were designed.

Engelberts and Klinkenberg (1951) observed another kind of instability that existed in scaled experiments on water displacing oil, where water moved through the oil in the form of fingers. Chouke et al. (1959) also showed that, at displacement rates greater than a critical value, instability will occur even in a highly uniform porous medium. They observed that the tendency to form viscous fingers increased at higher oil/water viscosity ratios. They introduced the concept of an effective interfacial tension for a porous medium which depends on the total surface area of the microscopic moving fluid—fluid interfaces in addition to the bulk interfacial tension to which it reduces for displacement between two parallel plates. Using these concepts, Peters (1979) has developed a dimensionless stability number which defines stable flow as occurring below a critical value of 13.56 and which is unstable beyond it. The dimensionless number depends on the viscosities of oil and water, superficial velocity, diameter of the core, effective interfacial tension, and absolute permeability. It was experimentally shown that, as velocity increased, the number of fingers increased thereby increasing instability. Rachford (1964) concluded from his numerical experiments, however, that the effect of flow velocity on the onset of instability should be small. He further observed that instabilities do not necessarily become more severe with increasing viscosity ratio. Demetre et al. (1981) have recently studied immiscible displacement at various superficial velocities in cylindrical cores having different diameters and lengths. They have concluded that neither linear nor scaling groups were adequate as correlating parameters in the unstable flow regime (stability number between 13.56 and 900). The flow became pseudo-stable when this number exceeded 900 and was stable below 13.56. In both stable and pseudo-stable flow regimes, geometrical similarity between the model and the prototype was not important.

Returning to the Buckley-Leverett frontal advance theory and assuming that displacement experiments are performed on scaled models and that no instability sets in, then it is possible to use Welge's method to calculate from displacement experiments (water or gas displacing oil) the ratio of two-phase relative permeabilities. Johnson et al. (1959) extended Welge's work which enabled them to calculate individual phase relative permeabilities as:

$$k_{ro,w} = \frac{f_{o,2}}{d\left(\dfrac{1}{Q_w I_r}\right)\Big/ d\left(\dfrac{1}{Q_w}\right)} \qquad (4\text{-}41)$$

and:

$$k_{rw,o} = \frac{f_{w,2}}{f_{o,2}} \cdot \frac{\mu_w}{\mu_o} \cdot k_{ro,w} \qquad (4\text{-}42)$$

These equations give relative permeabilities at the outlet face saturation, which is related to the average core saturation through equation (4-37). The term I_r is the relative injectivity ($= (q_{w1}/\Delta P)/(q_{w1}/\Delta P)_{\text{at start of injection}}$), and subscript "1" refers to the inlet end of the core. For gas drive, the same equations hold with subscript "w" being replaced by "g". Delclaud (1972) questioned the applicability of the Welge-Johnson et al. (JBN) method as applied in laboratory displacement tests on small cores to obtain relative permeability, because of capillary pressure effects.

Jones and Roszelle (1978) developed a graphical technique, which is an alternative to the JBN theory for the evaluation of individual relative permeabilities from displacement experiment data that are linearly scalable. These authors claim that their technique is easier to use and yields more accurate results. It must be pointed out that inasmuch as the above theory applies only to homogeneous porous media under conditions where capillary pressure is negligible compared to the applied gradient, it is similar to the Welge-JBN procedure. Huppler (1970) studied the effects of stratification, high- and low-permeability lenses, and vugs on waterflooding and found that well distributed heterogeneities did not significantly influence the results. A pronounced effect on relative permeability, however, was observed when heterogeneities became channel-like. Archer and Wong (1973) reported that the JBN method gave poor relative permeability results for strongly water-wet cores where displacement was piston-like, for mixed wettability systems, and for heterogeneous carbonate rocks. They obtained relative permeability curves by trial and error method: calculated oil recovery and injectivity curves matching the measured data. Inspired by these results, Sigmund and McCaffery (1979) developed a procedure for obtaining relative permeability curves by trial and error method, calculated oil recovery and injectivity fitting the oil recovery and pressure data in the least-square sense. For heterogeneous carbonate cores, their two-parameter model gave satisfactory relative permeabilities whereas the JBN method seemed to fail. Based on optimal control theory, Chavent et al. (1980) have reported the development of an automatic adjustment method for the determination of two-phase relative permeability and capillary pressure from two sets of displacement experiments—one at a fast flow rate and the other at a velocity representative of reservoir conditions.

Unsteady-state methods discussed above have primarily been used for obtaining two-phase relative permeabilities, but one can as easily extend them to three-phase systems by initially putting two fluids in the core instead of one and displacing them by the third phase. Sarem (1966) extended Welge's theory to include a third phase and was able to calculate relative permeabilities after making an additional simplifying assumption that the relative permeability to each phase depended only on its own saturation. While this assumption may be reasonably valid for water and gas phases in a water-wet system (Corey et al., 1956), its validity for an oil phase is certainly questionable. However, this is the only procedure currently available for obtaining three-phase relative permeabilities from displacement tests. The only other work in this area is that of Donaldson and Dean (1966) who do not seem to have made such a simplifying assumption and still obtained the requisite information. Their results indicate that relative permeabilities to all the phases depend rather strongly on all the saturations. A close look at their data reveals that they may have calculated their relative permeabilities at saturations averaged over the entire core, which may not be of much practical use in an unsteady flow condition where large saturation gradients develop as a consequence of displacement.

There is an urgent need to develop a procedure for calculating three-phase relative permeabilities without making assumptions of the kind Sarem (1966) made. A possibility also exists for extending the Sigmund and McCaffery-type parametric approach to obtain three-phase data. Molina (1980) highlighted difficulties in measuring relative permeabilities in the laboratory and in the applicability of such data to reservoirs. He has suggested the use of a simulation procedure where two-phase relative permeability functions assigned on an absolute permeability distribution basis can be adjusted to match reservoir performance data.

HANDLING OF TEST CORES

Reservoir cores, outcrop rock samples, and sandpacks have all been used in relative permeability and oil displacement studies. Reservoir rock is obviously preferred for specific field evaluations, although many situations can arise where preliminary tests with reasonably homogeneous samples of outcrop rock, such as Berea sandstone, are advisable. Tests with outcrop rock can, for example, be useful in preliminary studies of surfactant, polymer, or alkaline flooding to determine optimum chemical concentrations and to assess slug size effects on oil recovery efficiency.

The relative permeability properties of a porous medium depend on its pore geometry, wettability, saturations of wetting and non-wetting phases present and saturation history. Of these, wettability can be the hardest to maintain or re-establish at its reservoir condition for core flooding tests. Much past study has been devoted to the complex and often uncertain problem of obtaining, handling, preparing, and testing cores in order to conserve the reservoir wetting properties. For unconsolidated media, there is the added important problem of preserving the in-situ particle arrangement and being able to restore reservoir stress conditions for valid petrophysical measurements (Swanson and Thomas, 1980).

An accurate representation of reservoir wettability is important for obtaining realistic relative permeability and residual saturation data from core tests. According to Craig (1971), the majority of oil reservoirs seems to possess an intermediate wettability, with the rock surfaces exhibiting no strong wetting preference for either the oil or water phases. Thus, the oil—water flow characteristics of such reservoir core cleaned to a strongly water-wet condition could be quite misleading. Two distinct approaches to the core wettability question have emerged. In one case, a "native state" core is obtained through drilling with crude oil (Mungan, 1972) or a low-fluid-loss mud containing no surface active agents, followed by special packaging to prevent the core from drying and exposure to oxygen (Amott, 1959; Mungan, 1966; Morgan and Gordon, 1970; Craig, 1971). The second approach involves restoring reservoir wettability in a cleaned core by recontacting it with crude oil at the reservoir temperature (Cuiec, 1977). A sandstone water-flooding and relative permeability study reported by Mungan (1972) indicated that both procedures can give similar results. A brief discussion of the effect of rock wettability on relative permeability behavior is given in a later section of this chapter.

A variety of core cleaning methods has been used in the past. Grist et al. (1975), in their study on core cleaning methods, found that toluene extraction did not change the core wettability to any appreciable extent but, if it was followed by methanol extraction or brine soaking, the rock became water-wet. Cuiec (1977) observed that the nature of the solid surface (acidic or basic) governed the choice of solvents to be used in restoration of core samples to their natural states. For acidic surfaces (sandstones) benzene—benzoic acid or chloroform—acetic acid, or toluene—acetic acid—ethanol are good solvents, whereas, for carbonate rocks, chloroform or ethyl acetate, chloroform—ethanol, or sodium hydroxide are good solvents.

If the native reservoir is strongly water-wet, it is easier to restore wettability. Donaldson et al. (1966) found that cleaning with toluene followed by isopropyl alcohol and, finally, with water ensured restoration of water wettability. They reported that the addition of sodium tripolyphosphate to brine ensured the core to be water-wet even after exposure to crude oils that would make it oil-wet because of the presence of surface-active polar compounds (mostly organic acids) native within these crude oils. Jennings (1957) and Schneider and Owens (1980) found that heating sandstone cores at 600°C invariably made them water-wet. Owens and Archer (1971) and Lo and Mungan (1973) fired the cores at 870°C to make them water-wet. Firing, in addition to cleaning the core, stabilizes any clay minerals present in the pore space and ensures constant internal rock surface mineralogy (Owens and Archer, 1971; Parsons and Jones, 1977). Talash (1976) reported that the absolute permeability of fired Berea sandstone cores is two to three times that of unfired cores.

Schneider and Owens (1976) reported that solvent cleaning, particularly with polar compounds, had the tendency to make the cores water-wet even though they were initially oil-wet, and recommended the use of preserved cores particularly for water—gas flow through oil-wet rocks. Mungan (1972) prescribed the use of fresh, preserved core with reservoir fluids for relative permeability measurements.

Corey and Rathjens (1956) studied the effect of stratification on gas—oil two-phase relative permeabilities. They found both the critical gas saturation and the gas relative permeability to be more sensitive to slight stratification than was the oil relative permeability. Rosman and Simon (1976) have described a nitrogen—helium miscible displacement technique to measure microscopic flow heterogeneities for determining the extent of core damage. Earlier nitrogen breakthrough indicates a higher degree of heterogeneity. Huppler (1969) has suggested the use of composite cores for waterflood testing in order to reduce the importance of capillary end effects. By choosing the proper ordering of individual sections in preparing a composite core, it is possible to obtain relative permeabilities that are more representative of the reservoir.

Refined oils have been extensively used for measuring relative permeabilities, primarily because they are clean and can be chosen to have sufficiently high viscosities to ensure good subordinate production following breakthrough. Care should, however, be taken to ensure that they do not contain polar compounds which could alter the wettability of rock surface. Langnes et al. (1972) have reported a significant effect of oil polarity on oil/water relative permeabilities. Mungan (1964) suggested using an acid wash (H_2SO_4) and passing the oil through a silica gel column to remove surface-active impurities. Hirsch et al. (1972) were able to remove polar and aromatic compounds from crude oil using a dual packed adsorption column of silica gel and alumina gel.

FACTORS AFFECTING RELATIVE PERMEABILITY

Relative permeabilities have been found by most workers to depend on saturation of fluids, saturation history, wettability, temperature, and viscous, capillary and gravitational forces. The effects of all these parameters on relative permeabilities are discussed in the following section.

Fluid saturation and saturation history

Relative permeability is a direct consequence of the different proportions of each of the different fluids present in the porous medium and, as such, it

TABLE 4-1

Summary of three-phase relative permeability experimental results

Authors	Date	Porous medium	Experimental technique	Saturation measurement	End effect	Relative permeability is a function of:		
						k_{rw}	k_{ro}	k_{rg}
Leverett and Lewis	1941	unconsolidated sand	dynamic method	water by electrical resistivity: gas by P-V relation	neglected	S_w	S_o, S_w, S_g	S_g, S_o, S_w
Caudle et al.	1951	consolidated sandstone	dynamic method	weighing and vacuum distillation	neglected	S_w, S_o, S_g	S_o, S_w, S_g	S_g, S_o, S_w
Corey et al.	1956	Berea sandstone	Hassler capillary method	gravimetric	minimized	S_w	S_o, S_w, S_g	S_g
Reid	1956	unconsolidated sand	Hassler capillary method	oil by gamma-ray; water by electrical resistivity	eliminated	S_w, S_o, S_g	S_o, S_w, S_g	S_g, S_o, S_w
Snell	1961	unconsolidated sand	Hassler capillary method	water by RCL circuit; gas by neutron diffraction	eliminated	S_w, S_o, S_g	S_o, S_w, S_g	S_g, S_o, S_w
Sarem	1966	Berea sandstone	displacement	—	neglected	S_w	S_o	S_g
Donaldson and Dean	1966	Berea sandstone and Arbucle Limestone	displacement	—	minimized	S_w, S_o, S_g	S_o, S_w, S_g	S_g, S_o, S_w

TABLE 4-1 (continued)

Authors	Date	Porous medium	Experimental technique	Saturation measurement	End effect	Relative permeability is a function of:		
						k_{rw}	k_{ro}	k_{rg}
Saraf and Fatt	1967	Boise Sandstone	dynamic method	NMR	eliminated	S_w	S_o, S_w, S_g	S_g
Schneider and Owens	1970	Torpedo Sandstone	dynamic method	gas by X-ray; water by electrical resistivity	not discussed	S_w	S_o, S_w, S_g	S_g, S_o, S_w

is directly dependent upon saturations of wetting or non-wetting phases and it is always reported as a function of saturation. This dependence, however, does not seem uniform throughout the saturation range and diminishes towards the lower end, with the relative permeability becoming zero (i.e., independent of saturation) much before zero saturation of that phase is reached. An irreducible wetting-phase saturation is, therefore, normally defined as that which is always present and which cannot be displaced. As observed by Burdine (1953) and Ashford (1969), however, mobility to the wetting phase approaches zero at saturations higher than S_{wi}. This led Ashford (1969) to define a pseudo-irreducible saturation at which the wetting-phase relative permeability becomes practically zero. Since the mobility of the wetting phase is zero at this saturation, it could be considered a part of the solid matrix for all practical purposes, and one defines a reduced porosity $\phi^* = \phi \, (1 - S_{wi})$ and a reduced flowing wetting-phase saturation $S_{wf}^* = (S_w - S_{wi})/(1 - S_{wi})$. Relative permeability is then defined as a function of S_{wf}^*. For the non-wetting phase, a residual saturation is defined to be the saturation remaining after the non-wetting-phase-saturated porous medium has been flushed with a very large number of pore volumes of the wetting phase. For a water-wet matrix then, S_{or} or S_{gr} is the residual oil or gas saturation at which relative permeability to the corresponding phase drops to zero. The S_{or} and S_{gr} are not the irreducible minimum saturations and one could define S_{om} and S_{gm} as limiting quantities, respectively. For a water-wet system, these are quite small and at times taken to be zero. For a gas phase, however, one defines a critical gas saturation as a minimum gas saturation, which must be created before it starts flowing as a continuous phase.

In a three-phase flow, various relative permeabilities have been reported by different investigators to depend on the saturations of the three phases in different ways. Table 4-1 lists the published work in this area. Leverett and Lewis (1941), Corey et al. (1956), and Saraf and Fatt (1967) found that the relative permeability to water (wetting phase) depended only on water saturation. Corey et al. and Saraf and Fatt found that the gas relative permeability also depends on its own saturation. These authors reasoned that since oil has intermediate wettability (non-wetting with respect to water and wetting with respect to gas), its relative permeability is the only one influenced by saturations other than its own. Almost all investigators except Sarem (1966) have reported that the oil relative permeability depends on all of the saturations. Caudle et al. (1951), Reid (1956), Snell (1962), and Donaldson and Dean (1966), however, found relative permeabilities to all three phases to depend on the saturations of all the phases present.

The effect of saturation history in two-phase flow has long been recognized (Geffen et al., 1951). The terms commonly used are drainage (when saturation of wetting phase is decreasing) and imbibition (when wetting-phase saturation is increasing). It is generally believed that relative permeability to the wetting phase is only marginally affected by the saturation history, whereas imbibition relative permeability to the non-wetting phase is mark-edly lower compared to one during drainage in consolidated rocks. The hysteresis of relative permeability seems to be related directly to hysteresis of capillary presssure during drainage and imbibition cycles. The contact angle hysteresis observed during advancing and receding flows provides a partial explanation of hysteresis in the capillary pressure—saturation curve. During a two-phase flow in a glass capillary tube (Blake et al., 1967) it was observed that, when the wetting phase (water) was displacing the non-wetting phase (benzene), the contact angle in water was approximately 90°, whereas when the non-wetting phase was displacing the wetting phase, it was approxi-mately 43°. In a capillary, when two phases are flowing, the velocity profile for laminar flow is parabolic but the interface is spherical. This requires some additional flow in the vicinity of the moving boundary, causing some extra energy dissipation. This flow also deforms the interface, but differently during imbibition and drainage. During imbibition, the curvature of the velocity profile and interface oppose each other, whereas during drainage, they are in the same direction. Melrose (1965), using an ideal soil model for pore geometry, has shown the existence of hysteresis even when the contact angle is held at zero, which is not predictable from the capillary tube model of a porous medium. This type of hysteresis occurs because, during differ-ent directions of saturation change, different fluid—fluid interface configura-tions are achieved (Morrow and Harris, 1965). The other reason for hystere-sis in relative permeability is the trapping of non-wetting fluid during imbibi-tion in the pores, which is absent during drainage. Mohanty et al. (1980) have studied the physics of oil entrapment in water-wet rock on a square-pore, two-dimensional network model. It is hoped that the model will be ex-tended to include real three-dimensional reservoir pores with all their com-plexities so that it could predict the trapping phenomenon more accurately.

In three-phase flow, because of the presence of more than one non-wetting phase, the saturation history has to be more elaborate (Snell, 1962, 1963). Snell (1963) considered only the histories of liquid phase saturations to be important and defined four distinct cases: (a) imbibition of water, oil saturation increasing (II); (b) imbibition of water, oil saturation decreasing (ID); (c) drainage of water, oil saturation increasing (DI); and (d) drainage of water, oil saturation decreasing (DD). However, if one considers gas saturation history also to be important and recognizes that any given satura-tion may increase, decrease, or remain unchanged, then there are a total of 27 possible histories. Fortunately, several of them are not realizable in prac-tice because of mass balance requirements (e.g., saturation of all three phases decreasing, etc.), but still as many as 13 distinct histories are realizable under different flow conditions. How many of them will result in hysteresis of re-lative permeability is at present a matter of speculation, and only detailed experimentation can answer this satisfactorily. Snell (1962, 1963), while measuring three-phase relative permeabilities on unconsolidated sands, observed that of the four liquid saturation histories (II, ID, DI, and DD) only DD gave distinct oil relative permeabilities. For the other three, the results were indistinguishable. According to Snell (1963) and as observed by Hosain (1961), this effect also disappeared when the oil used was non-polar. Some of those engaged in reservoir simulation have a tendency to regard relative permeability as an adjustable parameter, which is entirely an unfortunate situation. A better approach would be to choose the appropriate relative permeability data obtained from mathematical models, experiments, or elsewhere, which are commensurate with the saturation history in question. Evrenos and Comer (1969) and Colonna et al. (1972) studied the evolution of capillary and relative permeability hysteresis during alternate displacement of water and gas in gas storage aquifers. Evrenos and Comer developed a simulator for multi-cycle drainage—imbibition relative permeability and capillarity. The simulator makes use of semi-empirical equations to calculate relative permeabilities at a given saturation depending on the saturation history. Colonna et al. (1972) made use of simple experimental data to calculate relative permeability with appropriate history. Killough (1976) developed a simulator with an interpolation scheme that would obtain history-dependent relative permeability.

Viscous, capillary, and gravitational forces

Yuster (1951) did not agree with the applicability of Darcy's law to multi-phase flow of fluids and theorized that, in addition to saturation, relative permeability should vary with viscosity ratio of the fluids. Odeh (1959) developed a theoretical expression for the dependence of relative permeability on viscosity ratio and made measurements to support his hypothesis. He observed that relative permeability to non-wetting phase increased with increasing viscosity ratio. Baker (1960), however, showed the incorrectness of Odeh's conclusions. Downie and Crane (1961) observed that oil viscosity could influence the effective permeability of some rocks to oil. They further qualified their statement by saying that once an increased permeability is obtained with high-viscosity oil, it may not be lost even after replacing this oil with one of low viscosity. They attributed these observations to movement of colloid particles at oil—water interfaces. Rose (1972) has argued on the lines similar to Yuster and Odeh that Darcy's law could not be extended to multi-phase flow, because a slip condition at the fluid—fluid interface will mean a non-zero velocity at that boundary. This brings in the dependence of relative permeability on viscosity ratio. Dullien's (1979) book seems to suggest that the presence of a low-viscosity wetting phase on grains may have a lubricating effect on high-viscosity non-wetting flow. Du Prey (1973), in his study on factors affecting two-phase liquid—liquid displacement in consolidated porous media, showed that the viscosity ratio had considerable influence on both relative permeabilities and on residual saturations. Also, unfavorable mobility ratio (related to viscosity ratio) may give rise to viscous fingering (Chouke et al., 1959; Peters and Flock, 1981), which in turn affects residual fluid saturations as well as relative permeabilities. Ehrlich and Crane (1969), based on their model of two-phase flow in consolidated rocks, concluded that relative permeability is independent of the viscosity ratio everywhere except near the irreducible wetting phase saturation, which itself decreases with increasing non-wetting to wetting-phase viscosity ratio. Leverett and Lewis (1941), Sandberg et al. (1958), and Donaldson et al. (1966) found relative permeability to be independent of flowing fluid viscosities. Donaldson et al. (1966) cautioned, however, that if care is not taken in the selection of oils of different viscosities to ensure similar wettability properties, then relative permeabilities may be affected because of variable wetting and not because of viscosity effects.

Rate of fluid flow has been shown (Leas et al., 1950; Geffen et al., 1951; Caudle et al., 1951; Richardson et al., 1952; and Sandberg et al., 1958) to have no effect on relative permeability so long as it does not create a saturation gradient caused by inertial effects (at higher flow rates) and by capillary end effects (at lower rates). Rose (1972), however, has pointed out that capillary pressure measured at static equilibrium is different from the capillary pressure present in dynamic situations, because of the influence of hydrodynamic forces. Recently, Labastie et al. (1980) have reported the effects of flow rate on capillary pressure to be significant, particularly in carbonate rocks, and have pointed out the difficulty in using static capillary pressure data in interpreting waterflood results. This hydrodynamic effect was found to be very small in sandstones. These authors found relative permeability to be independent of flow rate except near residual oil saturation, which itself changes with rate, usually decreasing with increasing velocity.

Dependence of relative permeabilities on capillary pressure has been amply demonstrated in the earlier section on mathematical models. Several workers (McEwen, 1959; Hovanessian and Fayers, 1961; Bentsen, 1977; Yortsos and Fokas, 1980; Batycky et al., 1980) have examined the effect of the capillary term in solution of the fractional flow equations. It has been recognized that the triple-valued saturation function obtained by Buckley and Leverett (1942) is an outcome of neglecting capillary pressure in Leverett's frontal advance formula. This problem can be eliminated by simply keeping that term (Hovanessian and Fayers, 1961).

An important property, which gives rise to capillary pressure in a porous medium, is the interfacial tension of the fluids used. A dimensionless group called the capillary number, which is a ratio of viscous to interfacial forces, has been used in the literature to demonstrate the dependence of ultimate recovery on viscous and capillary forces. For the case of water displacing oil:

$$N_c = \frac{\mu_w U_w}{\phi \sigma_{ow}} \qquad (4\text{-}43)$$

where N_c is the capillary number, U_w is water flow rate per unit area, and σ_{ow} is the oil—water interfacial tension. For an ordinary waterflood, its value is typically in the order of 10^{-6}.

Wagner and Leach (1966) studied the effect of interfacial tension on displacement efficiency and showed that increased recovery at breakthrough was obtained in both oil- and water-wet systems when σ was reduced to less than 0.07 mN/m. Du Prey (1973) and Foster (1973) were both able to reduce residual oil to near zero by increasing the capillary number to 10^{-2}. Up to about a value of 10^{-4}, there appeared to be no noticeable effect on recovery, but as N_c was increased beyond 10^{-4}, primarily by lowering interfacial tension to ultra-low values, significant improvement in recovery was achieved. Melrose and Brandner (1974) correlated the microscopic displacement efficiency, E_M, defined as:

$$E_M = \frac{1 - S_{or} - S_{wi}}{1 - S_{wi}} \tag{4-44}$$

to capillary number and obtained a critical value of N_c which, when exceeded, will permit E_M to increase to unity. Below this critical point, however, E_M was independent of N_c, as observed earlier by Du Prey (1973) and Foster (1973). Based on the premise that when viscous forces begin to compete with capillary forces, the critical value for N_c is reached, they calculated $N_{c,critical}$ to be approximately 10^{-3}, which is in the range of observed values of 10^{-4} to 3×10^{-2}. The lower critical value corresponds to displacement of large oil ganglia formed during original water displacement and occupying several pores, whereas the upper limit is associated with displacement of single pore oil droplets. According to them, it is the shape and size distribution of the pores rather than absolute size which controls the critical value of the capillary number.

The easiest way to significantly increase capillary number is to lower interfacial tension. Lo (1976), while studying the effect of capillary number on oil—water relative permeability in consolidated synthetic porous media comprised of Teflon, found displacement efficiency to be in general agreement with the correlation provided by Melrose and Brandner (1974). He further observed that imbibition relative permeability to wetting phase increased at wetting phase saturation higher than 50%, whereas non-wetting phase relative permeability decreased slightly when σ decreased from 50 to 0.01 mN/m. Bardon and Longeron (1980) measured two-phase relative permeabilities on the methane—n-heptane system and found qualitatively similar results as reported by Lo (1976) up to values of $\sigma > 0.04$—0.07 mN/m. They found relative permeability to the wetting phase to increase linearly with decrease in σ up to a value of 0.07 mN/m. Bardon and Longeron obtained markedly different results when σ was reduced below 0.04 mN/m compared to when σ was greater than 0.07 mN/m. The relative permeabilities to both wetting and non-wetting phases increased sharply, and the curvatures of the relative permeability—saturation curves decreased so as to approach straight lines when σ = 0.001 mN/m (cf. Foster, 1973). Whereas the wetting phase relative permeability increases monotonically with decreasing interfacial tension (increasing N_c), non-wetting phase relative permeability first decreases with decreasing σ (Lo, 1976; Bardon and Longeron, 1980) but begins to increase with further reduction in σ beyond 0.04 mN/m (Bardon and Longeron, 1980). Batycky and McCaffery (1978), while studying the effect of capillary number (changing σ_{ow}) on oil—water displacement behavior in unconsolidated sand, observed that hysteresis in the relative permeabilities decreases with increasing capillary number and disappears at the lowest interfacial tension of 0.02 mN/m. Amaefule and Handy (1981) have presented empirical correlations relating two-phase oil/water relative permeabilities and residual saturations with capillary number.

Buckley and Leverett (1942), in the derivation of their frontal advance equation, assumed that both gravity and capillary forces are negligible compared to viscous forces (equation (4-35)). Hovanessian and Fayers (1961) solved the one-dimensional displacement equation, which included a gravity function and a capillary function, using a finite difference scheme and obtained the effect of inclusion of these terms in a linear waterflood. The gravity term that was included in their treatment is a ratio of gravitational to viscous forces. They concluded that this ratio has significant effect on saturations, pressure, and fractional flow profiles of an oil-wet system. Catchpole and Fulford (1966) obtained a dimensionless group, the Bond number, which is a ratio of gravitational to interfacial forces and is involved in multiple-phase flow of fluids:

$$N_B = \frac{\Delta \rho g R^2}{\sigma} \qquad (4\text{-}45)$$

where N_B is the Bond number, $\Delta \rho$ is the fluid density difference, g is acceleration due to gravity, and R is the particle radius. For low-permeability porous media, this term is likely to have very little effect on relative permeability (Bardon and Longeron, 1980). However, in situations where interfacial tension is very low, such as during surfactant or microemulsion flooding or miscible displacement, Bond numbers may be high enough to cause vertical segregation of oil and water and permit tunneling of injection water through the lower portion of the core (Foster, 1973). This will affect relative permeabilities in an as yet undetermined way. Morrow and Songkran (1979) studied the effect of both capillary number and Bond number on trapping and mobilization of oil in random packings of equal spheres of glass. For a gas—oil system, they reported that gravity forces were important when the Bond number was in the range of 0.005—0.33. Below this range, these forces were negligible, whereas above 0.33, they dominated capillary forces.

Wettability

The relationship between capillary pressure and relative permeability has already been discussed. Contact angle, which is one measure of the wettability of a rock surface, and pore radius control the capillary pressure for a given fluid—fluid—rock system. These are believed to control the microscopic distribution of fluids, thus establishing a direct dependence of relative permeability on wettability.

It is believed (Craig, 1971) that preferential wettability of the reservoir rock to oil or water governs, to a great extent, the oil recovery in a waterflood. It is to be expected that the success of any enhanced oil recovery technique will be strongly influenced by this property. McCaffery (1973) has presented a comprehensive survey of previous work on this parameter and immiscible displacement. Despite numerous studies that have been conducted, its effects are not well understood. There are several different methods that investigators have used to measure wettability (McGhee et al., 1979) and most of them seem to give different results. Thus, the choice seems to be largely a matter of individual taste and opinion. Although one would expect contact angle measurement to be a direct and reliable method for characterizing wetting (Melrose and Brandner, 1974; Treiber et al., 1972; Craig, 1971), there is no way to measure it within a porous medium. Alternatively, contact angle has been measured on a flat polished crystal of the mineral type which predominates in a given rock and is taken to represent reservoir wettability. Treiber et al. (1972) have done extensive work in this direction studying a large number of reservoir rocks. However, one can easily realize the limitations of such an approach, because the actual rock consists of various other minerals, different cementing materials, and clays which are bound to influence the effective wettability. Brown and Fatt (1956) introduced a fractional wettability concept which may be of use when handling the complex situation found in a reservoir. They developed an NMR method to measure the fractional wettability (see also Kumar et al., 1969). Bobek et al. (1958) used a spontaneous imbibition method to measure the wetting properties of a porous medium. Later works in this area include those by Amott (1959), Mungan (1964), Donaldson et al. (1969), Morrow and Mungan (1971), Morrow et al. (1972), McCaffery and Bennion (1974), and Dullien (1979). McCaffery (1973) attempted to relate contact angle to imbibition rate.

Certain crude oils contain surface-active organic materials, causing the wetting character of the porous media to change when contacted by the oil for a sufficient length of time (Denekas et al., 1959; Johansen and Dunning, 1959). Also, temperature is known to have an influence on contact angle (Mungan, 1966). The effect of core handling procedures on wettability has already been discussed earlier. All of these complex effects make the wettability measurements and their influence on flow rather complicated.

Melrose (1965), with the help of the ideal soil model, showed that a wetting phase once displaced from a porous medium by a non-wetting phase will not re-imbibe if the contact angle of the non-wetting phase exceeds 40°. Morrow and Mungan (1971) observed that spontaneous imbibition only occurs over a fairly narrow range of contact angles. Richardson et al. (1952) found imbibition rate to be greater for extracted cores when compared to preserved cores, concluding that extraction had made the cores more water-wet. They also obtained higher oil recovery from preserved cores in water-flooding operations. Similar findings have been reported by Salathiel (1973) who attributed them to the mixed wettability character of the actual porous medium. Rathmell et al. (1973) have summarized the influence of wetta-bility on oil displacement. In water-wet reservoirs, waterflood seems to cause a piston-like displacement, with very little oil flowing after break-through has been reached. Kyte et al. (1961), Donaldson et al. (1969), Owens and Archer (1971), and several others have reported higher break-through recoveries in strongly water-wet systems. Amott (1959) and Rath-mell et al. (1973) pointed out that reservoirs with intermediate wettability give optimum recovery both at water breakthrough and in absolute terms. There is some evidence (Warren and Calhoun, 1955; Salathiel, 1973) that strongly oil-wet reservoirs exhibit the highest ultimate recovery. The appar-ent controversy seems to suggest that recovery is also dependent on some other rock and fluid properties in addition to wettability, such as pore geometry.

Batycky (1979) and Batycky and Singhal (1978) performed carefully designed experiments where they studied the force required to mobilize a ganglion of oil as a function of contact angle for varying degrees of surface roughness. According to Batycky, in addition to pore throat—pore body geometric contrast that must be overcome, there are two phenomena that are involved in the mobilization of a ganglion of oil: (a) a tangential dis-placement, and (b) a detachment of the rear end of the ganglion from the rock surface. The former is prevalent in a well-cemented rock where the pores are more nearly capillary shaped, whereas the latter will influence the recovery when sharp grains protrude into the pore spaces and the flow chan-nels are irregular. Since much greater energy is required for detachment (be-cause of adhesion) as compared to the tangential displacement, the recovery efficiency will be poorer in the latter case even though the wettability may have been nearly the same in the two cases (Batycky, 1980). Also, as point-ed out earlier, different workers using different methods to characterize wetting may arrive at different results for the same porous media. McCaffery and Bennion (1974), who have performed their experiments under well-controlled conditions to eliminate pore geometry and other effects, found that the most efficient recovery should be achieved with fluids capable of spontaneous imbibition. Singhal et al. (1976) studied the effect of hetero-geneous wettability on displacement efficiency and reported that the ultimate recovery increased with decrease in the fraction of surface wetting to the displaced phase.

Owens and Archer (1971) studied the effect of contact angle on relative permeability and found imbibition relative permeability to both oil and water to be dependent upon contact angle, with the oil relative permeability being highest for most water-wet conditions (contact angle $0°$) and that to water being lowest for the same conditions.* This seems to suggest that one can qualitatively infer the wetting properties of a rock from relative permeability measurements. Batycky et al. (1980) seem to support this hypothesis. Their findings that relative permeability to the wetting phase shows hysteresis are in agreement with those of Jones and Roszelle (1978). These findings, however, seem to be at variance with other studies (Geffen et al., 1951; Schneider and Owens, 1970), where it is reported that the relative permeability of the wetting phase is the same during imbibition and drainage. No reports have appeared in the literature on the effects of wettability on three-phase relative permeabilities.

Temperature

Davidson (1968), Poston et al. (1970), Ehrlich (1970), Weinbrandt and Ramey (1972), and Lo and Mungan (1973) have studied the effect of temperature on relative permeability, residual oil saturation, and irreducible water saturation. High temperature lowers the residual oil saturation and increases the irreducible water saturation; both of which affect relative permeability. The reason for these changes is believed to be related to the effect temperature has on lowering both interfacial tension and contact angle in oil—water—solid systems. The effect on contact angle is to make the matrix more water-wet at higher temperatures (McCaffery and Cram, 1971). Gas relative permeability may also be affected by temperature due to molecular slippage (Klinkenberg effect) in the gas phase (Davidson, 1968). At this stage, only general qualitative remarks can be made and more work is required before it is possible to quantify the effect of temperature on relative permeability, although Ehrlich (1970) has suggested a quantitative method to correct imbibition relative permeabilities for temperature effects.

CONCLUDING REMARKS

A critical review of two- and three-phase relative permeability studies has been presented. An attempt has been made to identify the various mathe-

---

* Opposite results were obtained by G.V. Chilingarian in drainage experiments (editorial comment).

matical models for estimation and experimental methods for the determination of multiphase relative permeabilities. The importance of core handling procedures in laboratory studies has been indicated. Several factors such as saturation history, wettability, capillary number, Bond number, and stability number have been shown to influence the relative permeability—saturation relationships.

ACKNOWLEDGEMENTS

Discussions by the senior author with Drs. J. Batycky, S. Cheshire, B. Hlavacek, and T. Okazawa, all of PRI, were very helpful. Ms. B. Moore and Ms. M. Porter typed the chapter.

The material in this chapter was originally prepared and issued as a report of the Petroleum Recovery Institute, Calgary, Canada, during the authors' employment at that organization.

REFERENCES

Amaefule, J.O. and Handy, L.L., 1981. The effect of interfacial tensions on relative oil-water permeabilities of consolidated porous media. *SPE/DOE 2nd Joint Symp., Tulsa, Okla., April 5—8*, SPE/DOE 9783.

Amott, E., 1959. Observations relating to the wettability of porous rock. *Trans. AIME*, 216: 156—162.

Archer, J.S. and Wong, S.W., 1973. Use of a reservoir simulator to interpret laboratory waterflood data. *Soc. Pet. Eng. J.*, December, pp. 343—347.

Ashford, F.E., 1969. Computed relative permeability drainage and imbibition. *44th Annu. Fall SPE Meet., Denver, Colo., September 28—October 1*, SPE 2582.

Aziz, K. and Settari, A., 1979. *Petroleum Reservoir Simulation*. Applied Science Publishers, London, 476 pp.

Baker, P.E., 1960. Dicussion of effect of viscosity ratio on relative permeability. *Tech. Note, J. Pet. Technol.*, 12: 65—66.

Bardon, C. and Longeron, D.G., 1980. Influence of very low interfacial tensions on relative permeability. *Soc. Pet. Eng. J.*, October, pp. 391—401.

Batycky, J.P., 1979. Dependence of residual oil mobilization on wetting and roughness. *PRI Res. Rep.*, RR-41.

Batycky, J.P., 1980. Towards understanding wettability effects on oil recovery. *Proc. Int. Energy Agency Workshop on EOR, Bartlesville, Okla., April 24*.

Batycky, J.P. and McCaffery, F.G., 1978. Low interfacial tension displacement studies. *PRI Res. Note*, RN-6.

Batycky, J.P. and Singhal, A.K., 1978. Mobilization of entrapped ganglia. *84th Natl. Meet. AIChE, Atlanta, Ga., February 26—March 1*.

Batycky, J.P., McCaffery, F.G., Hodgins, P.K. and Fisher, D.B., 1980. Interpreting capillary pressure and rock wetting characteristics from unsteady-state displacement measurements. *55th Annu. Fall SPE Meet., Dallas, September 21—24*, SPE 9403.

Bentsen, R.G., 1977. Conditions under which capillary term may be neglected. *28th Annu. Tech. Meet. Pet. Soc. CIM, Edmonton, Ont., May 30—June 3*.

Blake, T.D., Everett, D.H. and Haynes, J.M., 1967. Some basic considerations concerning the kinetics of wetting processes in capillary systems. In: *Wetting. SCI Monogr.*, 24: 164.

Bobek, J.E., Mattax, C.C. and Denekas, M.O., 1958. Reservoir rock wettability — its significance and evaluation. *Trans. AIME*, 213: 155—160.

Brooks, R.H. and Corey, A.T., 1964. Hydraulic properties of porous media. *Colo. State Univ. Hydrol. Paper*, 3.

Brooks, R.H. and Corey, A.T., 1966. Properties of porous media affecting fluid flow. *J. Irrig. Drain. Div., June*, pp. 61—87.

Brown, R.J.S. and Fatt, I., 1956. Measurement of fractional wettability of oil field rocks by the magnetic relaxation method. *Trans. AIME*, 207: 262—264.

Brownscombe, E.R., Slobod, R.L. and Caudle, B.H., 1950. Relative permeability, I and II. *Oil Gas J.*, February 9, pp. 68—69; February 16, pp. 98—102.

Burdine, N.T., Gournay, L.S. and Reichertz, P.P., 1950. Pore size distribution of reservoir rocks. *Trans. AIME*, 189: 195—204.

Burdine, N.T., 1953. Relative permeability calculations from pore-size distribution data. *Trans. AIME*, 198: 71—77.

Buckley, S.E. and Leverett, M.C., 1942. Mechanism of fluid displacement in sands. *Trans. AIME*, 146: 107—116.

Carman, P.C., 1937. *Trans. Inst. Chem. Eng.*, 15: 150.

Catchpole, J.P. and Fulford, G., 1966. Dimensionless groups. *Ind. Eng. Chem.*, 58 (3): 46—60.

Caudle, B.H., Slobod, R.L. and Brownscombe, E.R., 1951. Further developments in the laboratory determination of relative permeability. *Trans. AIME*, 192: 145—150.

Chavent, G., Cohen, G. and Espy, M., 1980. Determination of relative permeabilities and capillary pressures by an automatic adjustment method. *55th Annu. Fall SPE Meet., Dallas, Texas, September 21—24*, SPE 9237.

Chouke, R.L., van Meurs, P. and van der Poel, C., 1959. The instability of slow, immiscible, viscous liquid—liquid displacements in permeable media. *Trans. AIME*, 216: 188—194.

Colonna, J., Brissaud, F. and Millet, J.L., 1972. Evolution of capillarity and relative permeability hysteresis. *Soc. Pet. Eng. J.*, February, pp. 28—39.

Corey, A.T., 1954. The interrelation between gas and oil relative permeabilities. *Prod. Monthly*, November, pp. 38—41.

Corey, A.T., 1977. *Mechanics of Heterogeneous Fluids in Porous Media*. Water Resources Publications, Fort Collins, Colo.

Corey, A.T. and Rathjens, C.H., 1956. Effect of stratification on relative permeability. *J. Pet. Technol., Tech. Note*, 393: 69—71.

Corey, A.T., Rathjens, C.H., Henderson, J.H. and Wyllie, M.R.J., 1956. Three-phase relative permeability. *J. Pet. Technol., Tech. Note*, 375: 63—65.

Craig, F.F., Jr., 1971. *The Reservoir Engineering Aspects of Waterflooding. SPE-AIME Monogr. Ser.*, 3.

Croes, G.A. and Schwarz, N., 1955. Dimensionally scaled experiments and the theories on the water-drive process. *Trans. AIME*, 204: 35—52.

Cuiec, L., 1977. Study of problems related to the restoration of the natural state of core samples. *J. Can. Pet. Technol.*, October/December, pp. 68—80.

Darcy, H., 1856. *Les Fontaines Publiques de la Ville de Dijon*. Victor Dalmont, Paris.

Davidson, L.B., 1968. The effect of temperature on the permeability ratio of different fluid pairs in two-phase systems. *43rd Annu. Fall SPE Meet., Houston, Texas, September 29*, SPE 2298.

Delclaud, J.P., 1972. New results on the displacement of a fluid by another in a porous medium. *47th Annu. Fall SPE Meet., San Antonio, Texas, October 8—11*, SPE 4103.

Demetre, G.P., Bentsen, R.G. and Flock, D.L., 1981. A multi-dimensional approach to scaled immiscible fluid displacement. *32nd Annu. Technol. Meet., Pet. Soc. CIM, Calgary, Alta., May 3—6*, Paper 81—32—7.

Denekas, M.O., Mattax, C.C. and Davis, G.T., 1959. Effects of crude oil components on rock wettability. *Trans. AIME*, 216: 330—333.

Dietrich, J.K. and Bondor, P.L., 1976. Three-phase oil relative permeability models. *51st Annu. Fall SPE Meet., New Orleans, La., October 3—6*, SPE 6044.

Dietz, D.N., 1953. A theoretical approach to the problem of encroaching and bypassing edge water. *Verh. K. Ned. Akad. Wet.*, B56: 38.

Donaldson, E.C. and Dean, G.W., 1966. Two- and three-phase relative permeability studies. *U.S. Bur. Mines, Rep.*, RI6826.

Donaldson, E.C., Lorenz, B.P. and Thomas, R.D., 1966. The effects of viscosity and wettability on oil and water relative permeabilities. *41st Annu. Fall SPE Meet., Dallas, Texas, October 2—5*, SPE 1562.

Donaldson, E.C., Thomas, R.D. and Lorenz, P.B., 1969. Wettability determination and its effect on recovery efficiency. *Soc. Pet. Eng. J.*, March, pp. 13—20.

Downie, J. and Crane, F.E., 1961. Effect of viscosity on relative permeability. *Soc. Pet. Eng. J.*, June, pp. 59—60.

Dullien, F.A.L., 1979. *Porous Media: Fluid Transport and Pore Structure*. Academic Press, New York, N.Y.

Du Prey, E.J.L., 1973. Factors affecting liquid—liquid relative permeabilities of a consolidated porous medium. *Soc. Pet. Eng. J.*, February, pp. 39—47.

Ehrlich, R., 1970. The effect of temperature on water—oil imbibition relative permeability. *Eastern Regional SPE Meet., Pittsburgh, Pa., November 5—6*, SPE 3214.

Ehrlich, R. and Crane, F.E., 1969. A model for two-phase flow in consolidated materials. *Soc. Pet. Eng. J.*, June, pp. 221—231.

Engleberts, W.F. and Klinkenberg, L.J., 1951. Laboratory experiments on the displacement of oil by water from packs of granular materials. *Proc. 3rd World Pet. Congr., The Hague*, Part II, pp. 544.

Evrenos, A.I. and Comer, A.C., 1969. Sensitivity studies of gas—water relative permeability and capillarity in reservoir modeling. *44th Annu. Fall SPE Meet., Denver, Colo., September 28—October 1*, SPE 2668.

Fatt, I., 1957. The network model of porous media. *Trans AIME*, 207: 144—159.

Fatt, I. and Dykstra, H., 1951. Relative permeability studies. *Trans. AIME*, 192: 249—256.

Foster, W.R., 1973. A low-tension waterflooding process. *J. Pet. Technol.*, February, pp. 205—210.

Gates, J.I. and Lietz, W.T., 1950. Relative permeabilities of California cores by the capillary-pressure method. *Drill. Prod. Pract.*, pp. 285—301.

Geffen, T.M., Owens, W.W., Parish, D.R. and Morse, R.A., 1951. Experimental investigation of factors affecting laboratory relative permeability measurements. *Trans. AIME*, 192: 99—110.

Grist, D.M., Langley, G.O. and Neustadter, E.L., 1975. The dependence of water permeability on core cleaning methods in the case of some sandstone samples. *J. Can. Pet. Technol.*, April/June, pp. 48—52.

Handy, L.L. and Datta, P., 1966. Fluid distribution during immiscible displacements in porous media. *Trans. AIME*, 237: 261—280.

Happel, J. and Brenner, H., 1965. *Low Reynolds-Number Hydrodynamics*. Prentice-Hall, Englewood Cliffs, N.J.

Hassler, G.L., 1944. Methods and apparatus for permeability measurements. *U.S. Patent*, No. 2,345,935.

Hassler, G.L., Rice, R.R. and Leeman, E.H., 1936. Investigations on the recovery of oil from sandstones by gas drive. *Trans. AIME*, 118: 116—137.

Hirsch, D.E., Hopkins, R.L., Coleman, H.J., Cotton, F.O. and Thompson, C.G., 1972. Separation of high boiling petroleum distillates using gradient elution through dual-packed adsorption column. *Annu. Chem.*, 44: 915—919.

Hosain, A., 1961. M.Sc. Thesis, University of Birmingham, Birmingham.

Hovanessian, S.A. and Fayers, F.J., 1961. Linear waterflood with gravity and capillary effects. *Soc. Pet. Eng. J.*, March, pp. 32—36.

Huppler, J.D., 1969. Waterflood relative permeabilities in composite rocks. *J. Pet. Technol.*, May, pp. 539—540.

Huppler, J.D., 1970. Numerical investigation of the effects of core heterogeneities on waterflood relative permeabilities. *Trans. AIME*, 249: 160—171.

Jennings, H.Y. Jr., 1957. Surface properties of natural and synthetic porous media. *Prod. Monthly*, March, pp. 20—24.

Johansen, R.T. and Dunning, H.N., 1959. Relative wetting tendencies of crude oils by the capillarimetric method. *Prod. Monthly*, September, pp. 20—22.

Johnson, E.F., Bossler, D.P. and Naumann, V.O., 1959. Calculation of relative permeability from displacement experiments. *Trans. AIME*, 216: 370—372.

Jones, S.C. and Roszelle, W.O., 1978. Graphical techniques for determining relative permeability from displacement experiments. *J. Pet. Technol.*, May, pp. 807—817.

Josendal, V.A., Sandiford, B.B. and Wilson, J.W., 1952. Improved multiphase flow studies employing radioactive tracers. *Trans. AIME*, 195: 65—76.

Killough, J.E., 1976. Reservoir simulation with history-dependent saturation functions. *Soc. Pet. Eng. J.*, February, pp. 37—48.

King Hubbert, M., 1956. Darcy's law and the field equations of the flow of underground fluids. *Trans. AIME*, 207: 222—239.

Kozeny, J., 1927. *Sitzungsber. Akad. Wiss. Wien, Math.-Naturwiss. Kl., Abt. 2A*, 136: 271.

Kumar, J., Fatt, I. and Saraf, D.N., 1969. Nuclear magnetic relaxation time of water in a porous medium with heterogeneous surface wettability. *J. Appl. Phys.*, 40: 4165—4171.

Kyte, J.R. and Rapoport, L.A., 1958. Linear waterflood behavior and end effects in water-wet porous media. *Trans. AIME*, 213: 423—426.

Kyte, J.R., Naumann, V.O. and Mattax, C.C., 1961. Effect of reservoir environment on water—oil displacements. *J. Pet. Technol.*, June, pp. 579—582.

Labastie, A., Guy, M., Delclaud, J.P. and Iffly, R., 1980. Effect of flow rate and wettability on water—oil relative permeabilities and capillary pressure. *55th Annu. Fall SPE Meet., Dallas, Texas, September* 21—24, SPE 9236.

Land, C.S., 1968. Calculation of imbibition relative permeability for two- and three-phase flow from rock properties. *Soc. Pet. Eng. J.*, June, pp. 149—156.

Langnes, G.L., Robertson, J.O. and Chilingar, G.V., 1972. *Secondary Recovery and Carbonate Reservoirs*. Elsevier, New York, N.Y., Appendix F.

Leas, W.J., Jenks, L.H. and Russel, C.D., 1950. Relative permeability to gas. *Trans. AIME*, 189: 65—72.

Leverett, M.C., 1941. Capillary behavior in porous solids. *Trans. AIME*, 142: 152—169.

Leverett, M.C. and Lewis, W.B., 1941. Steady flow of gas—oil—water mixtures through unconsolidated sands. *Trans. AIME*, 142: 107—116.

Levine, J.S., 1954. Displacement experiments in a consolidated porous system. *Trans. AIME*, 201: 57—66.

Lo, H.Y., 1976. The effect of interfacial tension on oil—water relative permeabilities. *Pet. Recovery Inst., Res. Rep.*, PR-32.

Lo, H.Y. and Mungan, N., 1973. Temperature effect on relative permeabilities and residual saturations. *Pet. Recovery Inst., Res. Rep.*, PR-19.

Loomis, A.G. and Crowell, D.C., 1962. Relative permeability studies: gas—oil and water—oil systems. *U.S. Bur. Mines, Bull.*, 599.

McCaffery, F.G., 1973. *The effect of wettability on relative permeability and imbibition in porous media.* Ph.D. Thesis, University of Calgary, Calgary, Alta.

McCaffery, F.G. and Bennion, D.W., 1974. The effect of wettability on two-phase relative permeabilities. *J. Can. Pet. Technol.*, October/December, pp. 42—53.

McCaffery, F.G. and Cram, P.J., 1971. Wetting and adsorption studies of the *n*-dodecane—aqueous solution—quartz system. *Pet. Recovery Inst., Res. Rep.*, PR-13.

McEwen, C.R., 1959. A numerical solution of the linear displacement equation with capillary pressure. *Trans. AIME*, 216: 412—415.

McGhee, J.W., Crocker, M.E. and Donaldson, E.C., 1979. Relative wetting properties of crude oils in Berea sandstone. *U.S. Dep. Energy, Rep.*, BETC/RI-78/9.

Melrose, J.C., 1965. Wettability as related to capillary action in porous media. *Soc. Pet. Eng. J.*, September, pp. 259—271.

Melrose, J.C. and Brandner, C.F., 1974. Role of capillary forces in determining microscopic displacement efficiency for oil recovery by waterflooding. *J. Can. Pet. Technol.*, October, pp. 54—62.

Mohanty, K.K., Davis, H.T. and Scriven, L.E., 1980. Physics of oil entrapment in water-wet rock. *55th Annu. Fall SPE Meet., Dallas, Texas, September 21—24*, SPE 9406.

Molina, N.N., 1980. A systematic approach to the relative permeability problem in reservoir simulation. *55th Annu. Fall SPE Meet., Dallas, Texas, September 21—24*, SPE 9234.

Morgan, J.T. and Gordon, G.T., 1970. Influence of pore geometry on water—oil relative permeability. *J. Pet. Technol.*, October, pp. 199—208.

Morrow, N.R. and Harris, C.C., 1965. Capillary equilibrium in porous materials. *Soc. Pet. Eng. J.*, March, pp. 15—24.

Morrow, N.R. and Mungan, N., 1971. Mouillabilité et capillarité en milieux poreux. *Rev. Inst. Fr. Pét.*, 26: 629.

Morrow, N.R. and Songkran, B., 1979. Displacement mechanisms in oil recovery processes. *3rd Int. Conf. on Surface and Colloid Science, Stockholm, August 20—25.*

Morrow, N.R. and Chatiudompunth, S., 1980. Application of hydraulic radii concept to multiphase flow. *N. M. Pet. Recovery Res. Center, PRRC Rep.*, 80-48.

Morrow, N.R., Cram, P.J. and McCaffery, F.G., 1972. Displacement studies in dolomite with wettability control by octanoic acid. *47th Annu. Fall SPE Meet., San Antonio, Texas, October 8—11*, SPE 3993.

Morse, R.A., Terwilliger, P.L. and Yuster, S.T., 1947. Relative permeability measurements on small core samples. *Oil Gas J.*, 23: 109—125.

Mungan, N., 1964. Role of wettability and interfacial tension in waterflooding. *Soc. Pet. Eng. J.*, June, pp. 115—123.

Mungan, N., 1966. Interfacial effects on immiscible liquid—liquid displacement in porous media. *Soc. Pet. Eng. J.*, September, pp. 247—253.

Mungan, N., 1972. Relative permeability measurements using reservoir fluids. *Soc. Pet. Eng. J.*, October, pp. 398—402.

Naar, J. and Henderson, J.H., 1961. An imbibition model — its application to flow behavior and the prediction of oil recovery. *Soc. Pet. Eng. J.*, June, pp. 61—70.

Naar, J. and Wygal, R.J., 1961. Three-phase imbibition relative permeability. *Soc. Pet. Eng. J.*, December, pp. 254—258.

Naar, J., Wygal, R.J. and Henderson, J.H., 1962. Imbibition relative permeability in unconsolidated porous media. *Soc. Pet. Eng. J.*, March, pp. 13—17.

Odeh, A.S., 1959. Effect of viscosity ratio on relative permeability. *Trans. AIME*, 216: 346—353.

Osoba, J.S., Richardson, J.G. and Blair, P.M., 1951. Laboratory measurements of relative permeability. *Trans. AIME*, 192: 47—56.

Owens, W.W. and Archer, D.L., 1971. The effect of rock wettability on oil—water relative permeability relationships. *J. Pet. Technol.*, July, pp. 873—878.

Owens, W.W., Parrish, D.R. and Lamoreaux, W.E., 1956. An evaluation of a gas drive method for determining relative permeability relationships. *Trans. AIME*, 207: 275—280.

Parsons, R.W. and Jones, S.C., 1977. Linear scaling in slug-type processes — application to micellar flooding. *Soc. Pet. Eng. J.*, February, pp. 11—26.

Peters, E.J., 1979. *Stability theory and viscous fingering in porous media.* Ph.D. Thesis, University of Alberta, Edmonton, Alta.

Peters, E.J. and Flock, D.L., 1981. The onset of instability during two-phase immiscible displacement in porous media. *Soc. Pet. Eng. J.*, April, pp. 249—258.

Poston, S.W., Ysrael, S., Hossain, A.K.M.S., Montgomery, E.F., III and Ramey, H.J., Jr., 1970. The effect of temperature on irreducible water saturation and relative permeability of unconsolidated sands. *Soc. Pet. Eng. J.*, June, pp. 171—180.

Purcell, W.R., 1949. Capillary pressures — their measurement using mercury and the calculation of permeability therefrom. *Trans. AIME*, February, pp. 39—48.

Rachford, H.H., Jr., 1964. Instability in water flooding oil from water-wet porous media containing connate water. *Soc. Pet. Eng. J.*, June, pp. 133—148.

Rapoport, L.A. and Leas, W.J., 1951. Relative permeability to liquid in liquid—gas systems. *Trans. AIME*, 192: 83—98.

Rapoport, L.A. and Leas, W.J., 1953. Properties of linear waterfloods. *Trans. AIME*, 198: 139—148.

Rathmell, J.J., Braun, P.H. and Perkins, T.K., 1973. Reservoir waterflood residual oil saturation from laboratory tests. *J. Pet. Technol.*, 25: 175—185.

Reid, S., 1956. *The flow of three immiscible fluids in porous media.* Ph.D. Thesis, University of Birmingham, Birmingham.

Richardson, J.G., 1957. The calculation of waterflood recovery from steady state relative permeability data. *Trans. AIME*, 210: 373—375.

Richardson, J.G., Kerver, J.K., Hafford, J.A. and Osoba, J.S., 1952. Laboratory determination of relative permeability. *Trans. AIME*, 195: 187—196.

Rose, W., 1949. Theoretical generalizations leading to the evaluation of relative permeability. *Trans. AIME*, May, pp. 111—126.

Rose, W.D., 1972. Some problems connected with the use of classical description of fluid/fluid displacement processes. In: IAHR (Editor), *Fundamentals of Transport Phenomena in Porous Media.* Elsevier New York, N.Y.

Rose, W., 1980. Some problems in applying the Hassler relative permeability method. *J. Pet. Technol.*, July, pp. 1161—1163.

Rose, W. and Bruce, W.A., 1949. Evaluation of capillary character in petroleum reservoir rock. *Trans. AIME*, May, pp. 127—142.

Rosman, A. and Simon, R., 1976. Flow heterogeneity in reservoir rocks. *J. Pet. Technol.*, December, pp. 1427—1429.

Salathiel, R.A., 1973. Oil recovery by surface film drainage in mixed-wettability rocks. *J. Pet. Technol.*, October, pp. 1216—1224.

Sandberg, C.R., Gourney, L.S. and Sippel, R.F., 1958. The effect of fluid flow rate and viscosity on laboratory determinations of oil—water relative permeabilities. *Trans. AIME*, 213: 36—43.

Saraf, D.N., 1966. *Measurement of fluid saturations by NMR and its application to three-phase relative permeability studies.* Ph.D. Thesis, University of California, Berkeley, Calif.

Saraf, D.N., 1981. Methods, of in-situ saturation determination during core tests involving multiphase flow. *Pet. Recovery Inst., Rep.*, 1981—6.

Saraf, D.N. and Fatt, I., 1967. Three-phase relative permeability measurement using a NMR technique for estimating fluid saturation. *Soc. Pet. Eng. J.*, September, pp. 235—242.

Sarem, A.M., 1966. Three-phase relative permeability measurements by unsteady state method. *Soc. Pet. Eng. J.*, September, pp. 199—205.

Scheidegger, A.E., 1953. Theoretical models of porous matter. *Prod. Monthly*, 10: 17—23.

Scheidegger, A.E., 1974. *The Physics of Flow Through Porous Media*. University of Toronto Press, Toronto, Ont.

Schneider, F.N. and Owens, W.W., 1970. Sandstone and carbonate two- and three-phase relative permeability characteristics. *Soc. Pet. Eng. J.*, March, pp. 75—84.

Schneider, F.N. and Owens, W.W., 1976. Relative permeability studies of gas—water flow following solvent injection in carbonate rocks. *Soc. Pet. Eng. J.*, February, pp. 23—30.

Schneider, F.N. and Owens, W.W., 1980. Steady-state measurements of relative permeability for polymer—oil systems. *55th Annu. Fall SPE Meet., Dallas, Texas, September 21—24*, SPE 9408.

Schopper, J.R., 1966. A theoretical investigation on the formation factor/permeability/porosity relationship using a network model. *Geophys. Prospect.*, 14: 301—314.

Sigmund, P.M. and McCaffery, F.G., 1979. An improved unsteady state procedure for determining the relative permeability characteristics of heterogeneous porous media. *Soc. Pet. Eng. J.*, February, pp. 15—28.

Singhal, A.K., Mukherjee, D.P. and Somerton, W.H., 1976. Effect of heterogeneous wettability on flow of fluids through media. *J. Can. Pet. Technol.*, 15: 63—70.

Snell, R.W., 1962. Three-phase relative permeability in an unconsolidated sand. *J. Inst. Pet.*, 48: 80—88.

Snell, R.W., 1963. The saturation history dependence of three-phase oil relative permeability. *J. Inst. Pet.*, 49: 81—84.

Stone, H.L., 1970. Probability model for estimating three-phase relative permeability. *J. Pet. Technol.*, February, pp. 214—218.

Stone, H.L., 1973. Estimation of three-phase relative permeability and residual oil data. *J. Can. Pet. Technol.*, October/December, pp. 53—61.

Swanson, B.F. and Thomas, E.C., 1980. The measurement of petrophysical properties of unconsolidated sand cores. *Log Analyst*, September/October, pp. 22—31.

Talash, A.W., 1976. Experimental and calculated relative permeability data for systems containing tension additives. *SPE Symp. on Improved Oil Recovery, Tulsa, Okla., March, 22—24*, SPE 5810.

Terwilliger, P.L., Wilsey, L.E., Hall, H.N., Bridges, P.M. and Morse, R.A., 1951. An experimental and theoretical investigation of gravity drainage performance. *Trans. AIME*, 192: 285—296.

Treiber, L.E., Archer, D.L. and Owens, W.W., 1972. A laboratory evaluation of the wettability of fifty oil-producing reservoirs. *Soc. Pet. Eng. J.*, December, pp. 531—540.

Wagner, O.R. and Leach, R.O., 1966. Effect of interfacial tension on displacement efficiency. *Soc. Pet. Eng. J.*, December, pp. 335—344.

Warren, J.E. and Calhoun, J.C., Jr., 1955. A study of waterflood efficiency in oil-wet systems. *Trans. AIME*, 204: 22—29.

Weinbrandt, R.M. and Ramey, H.J., Jr., 1972. The effect of temperature on relative permeability of consolidated rocks. *47th SPE Annu. Meet., San Antonio, Texas, October 8—11*, SPE 4142.

Welge, H.J., 1952. A simplified method for computing oil recovery by gas or water drive. *Trans. AIME*, 195: 91—98.

Wyllie, M.R.J. and Gardner, G.H.F., 1958. The generalized Kozeny-Carman equation, I and II. *World Oil*, Part I, March, pp. 121—128; Part II, April, pp. 210—227.

Wyllie, M.R.J. and Spangler, M.B., 1952. Application of electrical resistivity measurements to problem of fluid flow in porous media. *Bull. Am. Assoc. Pet. Geol.*, 36: 359—403.

Yortsos, Y.C. and Fokas, A., 1980. An analytical solution for linear waterflood including the effects of capillary pressure. *55th Annu. Fall SPE Meet., Dallas, Texas, September 21—24*, SPE 9407.

Yuster, S.T., 1951. Theoretical considerations of multiphase flow in idealized capillary system. *Proc. 3rd World Pet. Congr., The Hague*, 2: 437—445.

Chapter 5

FLOW TESTS' ANALYSIS: LIQUID CASE

AZIZ S. ODEH

THE PHYSICAL SYSTEM AND ITS MATHEMATICAL DESCRIPTION

The system consists of a porous medium saturated with oil (a single phase liquid) and penetrated by a well. It is idealized or simulated mathematically by assuming that the drainage area around the well can be characterized by average properties. Moreover, the flow is radial.

The three fundamental equations (Matthews and Russell, 1967) which describe the above system are: (a) the mass balance equation, (b) the equation of motion, and (c) the equation of state.

The mass balance equation

If an element in the porous system is considered, the mass balance equation states that:

$$\text{mass of fluid entering the element} - \text{mass of fluid leaving it} = \text{change in the mass of fluid within the element} \tag{5-1}$$

Consider the element shown below:

$$\rho v A \, \Delta t \longrightarrow \boxed{\rho A \, \Delta x \, \phi} \longrightarrow [\rho v + \Delta (\rho v)] \, A \, \Delta t$$
$$\Delta x$$

(see Nomenclature for definition of symbols, p. 120).

The mass of fluid in during Δt time increment is $\rho v A \, \Delta t$, whereas the mass of fluid out is $[\rho v + \Delta (\rho v)] \, A \, \Delta t$. The fluid mass content of the element is $\Delta x A \phi \rho$, and the change in the mass content is $\Delta (\Delta x A \phi \rho)$. Thus:

$$\{\rho v - [\rho v + \Delta (\rho v)]\} \, A \, \Delta t = \Delta (\Delta x A \phi \rho) = A \, \Delta x \phi \Delta \rho \tag{5-2}$$

or:

$$-\frac{\Delta (\rho v)}{\Delta x} = \phi \frac{\Delta \rho}{\Delta t} \tag{5-3}$$

Writing the above equation in differential form, i.e., letting Δx go to dx and Δt go to dt gives:

$$-\frac{\partial(\rho v)}{\partial x} = \phi \frac{\partial \rho}{\partial t} \qquad (5\text{-}4)$$

Equation (5-4) is known as the continuity equation.

NOMENCLATURE

Symbols

A	cross sectional area, or drainage area of the well
A_x	cross sectional area of a fracture of half length equal to x
B	formation volume factor
c	compressibility
h	thickness
i	injection rate
k	permeability
k_r	relative permeability
m	slope
p	pressure
$p*$	the extrapolated pressure value at $(t + \Delta t)/\Delta t = 1$
q	flow rate
r	radius
R_s	gas in solution
S	saturation
s	skin
T	total time
t	time
Δt	time increment
Δx	length increment
v	velocity
β	non-Darcy flow constant
Δ	difference
λ	mobility
ρ	density
ϕ	porosity
μ	viscosity

Subscripts

av	average
g	gas
D	dimensionless
e	exterior
i	initial
o	oil
T	total
w	wellbore, water

The equation of motion

The transport of fluid through porous media is best described by a Forchheimer-type equation (Muskat, 1946, 1949), which is:

$$-\frac{\partial p}{\partial x} = \frac{\mu v}{k} + \beta \rho v^2 \tag{5-5}$$

where β is a constant determined by the physical properties of the media (Katz et al., 1959). The relative importance of the terms on the right-hand side of equation (5-5) for oil flow are examined below.

Consider a well producing 1000 reservoir bbl/day $\cong 1.84 \times 10^3$ cm³/s. Let $h = 25$ ft, $k = 100$ md, and $\phi = 15\%$. For this system (Katz et al., 1959), $\beta \simeq 3$ atm s²/g.

The flow velocity at the wellbore having a radius of 0.25 ft is:

$$v = \frac{q}{2\pi r_w h} = \frac{1.84 \times 10^3}{25 \times 30.5 \times 2 \times 3.14 \times 30.5 \times 0.25} = 0.05 \text{ cm/s.}$$

For $\rho = 0.8$, $\beta \rho v^2 = (0.05)^2 \times 3 \times 0.8 = 0.006$.

On the other hand for $\mu = 1$, $\mu v/k = 1 \times 0.05/0.1 = 0.5$. Thus, $\mu v/k \gg \beta \rho v^2$. For oil, therefore, flow equation (5-5) may be approximated by:

$$-\frac{\partial p}{\partial x} = \frac{\mu v}{k} \tag{5-6}$$

Equation (5-6) represents the well-known Darcy's law.

The equation of state

The equation of state relates pressure, p, to density, ρ. For a slightly compressible fluid, i.e., oil and water, the equation is (Muskat, 1946, 1949):

$$\rho = \rho_0 e^{c(p-p_0)} \tag{5-7}$$

where ρ_0 is the density at the pressure p_0 and c is the compressibility of the fluid.

Replacing equations (5-6) and (5-7) in equation (5-4) gives:

$$\frac{\partial}{\partial x}(\rho_0 e^{c(p-p_0)} \frac{k}{\mu} \frac{\partial p}{\partial x}) = \phi \frac{\partial}{\partial t}(\rho_0 e^{c(p-p_0)}) \tag{5-8}$$

Expanding and simplifying results in:

$$\frac{\partial^2 p}{\partial x^2} + c\left(\frac{\partial p}{\partial x}\right)^2 = \frac{\phi\mu c}{k}\frac{\partial p}{\partial t} \tag{5-9}$$

However, for oil flow $c\,(\partial p/\partial x)^2$ is very small and can be neglected to give:

$$\frac{\partial^2 p}{\partial x^2} = \frac{\phi\mu c}{k}\frac{\partial p}{\partial t} \tag{5-10}$$

In radial geometry equation (5-10) is:

$$\frac{\partial^2 p}{\partial r^2} + \frac{1}{r}\frac{\partial p}{\partial r} = \frac{\phi\mu c}{k}\frac{\partial p}{\partial t} \tag{5-11}$$

Equation (5-11) is identical to the diffusion equation (Muskat, 1946, 1949). It describes the flow of slightly compressible fluid in porous media.

In the derivation of equation (5-11) it was assumed that the porous rock is incompressible. In reality it is compressible. When this is accounted for, the resulting equation will be identical to (5-11), provided that the compressibility c is the sum of the fluid and rock compressibilities.

THE APPLICATION OF THE EQUATION OF FLOW TO THE PHYSICAL SYSTEM

Equation (5-11) relates the pressure to time and distance in a porous system filled with a slightly compressible fluid. Its solution is affected by the conditions imposed on the system.

Two conditions are recognized: (1) the boundary condition and (2) the initial condition. A radially symmetrical system with a well at the center may be viewed as two concentric cylinders, a small inner one representing the wellbore, and a larger outer one representing the drainage area. The conditions imposed at the inner and outer cylinders are called the boundary conditions. For instance, if the well produces at a constant rate, the condition is called a constant rate. The outer cylinder may be closed, thus giving a closed boundary condition. The boundary condition commonly used for the inner cylinder, i.e., at the wellbore, is constant rate.

The pressure distribution in the system at the beginning of the test is referred to as the initial condition. The one commonly used is uniform pressure.

All physical systems are finite. When a well is open to flow and a disturbance is created in the porous body, however, there is a period of time during which the boundary of the system is unaware of such an event, because the disturbance has not reached it. During this time, the boundary does not influence the behavior of the well, and the system acts as if it has no boundary or as infinite (Matthews and Russell, 1967; van Everdingen and Hurst, 1949). In flow tests, equation (5-11) is mainly solved for infinite systems.

Defining the following variables (van Everdingen and Hurst, 1949): $R = r/r_e$, and $t_D = kt/\phi\mu cr_e^2$, and substituting in equation (5-11) one obtains:

$$\frac{\partial^2 p}{\partial R^2} + \frac{1}{R}\frac{\partial p}{\partial R} = \frac{\partial p}{\partial t_D} \qquad (5\text{-}12)$$

Here r_e is the outside radius. The variable t_D is called dimensionless time, and equation (5-12) is the dimensionless form of (5-11). The advantage of equation (5-12) over equation (5-11) is that the former results in a general solution applicable to all systems of various k, ϕ, μ, and c. Sometimes, the area A is used in place of r_e^2 in t_D. In this case, t_D is normally written as t_{DA}.

STATES OF FLOW

Three states of flow are recognized (Matthews and Russell, 1967; Odeh and Nabor, 1966): unsteady, pseudo-steady, and transitional. During the time when the system acts as infinite, the flow is in an unsteady state. The pressure is a function of time and distance. This state is sometimes called early transient and its duration is given by:

$$t_D = \frac{kt}{\phi\mu cr_e^2} \simeq 0.1 \qquad (5\text{-}13)$$

where the units are darcy (d), seconds, fraction, centipoise (cp), vol/vol/atm, and cm^2. In practical engineering units (hours, md, cp, vol/vol/psi and ft), the duration is given by the following equation:

$$t = \frac{360\,\phi\mu cr_e^2}{k} \qquad (5\text{-}14)$$

When the boundary presence is felt, and if the system is closed, it will eventually behave like a tank. This means that the rate of pressure change, i.e., $\partial p/\partial t$, all over the system becomes constant, independent of location. When this happens, the system is in pseudo-steady state, and the relation between the well productivity and the average pressure is given by:

$$q = \frac{7.08 \times 10^{-3} \, kh \, (p_{av} - p_w)}{B\mu \left(\ln \dfrac{r_e}{r_w} - 0.75 \right)} \tag{5-15}$$

where q, the flow rate, is in stock tank barrels/day (STB/day); B, the formation volume factor, in reservoir bbl/STB; p_{av} is the average pressure within the system in psi; k is in md; h is in ft, and μ is in cp. Here the drainage area is assumed to be a circle. For non-circular drainage area see Table 5-1.

The start of pseudo-steady state is given by:

$$t = \frac{1440 \, \phi\mu c r_e^2}{k} \tag{5-16}$$

The period between t of equation (5-14) and t of equation (5-16) is called the transitional, or late transient. The flow is changing from unsteady to pseudo-steady.

In the literature, reference is made sometimes to steady-state. This condition only exists if the fluid is incompressible and is described by:

$$q = \frac{7.08 \times 10^{-3} \, kh \, (p_e - p_w)}{\mu \ln \dfrac{r_e}{r_w}} \tag{5-17}$$

where p_e is the pressure at the outside radius r_e. One should note the difference between equations (5-17) and (5-15).

FUNDAMENTAL SOLUTION AND SUPERPOSITION

The fundamental solution

A well-known solution to equations (5-11) or (5-12) for a well fully penetrating a homogeneous, productive sand and producing at a constant rate q in an infinite system, with the pressure being uniform at the start of the flow test, is the line source solution given by the Ei-function (Horner, 1951). Thus:

$$p_{r,t} = p_i + \frac{70.6 \, qB\mu}{kh} \, Ei \left(-\frac{948 \, \phi\mu c r^2}{kt} \right) \tag{5-18}$$

where the units are psi, STB/day, md, ft, vol/vol/psi, cp, and hours; $p_{r,t}$ is

TABLE 5-1

Factors for different shapes and well positions in drainage area[a] (after Odeh, 1978)

System	x	System	x
(circle, centered)	r_e/r_w	(rectangle 2×1, top-center)	$0.966A^{1/2}/r_w$
(square, centered)	$0.571A^{1/2}/r_w$	(rectangle 2×1)	$1.44A^{1/2}/r_w$
(hexagon, centered)	$0.565A^{1/2}/r_w$	(rectangle 2×1)	$2.206A^{1/2}/r_w$
(triangle, centered)	$0.604A^{1/2}/r_w$	(rectangle 4×1)	$1.925A^{1/2}/r_w$
(parallelogram 60°)	$0.61A^{1/2}/r_w$	(rectangle 4×1)	$6.59A^{1/2}/r_w$
(right triangle, 1/3)	$0.678A^{1/2}/r_w$	(rectangle 4×1)	$9.36A^{1/2}/r_w$
(rectangle 2×1)	$0.668A^{1/2}/r_w$	(square 1×1)	$1.724A^{1/2}/r_w$
(rectangle 4×1)	$1.368A^{1/2}/r_w$	(rectangle 2×1)	$1.794A^{1/2}/r_w$
(rectangle 5×1)	$2.066A^{1/2}/r_w$	(rectangle 2×1)	$4.074A^{1/2}/r_w$
(square 1×1)	$0.884A^{1/2}/r_w$	(rectangle 2×1)	$9.527A^{1/2}/r_w$
(square 1×1)	$1.485A^{1/2}/r_w$	(triangle)	$10.139A^{1/2}/r_w$

[a] A = drainage area of system shown; $A^{1/2}/r_w$ is dimensionless.

$$q = \frac{7.08 \times 10^{-3}\, kh\,(p_{av} - p_w)}{\mu B\,(\ln x - 0.75 + s)}$$

the pressure at any point r; and p_i is the pressure at the start of the test. The Ei-function is defined by:

$$Ei\,(-x) = -\int_x^\infty \frac{e^{-x}\,dx}{x} \qquad (5\text{-}19)$$

The pressure at the wellbore (r_w) is:

$$p_w = p_i + \frac{70.6\,Bq\mu}{kh}\,Ei\,\left(-\frac{948\,\phi\mu cr_w^2}{kt}\right) \qquad (5\text{-}20)$$

However, $Ei\,(-x) \simeq \ln x + 0.5772$ for $x < 10^{-2}$. In equation (5-20), $948\,\phi\mu cr_w^2/kt < 10^{-2}$ for very small t (a few seconds). Thus:

$$p_w = p_i - \frac{162.6\,Bq\mu}{kh}\,\left(\log t + \log \frac{k}{\phi\mu cr_w^2} - 3.23\right) \qquad (5\text{-}21)$$

Equation (5-21) is the basic equation used in flow tests' analyses. It is important to remember the main assumptions employed to derive it:

(1) The well is producing at a constant rate q.

(2) The medium is homogeneous.

(3) The flow is in unsteady state, i.e., the reservoir is infinite.

(4) The pressure at the beginning of the test is uniform all over the reservoir.

(5) The well completely penetrates the productive sand.

The above conditions indicate why requirements such as shutting the well and letting the pressure stabilize prior to the flow test are imposed.

Although conditions (1) to (5) were imposed to obtain equations (5-18), (5-20), and (5-21), it will be seen later how the variations in conditions (1) to (5) in testing procedures and interpretations, are accounted for.

If the well produces at a constant pressure p, and assumptions (2), (3), (4), and (5) are still applicable, the solution is (Earlougher, 1977):

$$\frac{1}{q} \simeq \frac{162.6\,\mu B}{kh\,(p_i - p_w)}\,\left(\log t + \log \frac{k}{\phi\mu cr_w^2} - 3.48\right) \qquad (5\text{-}22)$$

Superposition

Superposition is a very powerful tool used to generate solutions to complicated problems from simple ones (Matthews and Russell, 1967).

The superposition principle applies when the equation is linear; i.e., the dependent variable, which is p in our case, appears raised to the same power in all the terms. The use of superposition is illustrated by the following example:

Let a well A produce at a rate q_1 in an infinite reservoir. Let a second well B be put on production at a rate of q_2, Δt hours after well A. Calculate the pressure at time T at a point r_1 and r_2 away from A and B, respectively.

Superposition allows solution of this problem as follows:

(1) Calculating the pressure drop due to well A as if B does not exist.
(2) Calculating the pressure drop due to well B ignoring A.
(3) Obtaining the total pressure drop by summing (1) and (2).

The pressure drop due to A using equation (5-18) is:

$$p_i - p_{r,t} = \Delta p_A = - \frac{70.6 \, q_1 \, B\mu}{kh} \, Ei \left(- \frac{948 \, \phi\mu cr_1^2}{kT} \right) \tag{5-23}$$

The pressure drop due to well B which produced $(T - \Delta t)$ hours is:

$$\Delta p_B = - \frac{70.6 \, q_2 B\mu}{kh} \, Ei \left(- \frac{948 \, \phi\mu cr_2^2}{k(T - \Delta t)} \right) \tag{5-24}$$

Thus:

$$\Delta p_T = - \frac{70.6 \, B\mu}{kh} \left[q_1 Ei \left(- \frac{948 \, \phi\mu cr_1^2}{kT} \right) + q_2 Ei \left(- \frac{948 \, \phi\mu cr_2^2}{k(T - \Delta t)} \right) \right] \tag{5-25}$$

The same procedure is used to treat variable rate cases and will be illustrated fully in the next sections.

CONSTANT-RATE DRAWDOWN AND BUILDUP

Drawdown

When the well is opened to flow from a shut-in condition, and the flowing bottomhole pressure and rate (as a function of time) are recorded, the well is undergoing a drawdown test. If the assumptions stated with equation (5-21) are met, the equation will describe the pressure behavior with time. A plot of p_w versus t on semilog paper, where t is plotted on the semilog scale results in a straight line with the slope m given by:

$$m = - \frac{162.6 \, Bq\mu}{kh} \tag{5-26}$$

from which:

$$k = -\frac{162.6\,Bq\mu}{mh} \tag{5-27}$$

where m is the slope in psi/cycle, and the rest of the symbols are as previously defined.

Buildup

In a buildup test, the well is produced for t hours and then is shut in for Δt hours. The equation which describes the pressure behavior of the well during the shut-in time, Δt, may be derived from equation (5-21) using superposition. The buildup test can be viewed in the following manner. A well is produced at a constant rate q for t hours, at which time injection is started at a rate q through a well located in the same wellbore for Δt hours. The wellbore pressure behavior during the Δt hours is to be calculated.

As discussed previously, the net pressure drop is calculated by summing the pressure drop due to the producing well and the pressure drop due to the injection well. The pressure drop due to the producing well at $(t + \Delta t)$ hours, by equation (5-21) is:

$$\Delta p_{(1)} = \frac{162.6\,Bq\mu}{kh} \left[\log\,(t + \Delta t) + \log\,\frac{k}{\phi\mu cr_w^2} - 3.23 \right] \tag{5-28}$$

The pressure drop due to the second well is equal to:

$$\Delta p_{(2)} = \frac{162.6\,B\,(-q)\mu}{kh} \left[\log\,\Delta t + \log\,\frac{k}{\phi\mu cr_w^2} - 3.23 \right] \tag{5-29}$$

In the second case, q is negative representing injection; also the time is Δt because the well was not injecting during t. Adding the two solutions gives:

$$\Delta p_1 + \Delta p_2 = \Delta p_T = p_i - p_{w\Delta t} = \frac{162.6\,Bq\mu}{kh} \left(\log\,\frac{t + \Delta t}{\Delta t} \right) \tag{5-30}$$

Equation (5-30) is the buildup equation derived by Horner (1951). This equation shows that a plot of $p_{w\Delta t}$ versus $(t + \Delta t)/\Delta t$ on semilog paper results in a straight line. Its slope m enables calculation of k as follows:

$$k = -\frac{162.6\,Bq\mu}{mh} \tag{5-31}$$

The buildup is preferred to the drawdown, because the results obtained from the drawdown are very sensitive to a small variation in rate, whereas those of the buildup are not.

SKIN

The skin (see Hurst, 1953; van Everdingen, 1953) is defined as any factor at or around the wellbore which causes a change in the pressure drop, as compared to the pressure drop that occurs when the stratum is homogeneous, and the well fully penetrates the sand. Thus, if drilling fluid filtrate causes reduction in the permeability around the wellbore, it causes skin. If the well partially penetrates the sand, it causes skin (Brons and Marting, 1961; Odeh, 1968, 1980). If the well is fractured, the result is a negative skin, indicating improvement in the permeability. The most widely discussed skin is the one due to change in the permeability around the wellbore. The physical representation of this skin is illustrated below (Hawkins, 1956).

Consider a well where the permeability up to a radius r_1 has been changed to a value of k_1 due to mud damage, for example. The flow equation around the wellbore prior to the permeability damage is:

$$q = \frac{7.08 \times 10^{-3}\, kh\, (p_{r_1} - p_w)}{B\mu\, \ln \dfrac{r_1}{r_w}}$$

(5-32)

or:

$$p_{r_1} - p_w = \Delta p_1 = \frac{qB\mu\, \ln (r_1/r_w)}{7.08 \times 10^{-3}\, kh}$$

(5-33)

where k is the original permeability. The pressure drop after damage is:

$$\Delta p_2 = \frac{qB\mu\, \ln (r_1/r_w)}{7.08 \times 10^{-3}\, k_1 h}$$

(5-34)

Thus:

$$\Delta p_2 - \Delta p_1 = \Delta p_{skin} = \frac{141.2\, qB\mu\, \ln (r_1/r_w)}{h} \left(\frac{1}{k_1} - \frac{1}{k} \right) =$$

$$= \frac{141.2\, qB\mu\, \ln (r_1/r_w)}{kh} \left(\frac{k - k_1}{k_1} \right)$$

(5-35)

If:

$$\ln \frac{r_1}{r_w} \left(\frac{k - k_1}{k_1} \right) = s$$

then:

$$\Delta p_{skin} = \frac{141.2 \, qB\mu s}{kh} = \frac{162.6 \, qB\mu}{kh} \, (0.87s) = -0.87 \, ms \tag{5-36}$$

When skin is present, therefore, the total pressure drop in a flow test is obtained by adding equation (5-36) to (5-21) giving:

$$p_i - p_{wt} = \frac{162.6 \, qB\mu}{kh} \left[\log t + \log \frac{k}{\phi\mu cr_w^2} - 3.23 + 0.87s \right] \tag{5-37}$$

Skin may be calculated from drawdowns or buildups, but the common method is to calculate it from buildup plots.

Calculation of skin from drawdown. Equation (5-37) is used directly to solve for s. Thus:

$$s = 1.15 \left(\frac{p_i - p_{wt}}{-m} - \log \frac{kt}{\phi\mu cr_w^2} + 3.23 \right) \tag{5-38}$$

The value of p_{wt} should be read on the straight line or its extrapolation.

Calculation of skin from buildup. If equation (5-30) is subtracted from equation (5-37) the result is:

$$p_{w\Delta t} - p_{wt} = \frac{162.6 \, q\mu}{kh} \left(\log t - \log \frac{t + \Delta t}{\Delta t} + \log \frac{k}{\phi\mu cr_w^2} - 3.23 + 0.87s \right)$$

$$\tag{5-39}$$

If $t \gg \Delta t$, then $\log (t + \Delta t) \simeq \log t$. Substituting in equation (5-39) and solving for s gives:

$$s = 1.15 \left(\frac{p_{w\Delta t} - p_{w\Delta t = 0}}{-m} - \log \frac{k\Delta t}{\phi\mu cr_w^2} + 3.23 \right) \tag{5-40}$$

The value of $p_{w\Delta t}$ should be read on the straight line plot or its extrapolation; it is normally chosen for $\Delta t = 1$ hour.

Thus, to calculate s, a buildup plot is constructed; p is read on the straight line or its extrapolation corresponding to $\Delta t = 1$ hour, i.e., $(t + \Delta t)/\Delta t = t + 1$. Also, the slope is determined as well as the flowing bottomhole pressure of the well prior to shut in, i.e., $p_{w\Delta t = 0}$. Then s is calculated using equation (5-40).

The skin factor is composed of two major components. One is due to the physical change of the formation and the other is due to restricted entry to flow at the wellbore. The latter is caused by opening only part of the sand to flow or by plugged perforations. Thus:

$$s_t = s_1 + s_2 \tag{5-41}$$

The skin due to restricted entry to flow can be calculated by the methods presented by Brons and Marting (1961), and Odeh (1968, 1980).

Since s may or may not be present, the pseudo-steady state equation (5-15) should be modified to read:

$$q = \frac{7.08 \times 10^{-3} \, kh \, (p_{av} - p_w)}{\mu \, [\ln \, (r_e/r_w) - 0.75 + s]} \tag{5-42}$$

THE TWO-RATE FLOW TEST

The two-rate flow test (see Matthews and Russell, 1967; Russell, 1963; Earlougher, 1977) is a special case of the buildup with the second flow rate not equal to zero. In a buildup, the second rate is zero. The procedure consists of stabilizing the well at a constant rate q_1 for a period of time such as 24 hours, then changing the rate to q_2 and measuring the flowing bottomhole pressure versus time. The test had the advantage of not requiring a shut-in period; however, it has the disadvantage of being sensitive to small fluctuations in rate like the drawdown.

The equation for the two-rate flow test is derived following the same procedure used to derive the buildup equation. Assume a well producing at q_1 for time t; at that time another producing well in the same wellbore is placed on production at a rate $(q_2 - q_1)$. The net flow rate of the two wells is $q_1 + q_2 - q_1 = q_2$. The pressure behavior after the second well has been producing for Δt time is calculated as follows:

$$\Delta p_1 = \frac{162.6 \, Bq_1\mu_1}{kh} \left[\log \, (t + \Delta t) + \log \, \frac{k}{\phi\mu cr_w^2} - 3.23 + 0.87s\right] \tag{5-43}$$

$$\Delta p_2 = \frac{162.6 \, B \, (q_2 - q_1) \, \mu_1}{kh} \left[\log \Delta t + \log \frac{k}{\phi \mu c r_{\mathrm{w}}^2} - 3.23 + 0.87 s \right] \qquad (5\text{-}44)$$

Adding and rearranging gives:

$$\Delta p_T = p_i - p_w = \frac{162.6 \, B \mu q_1}{kh} \left[\log \frac{t + \Delta t}{\Delta t} + \frac{q_2}{q_1} \left(\log \Delta t + \right.\right.$$

$$\left.\left. + \log \frac{k}{\phi \mu c r_{\mathrm{w}}^2} + 0.87 s - 3.23 \right) \right] \qquad (5\text{-}45)$$

Equation (5-45) shows that a plot of p_w versus log $\{[(t + \Delta t)/\Delta t] + (q_2/q_1)$ log $\Delta t\}$ on rectangular coordinate paper gives a straight line whose slope m is equal to:

$$m = - \frac{162.6 \, B q_1 \mu}{kh} \qquad (5\text{-}46)$$

The skin s may be calculated by:

$$s = 1.15 \left[\frac{q_1}{q_1 - q_2} \left(\frac{p_{w \, \Delta t = 1} - p_{w \, \Delta t = 0}}{-m} \right) - \log \frac{k}{\phi \mu c r_{\mathrm{w}}^2} + 3.23 \right] \qquad (5\text{-}47)$$

If the well has been producing at a variable rate prior to stabilization, then t is calculated by:

$$t = \frac{\text{total production}}{q_1} \qquad (5\text{-}48)$$

If $t \gg \Delta t$, such that log $(t + \Delta t) \simeq$ log t, equations (5-45) and (5-47) become:

$$p_i - p_w = \frac{162.6 \, B \mu}{kh} (q_2 - q_1) \log \Delta t + \frac{162.6 \, B \mu}{kh} \left[q_1 \log t + \right.$$

$$\left. + q_2 \left(\log \frac{k}{\phi \mu c r_{\mathrm{w}}^2} + 0.87 s - 3.23 \right) \right] \qquad (5\text{-}49)$$

and:

$$s = 1.15 \left(\frac{p_{w \, \Delta t = 1} - p_{w \, \Delta t = 0}}{-m} - \log \frac{k}{\phi \mu c r_{\mathrm{w}}^2} + 3.23 \right) \qquad (5\text{-}50)$$

Equation (5-49) which is applicable for the majority of cases with minor error, shows that a plot of p_w versus Δt on semilog paper gives a straight line whose slope m is equal to

$$m = -\,\frac{162.6\,B\mu\,(q_2 - q_1)}{kh} \qquad (5\text{-}51)$$

When calculating the skin, the analyst must remember to read $p_{w\,\Delta t=1}$ at the straight line or its extension.

MULTIPLE FLOW RATE, DRAWDOWN

Multiple flow rate or variable rate drawdown is easy to run, but difficult to analyze by hand. Inasmuch as drawdowns' results are very sensitive to small fluctuations in rate and because it is difficult to maintain a constant rate during the test, most, if not all, drawdowns should be considered as variable rate drawdowns and analyzed as such. It is easy to program the calculations and eliminate the tedious hand calculations (Odeh and Jones, 1965).

In multiple rate tests, one keeps a record of flow rate, time, and pressure. After plotting rate versus time, the resulting curve is divided into a number of constant rate increments as shown in Fig. 5-1.

The equations applicable to variable rate tests are derived from the fundamental solution (equation 5-37) using superposition.

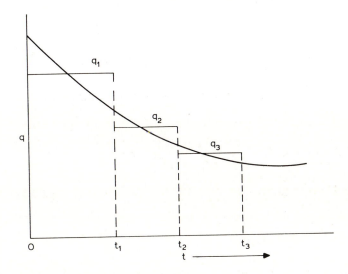

Fig. 5-1. Flow rate versus time plot.

To illustrate the method of derivation, consider a well whose rate of flow versus time is as given in Fig. 5-1. The pressure behavior, as a function of time, is to be calculated as was done previously. In this case there are multiple wells producing from the same wellbore, in place of one well producing at multiple rates. With this in mind, the pressure at time t_3 is calculated as follows: from 0 to t_1 there is one well producing at a q_1 rate. At time t_1 another well starts at a rate equal to $q_2 - q_1$, and at time t_2 a third well starts at a rate $q_3 - q_2$. Thus:

$$\Delta p_1 = \frac{162.6\, Bq_1\mu}{kh} \,(\log t_3 + C) \tag{5-52}$$

$$\Delta p_2 = \frac{162.6\, B\,(q_2 - q_1)\mu}{kh} \,[\log (t_3 - t_1) + C] \tag{5-53}$$

and:

$$\Delta p_3 = \frac{162.6\, B\,(q_3 - q_2)\mu}{kh} \,[\log (t_3 - t_2) + C] \tag{5-54}$$

where $C = \log (k/\phi\mu cr_w^2) - 3.23 + 0.87s$.

Adding to obtain the total pressure drop gives:

$$\Delta p_T = p_i - p_w = \frac{162.6\, B\mu}{kh} \,[q_1 \log t_3 + (q_2 - q_1) \log (t_3 - t_1) +$$

$$+ (q_3 - q_2) \log (t_3 - t_2) + C\,(q_1 + q_2 - q_1 + q_3 - q_2)] \tag{5-55}$$

which may be written in a compact notation as:

$$p_i - p_{wn} = \frac{162.6\, B\mu}{kh} \sum_{j=1}^{n} \Delta q_j \log (t_n - t_{j-1}) + q_n C \tag{5-56}$$

or:

$$\frac{p_i - p_{wn}}{q_n} = \frac{162.6\, B\mu}{kh} \sum_{j=1}^{n} \frac{\Delta q_j}{q_n} \log (t_n - t_{j-1}) + C \tag{5-57}$$

where $\Delta q_j = q_j - q_{j-1}$, $q_0 = 0$, and $t_0 = 0$.

Thus a plot of:

$$\frac{p_i - p_{wn}}{q_n} \quad \text{versus} \quad \sum_{j=1}^{n} \frac{\Delta q_j}{q_n} \log (t_n - t_{j-1})$$

should result in a straight line whose slope m is given by $m = 162.6 \, B\mu/kh$. The calculations are made as follows:

After the flow—time history is divided into a number of constant rate increments, one starts with one increment and calculates the left-hand side of equation (5-57) from the given pressure data and then the corresponding right-hand side. Two increments are then taken, i.e., $n = 2$, and $(p_i - p_{w2})/q_2$ is calculated, as well as the $\sum_{j=1}^{2} (\Delta q_j/q_2) \log (t_2 - t_{j-1})$. After that, 3, 4, etc., increments are chosen and the corresponding points are calculated, giving a table of $(p_i - p_{wn})/q_n$ versus Σ. These values are plotted on rectangular coordinate paper to obtain a straight line. It should be noted that the pressure at the beginning of the test, p_i, must be known for this method of analysis.

To illustrate the calculations for one pair of points, let q_1, q_2, and q_3 be 500, 400, and 600 STB/day; t_1, t_2, and t_3 be 1, 5, and 10 hours; and p_i, p_3 be 2000 and 1800 psi. The values of $(p_i - p_w)/q_n$ and the Σ at time t_3 are to be calculated. Here $n = 3$. Thus $(p_i - p_{wn})/q_n = (2000 - 1800)/600 = 0.33$. The Σ is:

$$\sum_{j=1}^{3} \frac{\Delta q_j}{q_n} \log (t_3 - t_{j-1}) = \frac{q_1 - q_0}{q_3} \log (t_3 - t_0) + \frac{q_2 - q_1}{q_3} \log (t_3 - t_1) +$$

$$+ \frac{q_3 - q_2}{q_3} \log (t_3 - t_2) = \frac{500 - 0}{600} \log (10 - 0) + \frac{400 - 500}{600} \log (10 - 1) +$$

$$+ \frac{600 - 400}{600} \log (10 - 5) = 0.91$$

The skin s is calculated using the following equation:

$$s = 1.15 \left(\frac{p_i - p_w}{q_n m} - \Sigma + 3.23 - \log \frac{k}{\phi \mu c r_w^2} \right) \tag{5-58}$$

where $(p_i - p_w)/q_n$ corresponds to the Σ. The easiest way to do this is to select any value of the Σ on the x-axis and read the corresponding $(p_i - p_w)/q_n$ value which falls on the straight line or its extension.

VARIABLE-RATE BUILDUP

Approximate method

The most widely used of the approximate methods is Horner's (1951). It is based on obtaining a corrected time, t_c, to reflect the variation in rate. The t_c is calculated by dividing the total cumulative production by the last stabilized rate. The total cumulative production is measured from the last time when the well was shut-in and achieved pressure buildup. After calculating t_c, the analysis proceeds as indicated by equation (5-30). (See also Matthews and Russell, 1967; Earlougher, 1977).

Odeh and Nabor (1966) found that the above analysis could result in considerable error if the last stabilized flow period is not long enough. They found that if the stabilized flow period is $< 1440 \phi \mu c r_e^2 / k$, which is the time required for the start of pseudo-steady state flow, then the part of the buildup that can be trusted to yield accurate results is the early part and is given by:

$$\Delta t < \frac{t_L}{4} \tag{5-59}$$

where Δt is the time of the buildup when the data can be trusted and t_L is the last stabilized flow period prior to the buildup.

The t*q* method

The t^*q^* method (Odeh and Selig, 1963) is intended to preserve the simple form of Horner's buildup equation. A modified flow time, t^*, and flow rate, q^*, are calculated and used in equation (5-30). The time t^* is given by:

$$t^* = 2 \left[t - \frac{1}{V} \sum_{i=0}^{n-1} \bar{q}_i B \left(\frac{t_i + t_{i+1}}{2} \right) (t_{i+1} - t_i) \right] \tag{5-60}$$

where $\bar{q}_i$ is the average flow rate between t_i and t_{i+1}, n is the number of flow rates, i.e., time increments, t is the actual flow time in days, and V is the total production in reservoir barrels. Also:

$$q^* B = \frac{V}{t^*} \tag{5-61}$$

To illustrate the use of the t^*q^* method, one can consider a test with the following data:

Time (hours)	qB (res. bbl/day)
0—3	478.5
3—6	319.0
6—9	159.5

Solving:

$$t^* = 2 \left[\frac{9}{24} - \frac{478.5 \left(\frac{0+3}{2 \times 24}\right)\left(\frac{3-0}{24}\right) + 319\left(\frac{3+6}{2 \times 24}\right)\left(\frac{6-3}{24}\right) + 159.5\left(\frac{6+9}{2 \times 24}\right)\left(\frac{9-6}{24}\right)}{\dfrac{478.5 \times 3 + 319 \times 3 + 159.5 \times 3}{24}} \right]$$

$$= 2 \left[\frac{9 - 3.5}{24} \right] = \frac{11}{24} \text{ days or 11 hours}$$

and:

$$q^*B = \frac{478.5 \times 3 + 319 \times 3 \times 159.5 \times 3}{24 \times \dfrac{11}{24}} = 261 \text{ res. bbl/day}$$

The t^*q^* is applicable to a test in which the buildup time Δt is greater than the drawdown time t, since the method is accurate when $\Delta t \geqslant t$. In drillstem testing, Δt is usually greater than t and the t^*q^* method is preferred to Horner's.

The exact method — superposition

The superposition method (Horner, 1951) is the most accurate method, derived by applying the superposition principle and should be used whenever possible; the calculations can be programmed easily. The equation used is:

$$p_{w \Delta t} = p_i - m \sum_{j=1}^{n} q_j \log \frac{t_n - t_{j-1} + \Delta t}{t_n - t_j + \Delta t} \tag{5-62}$$

where $m = 162.6 \, \mu/kh$, $t_0 = 0$, and n is the number of constant flow periods into which the drawdown was divided.

Equation (5-62) shows that a plot of p_w versus Σ on rectangular coordinate paper should result in a straight line. The procedure is to select several values of Δt, calculate the summation, and plot it versus the corresponding buildup pressure $p_{w \Delta t}$ on rectangular coordinate paper.

A problem common to drawdowns and buildups is the wellbore storage. It is known as after-flow in case of buildups and unloading in case of drawdowns (Agrawal et al., 1970; Earlougher, 1977). The problem is caused by the fact that when the well is shut in at the surface, fluid migration from the formation into the wellbore continues for a period of time. During this time, q is not zero as the mathematics of buildups dictate. In case of drawdowns, early surface production comes mainly from the expansion of the fluids in the wellbore and should be accounted for in the analysis. This is treated in the section "Wellbore storage".

AVERAGE RESERVOIR PRESSURE

If the flow disturbance has not reached the drainage boundary of the well, i.e., the reservoir is infinite, then extrapolation of the straight-line portion of the buildup plot to $(t + \Delta t)/\Delta t = 1$ gives p_i, the initial pressure. For this case, the average pressure is taken to be equal to p_i.

If the effect of the boundary has been felt, extrapolation of the straight line gives $p*$ where $p_{av} < p* < p_i$. Several methods are available to relate $p*$ to p_{av}:

The Horner method. The Horner method (Horner, 1951) requires radial symmetry and the knowledge of p_i. It enables determination of p_{av} and an estimate for the drainage volume.

The method is based on an approximate solution of the equation for the bounded system:

$$p_i - p* = \frac{70.6 \, qB\mu}{kh} \, Y(u) \tag{5-63}$$

From the extrapolated straight line of the buildup plot, one reads $p*$ and calculates $Y(u) = (p_i - p*)/(70.6 \, qB\mu/kh)$. From a plot of $Y(u)$ versus u given by Horner, u is determined. The symbol u is equal to $948 \, \phi\mu cr_e^2/kt$. From material balance:

$$p_{av} = p_i - \frac{0.0745 \, qBt}{\phi hcr_e^2} = p_i - \frac{70.6 \, qB\mu}{kh} / u \tag{5-64}$$

Thus, p_{av} is calculated. From $u = 948 \, \phi\mu cr_e^2/kt$ and the slope of the buildup plot, one can estimate $\pi\phi hr_e^2$, which is the drainage volume.

The MBH method. The MBH method (Matthews et al., 1954; Matthews and Russell, 1967) requires knowledge of the drainage area of the well and its geometric shape. After $p*$ is determined, the dimensionless flow time t_D is calculated by:

$$t_D = \frac{2.64 \times 10^{-4} \, kt}{\phi\mu cA} \tag{5-65}$$

where t is time in hours; and A is the drainage area in ft^2. From figures provided by Matthews et al. (1954) one reads for the calculated t_D a correction factor $= (p* - p_{av})/(70.6\ qB\mu/kh)$. This allows the calculation of p_{av}.

The Dietz method. The Dietz method (Dietz, 1965) is a graphical solution to the MBH method. It applies when the flow prior to the buildup is in pseudo-steady state. In this case, one can show that:

$$p_{av} - p_{w\,\Delta t} = \frac{162.6\ qB\mu}{kh} \left(\log \frac{0.000264\ c_A\,k\,\Delta t}{\phi\mu c A} \right) \tag{5-66}$$

where c_A is a shape factor given by Dietz (1965) for several drainage areas and well locations. The above equation shows that $p_{av} = p_{w\,\Delta t}$, when $\Delta t = (3.8 \times 10^3\ \phi\mu cA/c_A k)$ (units are in hours, cp, inch/psi, ft^2, and md). Thus the straight line of the buildup plot is extended to $\Delta t = 3.8 \times 10^3\ \phi\mu cA/c_A k$, and the corresponding pressure value is read. This gives p_{av}.

Odeh and Al-Hussainy's method. The method of Odeh and Al-Hussainy (1971) is a combination of the Horner and MBH methods. It requires the original pressure p_i, but not the drainage area. It is applicable to all the geometries where the MBH method applies. The principal equation is:

$$\frac{p_i - p*}{|slope|} = \frac{p_i - p_{av}}{|slope|} - \log \frac{p_i - p_{av}}{|slope|} - \log c_A + 0.74 \tag{5-67}$$

From the straight-line portion of the buildup and its extrapolation $p*$, $(p_i - p*)/|slope|$ is determined. The corresponding $(p_i - p_{av})/|slope|$ is read from the plots given in Odeh and Al-Hussainy (1971). This allows the calculation of p_{av}.

The MDH method. In the MDH method (Miller et al., 1950) two plots (one corresponding to a closed outside boundary and the other to a constant pressure at the boundary) of dimensionless time versus dimensionless pressure drop are given. First, one calculates Δt_D by using the following equation:

$$\Delta t_D = \frac{2.64 \times 10^{-4}\ k\,\Delta t}{\phi\mu c r_e^2} \tag{5-68}$$

and then from the plots reads the value of the corresponding pressure drop given by:

$$\frac{1.15\ (p_{av} - p_{\Delta t})}{-m} = \frac{7.08 \times 10^{-3}\ kh\ (p_{av} - p_{\Delta t})}{q\mu B} \tag{5-69}$$

Since $p_{\Delta t}$ and m are known, p_{av} can be calculated easily. The value of $p_{\Delta t}$ should be read off the straight line of the buildup corresponding to the selected Δt in the t_D equation.

Pitzer (1964) extended this method to various drainage geometries.

The Muskat method. The Muskat method (see Larson, 1963) is useful in the case of a stratified reservoir. The basic equation, which is applicable for $\Delta t_D = 2.64 \times 10^{-4}\, \Delta tk/\phi\mu cr_e^2 > 0.1$, is:

$$\log\,(p_{av} - p_{w\,\Delta t}) = \log\,\frac{118.6\,q\mu}{kh} - 0.00168\,\frac{k\Delta t}{\phi\mu cr_e^2} \tag{5-70}$$

Thus, a plot of $(p_{av} - p_{w\,\Delta t})$ versus Δt on semilog paper, where $(p_{av} - p_{w\,\Delta t})$ occupies the semilog scale, should result in a straight line if the correct p_{av} is chosen. One guesses p_{av}, calculates $(p_{av} - p_{w\,\Delta t})$, and prepares the plot. The p_{av} which results in a straight line plot is the correct one. The slope of the line is equal to $-0.00168k/\phi\mu cr_e^2$. The intercept at $\Delta t = 0$ gives $118.6\,q\mu/kh$. This allows the calculation of k and $h\phi r_e^2$. The straight-line portion should be the one given by $\Delta t_D > 0.1$, where $\Delta t_D = 2.64 \times 10^{-4}\,kt/\phi\mu cr_e^2$. This method is difficult to apply sometimes, because it may not be easy to recognize when the straight line occurs. Usually one assumes too small a value for p_{av}, which results in the plot curving downward and increases the assumed value until the first straight line occurs.

HETEROGENEITIES AND FLOW TESTS' INTERPRETATION

The previous analysis was based on the assumption of a homogeneous reservoir. Gross heterogeneity results in modification to the shape of the normal plots. It should be noted that, in general, the modified plots are not unique, which means that more than one heterogeneity gives the same general shape. Thus, flow analysis results must be combined with all geological and geophysical data for a proper characterization of the reservoir. The general effect on the normal plots of some heterogeneities is discussed below.

Faults. The mathematical representation of a fault (Horner, 1951; Matthews and Russell, 1967) may be arrived at by placing a well image at a distance $2d$ from the tested well, where d represents the actual distance to the fault. Application of superposition to equation (5-18) shows that both the drawdown and the buildup plots in the presence of a fault show two straight lines. The slope of the second straight line is twice the slope of the first.

Horner (1951) gave an exact method for calculating the distance to the fault. However, the following approximate formulas may be used (Gray, 1965):

$$d = \sqrt{\frac{1.48 \times 10^{-4} \, \Delta t_x \, k}{\phi \mu c}}, \quad \text{when} \quad \frac{t + \Delta t}{\Delta t} \geqslant 30 \qquad (5\text{-}71)$$

and:

$$d = 0.0122 \left\{ \left[1 + 0.4 \left(\frac{\Delta t}{t + \Delta t} \right)_x \right] \sqrt{\frac{k \, \Delta t_x}{\phi \mu c}} \right\}, \quad \text{when} \quad \frac{t + \Delta t}{\Delta t} < 30 \quad (5\text{-}72)$$

where the subscript x refers to the intersection of the two straight lines.

Contacts. The presence of an oil—gas or an oil—water contact in the drainage area of the tested well results in a drawdown or a buildup plot of two straight lines connected by a curve (Bixel et al., 1963; Odeh, 1969). The ratio of the slope of the second line to the slope of the first gives the ratio of kh/μ of the area around the well to the kh/μ of the water or gas zone. The presence of any concentric zone around the drainage area of the well having different kh/μ from that of the drainage area, results in similar plots. Two basic methods are available in the literature to estimate the distance to the discontinuity. One method requires an overlay technique (Bixel et al., 1963), whereas the other uses the intersection of the two lines (Odeh, 1969).

Layered reservoirs with communication. A layered reservoir with communication behaves like a homogeneous reservoir with a total kh and ϕh given by (Matthews and Russell, 1967):

$$kh = \sum_{i=1}^{h} k_i h_i \qquad (5\text{-}73)$$

and:

$$\phi h = \sum_{i=1}^{n} \phi_i h_i \qquad (5\text{-}74)$$

The previous analysis is thus applicable to this reservoir.

Layered reservoirs without communication. In layered reservoirs without communication, the transient behavior is described by the following equation (see Matthews and Russell, 1967; Lefkovits et al., 1961):

$$p_i - p_w = \frac{162.6 \, q_T B \mu}{(kh)_T} \left(\log t + \frac{\sum\limits_{i=1}^{n} k_i h_i \log \dfrac{k_i}{\phi_i \mu c_i r_w^2}}{(kh)_T} - 3.23 \right) \qquad (5\text{-}75)$$

where $(kh)_T = \sum\limits_{i=1}^{n} k_i h_i$.

The buildup or the drawdown plot has the classical shape of an early straight line, followed by a leveling off, and then a rise. The slope of the first straight line gives $(kh)_T$. The average reservoir pressure is best obtained by using the Muskat method (see Larson, 1963).

Fractured reservoirs. The buildup plot of fractured reservoirs may exhibit an "S"-shaped plot (Warren and Root, 1963; Matthews and Russell, 1967). If this happens, the separation between the two straight lines may be used to calculate the pore volume of the fractures. On the other hand, the first part of the "S" may be missing, thus giving a plot identical to the homogeneous reservoir (Odeh, 1965).

Hydraulically fractured wells. Russell and Truitt (1964) showed that for a vertical fracture the normal plots yield permeabilities higher than the true ones. They published correlations relating the true permeability to the calculated value as a function of the fracture length (see also Matthews and Russell, 1967).

Since the publication of the Russell and Truitt paper, and because of the interest in massive fracturing, many papers have been published on this subject. Some of the papers are based on the assumption that the conductivity of the fracture is infinite, whereas others consider a finite conductivity fracture (Gringarten et al., 1974; Cinco, 1978).

For infinite conductivity fracture, the early flow is linear, and is described by:

$$p_i - p_{wt} = \frac{4.1 \, qB\mu \, t^{1/2}}{A_x \, (k\phi\mu c)^{1/2}} \qquad (5\text{-}76)$$

Analysis of test data of hydraulically fractured wells is beyond the scope of this chapter, and the reader is advised to refer to the above cited references.

Restricted entry to flow. Restricted entry to flow causes additional pressure drop, thus appearing as a positive skin (Brons and Marting, 1961; Odeh, 1968, 1980). This was discussed in the section "Skin".

WELLBORE STORAGE

The methods of analysis given previously for drawdowns and buildups assume that when the well is open to flow, in the case of a drawdown, the first drop of production comes out of the formation, whereas when the well is closed, in the case of the buildup, the flow from the formation stops instantaneously. In reality, this situation does not occur. The fluid in the wellbore when the well is open to flow expands and contributes to the early production. In fact, the real early production may be due to the expansion of the fluid in the wellbore. In case of buildups, the flow into the wellbore may continue for a while after the well is closed at the surface. During these periods, Horner-type methods of analysis do not apply.

The engineer needs to recognize when wellbore storage effects die out and the Horner methods of analysis become applicable. For example, on considering the wellbore to be a tank full of fluid under pressure, when the well is opened, the fluid expands, resulting in a production rate q:

$$q = V_w \, c \, \frac{p_i - p_w}{t} \tag{5-77}$$

where V_w is the fluid volume in the wellbore, c is the compressibility, t is flow time, and q is the flow rate (constant).

The previous equation may be written as:

$$t = \frac{V_w c}{q} \, (p_i - p_w) = C \, (p_i - p_w) \tag{5-78}$$

Taking the log of both sides gives:

$$\log t = \log \frac{V_w c}{q} + \log (p_i - p_w) \tag{5-79}$$

Thus a plot of (1) flow time versus $(p_i - p_w)$ in the case of a drawdown and of (2) shut-in time, Δt, versus $(p_{w \, \Delta t} - p_{w \, \Delta t = 0})$ in the case of a buildup on a log-log paper should result in a straight line of unit slope if the flow is dominated by wellbore storage. This will occur in the case of the buildup only if $t \gg \Delta t$, so that $\log (t + \Delta t) \simeq \log t$. The analysis proceeds as follows:

(1) Plot $(p_i - p_w)$ versus flow time in case of a drawdown or $(p_{w \, \Delta t} - p_{w \, \Delta t = 0})$ versus shut-in time in case of a buildup on log-log paper using the earliest data possible. Time is plotted on the x-axis.

(2) Examine the slope of the plot using the early data. If the slope is unity, wellbore storage dominates. If the slope $\ll 1$, the wellbore storage effect is negligible.

(3) If the slope is one, find out where the data start to deviate from the slope of unity, and advance at least one log cycle to the right. This gives the start of the data that can be used safely in the Horner-type analysis.

MULTIPHASE FLOW

Martin (1959) showed that multiphase flow can be described by an equation similar to equation (5-11). The equation is:

$$\frac{\partial^2 p}{\partial r^2} + \frac{1}{r}\frac{\partial p}{\partial r} = \frac{\phi c_t}{\lambda_t}\frac{\partial p}{\partial t} \tag{5-80}$$

where $c_t = c_o S_o + c_w S_w + c_g S_g + c_r$, and:

$$\lambda_t = k\left(\frac{k_{r_o}}{\mu_o} + \frac{k_{r_w}}{\mu_w} + \frac{k_{r_g}}{\mu_g}\right)$$

The assumption is made that:

$$\left(\frac{\partial p}{\partial r}\right)^2 = \frac{\partial p}{\partial r}\frac{\partial S_o}{\partial r} = \frac{\partial p}{\partial r}\frac{\partial S_w}{\partial r} = \frac{\partial p}{\partial r}\frac{\partial S_g}{\partial r} = 0$$

The slopes of the straight lines of the normal buildups and drawdown plots give:

$$\lambda_o = \frac{kk_{r_o}}{\mu_o} = \frac{162.6\,B_o q_o}{-mh} \tag{5-81}$$

$$\lambda_w = \frac{162.6\,B_w q_w}{-mh} \tag{5-82}$$

and:

$$\lambda_g = \frac{162.6\,B_q}{-mh}(q_g - q_o R_s) \tag{5-83}$$

The rest of the analysis is identical to that of a single-phase flow, except for substituting c_T for c_i and λ_T for λ_i.

INTERFERENCE AND PULSE TESTING

Interference tests (Matthews and Russell, 1967) and pulse testing (Johnson et al., 1966; Brigham, 1970) require two wells: the tested well and the responding well. Basically both tests are the same. The point source solution, i.e., the *Ei*-function, is the basis for the interpretive theory. The method of interpretation, however, is different. In interference tests, a normal constant-rate drawdown is run on a well and the pressure behavior at an observation well is also recorded. The kh/μ is calculated first by the normal drawdown or buildup analysis. Then the pressure drop recorded on the observation well is used to calculate the argument of the *Ei*-function from the following equation:

$$p_i - p = \Delta p = \frac{70.6\,Bq\mu}{kh}\,Ei\left(-\frac{948\,\phi h\mu cr^2}{kht}\right) \tag{5-84}$$

where r is the distance between the two wells in ft, Δp is the measured pressure drop on the observation well in psi, t is time in hours, and the rest of the symbols are in engineering units. This allows the calculation of ϕhc.

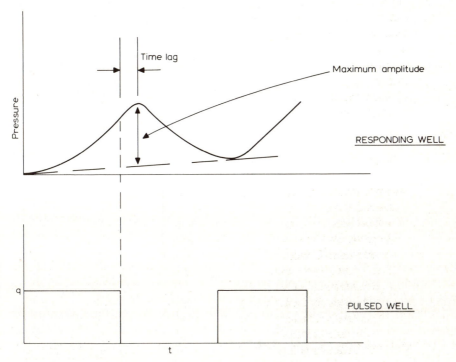

Fig. 5-2. Pressure versus time plot and responding wells.

In pulse testing one well is pulsed, i.e., produced at a constant rate for Δt, shut in for Δt, and so on. The pressure behavior in an observation well called the responding well is recorded.

The dimensionless time lag and the dimensionless maximum pressure amplitude are used to calculate kh/μ and ϕhc. These two quantities are illustrated in Fig. 5.2.

The calculation procedure has been simplified and presented in figure form by Brigham (1970). One of the advantages claimed by the advocates of pulse testing is that it requires considerably shorter time than interference testing. From the theoretical point of view, both tests require comparable time. However, very sensitive equipment is used in pulse testing, allowing the measurements ot hundredths of psi and thus reducing testing time.

INJECTION WELLS

Because of the complexity of the physical system when a fluid is injected into the reservoir, the methods of pressure analysis in injection wells are approximate. Basically the methods fall under two categories:

Unit mobility ratio. The unit mobility ratio method (Nowak and Lester, 1955; Matthews and Russell, 1967) is applicable to waterfloods which achieved fillup and in which the water and oil mobilities are approximately equal. In this case, the previously developed pressure analysis techniques for producing wells apply. One must remember, however, that for an injection well, its pressure is higher than the reservoir pressure. The fundamental solution of equation (5-21) becomes:

$$p_{\mathrm{w}} = p_i + \frac{162.6\, i\mu}{kh} \left(\log t + \log \frac{k}{\phi\mu c r_{\mathrm{w}}^2} - 3.23 \right) \tag{5-85}$$

where i is the injection rate in reservoir bbl/day.

Non-unit mobility ratio. The non-unit mobility ratio method applies prior to fillup and after fillup when the mobilities of the injected and the produced fluids differ. An approximate mathematical solution for this case was developed by Hazebroek et al. (1958). They presented the appropriate figures which simplify the analysis considerably. The analyst is advised to refer to Matthews and Russell (1967) for an explanation of the method.

In addition to the mobility ratio, the behavior of the surface pressure of the injection well after the well is closed in affects the method of analysis as shown by Matthews and Russell (1967).

CURVE MATCHING

The classical way of analyzing test data was presented here. In the early seventies, the curve matching method made its inroad into the petroleum industry. This method has been used in hydrology. To illustrate the method, the case of a well producing with skin and wellbore storage effect is selected as an example.

A set of curves is prepared relating dimensionless pressure drop versus dimensionless time. Each curve is for an assumed value of skin and wellbore storage constant. Dimensionless pressure drop, p_D, is usually defined as:

$$p_D = \frac{2\pi kh (p_i - p_w)}{Bq\mu} \qquad (5\text{-}86)$$

whereas dimensionless time, t_D, is usually defined as:

$$t_D = \frac{kt}{\phi\mu cr_w^2} \qquad (5\text{-}87)$$

To use the set of curves for analyzing the flow test data, one proceeds as follows:

(1) A piece of tracing paper is placed and taped on top of the set of curves, and the x and y lines of the set of curves are drawn on the tracing paper.

(2) The actual flow test data, i.e., $p_i - p_w$ versus t, are plotted on the tracing paper using the grid lines of the set of curves, and any appropriate scale.

(3) The tracing paper is untaped and is moved up, down, or sideways, but always keeping the x and y lines of the tracing paper parallel to the original x and y lines, until a curve in the orginal set is found on which the actual field data fit.

(4) A point is selected on the matched data and the corresponding values of $p_i - p_w$ of the tracing paper, p_D of the original set, t of the tracing paper and t_D of the original set are read.

(5) Knowing the value of p_D and the corresponding $p_i - p_w$, the value of kh/μ is given by:

$$\frac{kh}{\mu} = \frac{p_D qB}{2\pi (p_i - p_w)} \qquad (5\text{-}88)$$

In addition:

$$\frac{k}{\phi} = \frac{t_D \mu cr_w^2}{t} \qquad (5\text{-}89)$$

Theoretically, this method should be easy to apply and should result in much useful information. However, because of the many curves in a set with about the same shape, and because one deals with real data with some scatter, it is possible to find more than one curve which matches. This results in many answers which presents a serious problem with curve matching.

There are many varieties of curves by many authors for various physical models such as fractures, drillstem test, etc. A good exposure of these curves is given by Earlougher (1977).

REFERENCES

Agrawal, R.G., Al-Hussainy, R. and Ramey, H.J., Jr., 1970. An investigation of wellbore storage and skin effect in unsteady liquid flow. *Trans. AIME*, 249: 279—290.

Bixel, H.C., Larkin, B.K. and van Poollen, H.K., 1963. Effect of linear discontinuities on pressure buildup and drawdown behavior. *Trans. AIME*, 228: 885—895.

Brigham, W.E., 1970. Planning and analysis of pulse tests. *J. Pet. Technol.*, 22 (5): 618—624.

Brons, F. and Marting, V.E., 1961. The effect of restricted fluid entry on well productivity. *Trans. AIME*, 222: 172—174.

Cinco, H., Samaniego, L.F. and Domingues, A., 1978. Transient pressure behavior for a well with finite conductivity vertical fracture. *Soc. Pet. Eng. J.*, 18 (4): 253—264.

Dietz, D.N., 1965. Determination of average reservoir pressure from build-up surveys. *Trans. AIME*, 234: 955—959.

Earlougher, R., Jr., 1977. *Advances in Well Test Analysis. SPE Monogr.*, 2: 264 pp.

Gray, K.E., 1965. Approximating well-to-fault distance from build-up tests. *J. Pet. Technol.*, 17 (7): 761—767.

Gringarten, A., Ramey, H.J. and Raghavan, R., 1974. Unsteady-state pressure distribution created by a well with single infinite conductivity vertical fracture. *Trans. AIME*, 257: 347—360.

Hawkins, M.F., Jr., 1956. A note on the skin effect. *Trans. AIME*, 207: 356—357.

Hazebroek, P., Rainbow, H. and Matthews, C.S., 1958. Pressure fall-off in water injection wells. *Trans. AIME*, 213: 250—260.

Horner, D.R., 1951. Pressure buildup in wells. *Proc. 3rd World Pet. Congr., The Hague*, Part II, pp. 503—521.

Hurst, W., 1953. Establishment of the skin effect and its impediment to fluid flow into a wellbore. *Pet. Eng. J.*, 25 (11): B6—B16.

Johnson, C.R., Greenkorn, R.A. and Woods, E.G., 1966. Pulse testing: a new method for describing reservoir flow properties between wells. *Trans. AIME*, 237: 1599—1604.

Katz, D.L., Cornell, D., Vary, J.A., Kobayashi, R., Poettmann, F.H., Elenbaas, J.R. and Weinaug, C.F., 1959. *Handbook of Natural Gas Engineering*. McGraw-Hill, New York, N.Y., 802 pp.

Larson, V.C., 1963. Understanding the Muskat method of analyzing pressure buildup curves. *J. Can. Pet. Technol.*, 2 (Fall): 136.

Lefkovits, H.C., Hazebroek, P., Allen, E.E. and Matthews, C.S., 1961. A study of the behavior of bounded reservoirs composed of stratified layers. *Trans. AIME*, 222: 43—58.

Martin, J.C., 1959. Simplified equations of flow in gas drive reservoirs and the theoretical foundation of multiphase pressure buildup analysis. *Trans. AIME*, 216: 309—311.

Matthews, C.S., Brons, F. and Hazebroek, P., 1954. A method for determining average reservoir pressure in bounded reservoir. *Trans. AIME*, 201: 182—191.

Matthews, C.S. and Russell, D.G., 1967. *Pressure Buildup and Flow Tests Analysis. SPE Monogr.*, 1: 176 pp.

Miller, C.C., Dyes, A.B. and Hutchinson, C.A., Jr., 1950. The estimation of permeability and reservoir pressure from bottom-hole pressure buildup characteristics. *Trans. AIME*, 189: 91—104.

Muskat, M., 1946. *Flow of Homogeneous Fluids Through Porous Media.* J.W. Edwards, 763 pp.

Muskat, M., 1949. *Physical Principles of Oil Production.* McGraw-Hill, New York, N.Y., 922 pp.

Nowak, T.J. and Lester, C.W., 1955. Analysis of pressure fall-off curves obtained in water injection well to determine injective capacity and formation damage. *Trans. AIME*, 204: 96—102.

Odeh, A.S., 1965. Unsteady state behavior of naturally fractured reservoirs. *Trans. AIME*, 234: 60—66.

Odeh, A.S., 1968. Steady-flow capacity of wells with limited entry to flow. *Trans. AIME*, 243: 43—51.

Odeh, A.S., 1969. Flow test analysis for a well with radial discontinuity. *J. Pet. Technol.*, 21 (2): 207—211.

Odeh, A.S., 1978. Pseudo steady-state flow equation and productivity index for a well with noncircular drainage area. Forum Article. *J. Pet. Technol.*, 30 (11): 1630—1632.

Odeh, A.S., 1980. An equation for calculating skin factor due to restricted entry. *Trans. AIME.*, 269: 964—965.

Odeh, A.S. and Al-Hussainy, R., 1971. A method for determining the static pressure of a well from buildup data. *Trans. AIME*, 251: 621—624.

Odeh, A.S. and Jones, L.G., 1965. Pressure drawdown analysis, variable-rate case. *Trans. AIME*, 234: 960—964.

Odeh, A.S. and Nabor, G.W., 1966. The effect of production history on determination of formation characteristics from flow tests. *Trans. AIME*, 237: 1343—1350.

Odeh, A.S. and Selig, F., 1963. Pressure buildup analysis, variable rate case. *Trans. AIME*, 228: 790—794.

Odeh, A.S., Moreland, E.E. and Schueler, S., 1975. Characterization of a gas well from one flow test sequence. *Trans. AIME*, 259: 1500—1504.

Pitzer, S.C., 1964. Uses of transient pressure tests. *Drill. Prod. Pract. API*, 115—130.

Ramey, H.J., Jr., 1965. Non-Darcy flow and wellbore storage effects in pressure buildup and drawdown of gas wells. *Trans. AIME*, 234: 223—233.

Ramey, H.J., Jr., 1970. Short term well test data interpretation in the presence of skin effect and wellbore storage. *Trans. AIME*, 249: 97—106.

Russell, D.G., 1963. Determination of formation characteristics from two-rate flow tests. *Trans. AIME*, 228: 1347—1355.

Russell, D.G. and Truitt, N.E., 1964. Transient pressure behavior in vertically fractured reservoirs. *Trans. AIME*, 231: 1159—1170.

van Everdingen, A.F., 1953. The skin effect and its influence on the productive capacity of a well. *Trans. AIME*, 198: 171—176.

van Everdingen, A.F. and Hurst, W., 1949. The application of the Laplace transformation to flow problems of reservoirs. *Trans. AIME*, 186: 305—324.

Warren, J.E. and Root, P.J., 1963. The behavior of naturally fractured reservoirs. *Trans. AIME*, 228: 245—255.

Chapter 6

ENHANCED OIL RECOVERY INJECTION WATERS

A. GENE COLLINS and CHARLES C. WRIGHT

INTRODUCTION

Oilfield waters must be considered in all enhanced oil recovery (EOR) operations. This is true for several reasons, most of which will be elaborated upon later. The ultimate truth, however, is related to the origin of the petroleum or the chicken and the egg philosophy. Water preceded petroleum and many petroleum-bearing rocks are water-wet. Water is ubiquitous in Nature and in sedimentary rocks.

There are about seven major EOR techniques: (1) steam injection, (2) in-situ combustion, (3) carbon dioxide injection, (4) surfactant—polymer injection, (5) polymer injection, (6) alkaline (caustic) injection, and (7) injection of petroleum miscible hydrocarbons.

The importance of water in EOR technology becomes obvious when one considers the amount of water necessary to recover one barrel of oil. Table 6-1 illustrates the approximate number of barrels of water that are required to recover a barrel of oil for some EOR processes. The water quality required may vary from excellent to poor. Steam, for example, requires a high quality water for use in boilers.

Table 6-1 also presents some information on the estimated percent recovery of the residual oil-in-place for some EOR processes.

TABLE 6-1

Estimated water usage for EOR process and estimated percent recovery of residual oil-in-place

Process	Water usage,[a] bbl water/bbl oil	Percent recovery of residual oil-in-place
Surfactant—polymer	10—15	30—43
Polymer	16—50	4
Alkaline (caustic)	22—33	6—13
Carbon dioxide (CO_2)	1—3	15—19
In-situ combustion, wet	0.5—1	28—38
Steam	2—5	25—45

[a] Department of Energy (DOE), 1981.

The objective of EOR technology is to recover oil that cannot be recovered by primary and/or secondary recovery operations. However, more specifically or in more practical terms, EOR refers to technology that can be applied to increase the production of oil above the amounts that could be recovered using primary reservoir energy or secondarily-applied artificial reservoir energy. Essentially, EOR technology strives for the ultimate, i.e., to recover all the oil in a given reservoir. The EOR technique applied should be tailored to the reservoir.

A major problem in oil recovery is overcoming the interfacial tension forces between the oil and the rock (Dickey, 1979). In a water-wet rock surface, tension forces tend to create bubbles of oil which block pore openings. In an oil-wet reservoir, if it is truly oil wet, the interfacial tension forces tend to cause the oil to bind to the rock. These two types of interfacial tension forces are a prime reason why oil becomes increasingly more difficult to recover from a reservoir as water saturation increases and oil saturation decreases.

Reduction of these interfacial tension forces is a major objective of EOR. This relates to oilfield waters because oilfield waters are present in all oil reservoirs. Further, these same oilfield waters, in the form of produced water, are employed in some EOR techniques. For example, they are reinjected into the oil reservoir and serve as an oil-driving medium.

These same oilfield waters are a detriment to some EOR techniques. For example, some oilfield waters contain relatively large quantities of dissolved multivalent cations such as calcium, magnesium, strontium, and barium. These particular cations will react with some EOR chemicals such as surfactants and polymers. These reactions cause the EOR chemicals to become inactive in the EOR operation. In fact, the new chemicals formed by the reactions can hinder or prevent the additional recovery of oil.

Oilfield waters include all waters or brines found in oilfields. Such waters have certain distinct chemical characteristics.

About 70% of the world petroleum reserves are associated with waters containing more than 100 g/l of dissolved solids. A water containing dissolved solids in excess of 100 g/l can be classified as a brine. Waters associated with the other 30% of petroleum reserves contain less than 100 g/l of dissolved solids. Some of these waters are almost fresh. The presence of fresher waters, however, usually is attributed to water invasion after the petroleum accumulated in the reservoir trap.

Examples of some of the low-salinity waters can be found in the Rocky Mountain areas of the United States in Wyoming in fields such as Enos Creek, South Sunshine, and Cottonwood Creek. The Douleb oilfield in Tunisia is another example.

The composition of dissolved solids found in oilfield waters is dependent upon several factors. Some of these factors are the composition of the water in the depositional environment of the sedimentary rock, subsequent changes by rock—water interaction during sediment compaction, changes by rock—water interaction during water migration (if migration occurs), and changes by mixing with other waters, including infiltrating younger waters such as meteoric waters. The following are definitions of some different types of water.

Meteoric water. Water that was recently involved in atmosphere circulation; furthermore, "the age of meteoric groundwater is slight when compared with the age of the enclosing rocks and is not more than a small part of a geologic period" (White, 1957).

Seawater. The composition of seawater varies somewhat, but in general will have a composition relative to the following (in mg/l): chloride, 19,375; bromide, 67; sulfate, 2712; potassium, 387; sodium, 10,760; magnesium, 1294; calcium, 413; and strontium, 8.

Interstitial water. Interstitial water is the water contained in the small pores or spaces between the minute grains or units of rock. Interstitial waters are: (1) syngenetic (formed at the same time as the enclosing rocks); or (2) epigenetic (originated by subsequent infiltration into rocks).

Connate water. The term connate implies born, produced or originated to-gether, and connascent. Connate water , therefore, probably should be con-sidered to be an interstitial water of syngenetic origin. Connate water of this definition is fossil water that has been out of contact with the atmosphere for at least a large part of a geologic period. The implication that connate waters are only those "born with" the enclosing rocks, however, is an un-desirable restriction (White, 1957).

Diagenetic water. Diagenetic waters are those waters that have changed chemically and physically before, during and after sediment consolidation. Some of the reactions that occur in or to diagenetic waters include bacterial, ion exchange, replacement (dolomitization), infiltration by permeation, and membrane filtration.

Formation water. Formation water, as here defined, is water that naturally occurs in the rocks and is present in them immediately before drilling.

Juvenile water. Water that is present in primary magma or is derived from primary magma is juvenile water (White, 1957).

Condensate water. Water associated with gas sometimes is carried as vapor to the surface of the well where it condenses and precipitates because of temperature and pressure changes. More of this type of water occurs in the winter months and in colder climates and only in gas-producing wells. This type of water is easy to recognize because it contains a relatively small amount of dissolved solids, most derived from reactions with chemicals in or on the well casing or tubing.

OCCURRENCE, ORIGIN AND EVOLUTION OF OILFIELD WATERS

The sedimentary rocks, which now consist of stratified deposits, originally were laid down as sediments in oceans, seas, lakes, and streams. Naturally, these sediments were filled with water. This water is still present in the stratified sediments and millions of years later would be considered truly connate water.

Many large sedimentary strata were originally associated with oceans and seas. The original associated water, therefore, was marine in such sediments. Sediments laid down by lakes and streams would not contain a marine water during their initial deposition. However, with time and tectonic events, plus transgression and regression of oceans and seas, even these sediments probably were exposed to marine waters by infiltration.

In any event, the petroleum, which formed from organic matter deposited with the sediments, migrated from what is usually called the source rock into more porous and permeable sedimentary rock (reservoir rock). Petroleum, i.e., oil and gas, is less dense than water; therefore, it tends to float to the top of a water body regardless of whether the water is on the surface or in the subsurface.

Water associated with petroleum in a subsurface reservoir is called an oilfield water. By this definition, any water associated with a petroleum deposit is an oilfield water.

The question of the origin of oilfield brines is difficult to answer in a general manner. The water involved and the constituents dissolved in the water to form the brine can involve divergent histories. Subsurface water is there either because it originally was there or because it infiltrated to the subsurface from the surface. If it was there originally, it would be endogenetic, whereas if it infiltrated from the surface and/or penetrated from above with sediment accumulations, it would be exogenetic.

Obviously these two types of waters could meet and mix in the subsurface, and thus the mixture would contain water of two separate origins. The problem could multiply if more than one exogenetic water were involved. In addition, there is water of compaction, which moves upward due to the compaction of sediments.

The chemical composition of an oilfield brine is an end-product of several variables. These variables include the dissolved ions, salts, gases and organic matter; reactions between these dissolved and dispersed constituents; and interaction of the brine with the surrounding rocks, petroleum, etc. There are a number of pertinent reactions which could cause the composition of a subsurface oilfield brine to change, including leaching of the rocks, ion exchange between water and rock, oxidation—reduction, mineral hydration, mineral formation and/or dissolution, ion diffusion, gravitational segregation of ions, and membrane filtration or other osmotic effects.

It is rather difficult to rank the factors which might be more important for general consideration. However, two of the more important factors probably are: (1) the original composition of the water, and (2) interaction of that water with the rocks. If one assumes that the original water was a marine water and that the associated sediments (subsequently sedimentary rocks) were marine, then the original composition of the marine water could be an important factor.

However, even the salinity in the various oceans and seas is not constant. For example, the salinity of the waters in the major oceans ranges from 33,000 to 38,000 mg/l, about 40,000 mg/l in the Mediterranean Sea, up to 70,000 mg/l in the Red Sea, about 18,000 to 22,000 mg/l in the Black Sea, and only about 1000 mg/l in the Baltic Sea. Some land-locked waters, such as the Dead Sea, Great Salt Lake, etc., contain waters that are nearly saturated with dissolved solids.

Studies of formation waters in the western Canada sedimentary basin indicate that 85% of the strata were deposited under marine conditions, whereas 15% were deposited under brackish water and, possibly, under fresh-water conditions (Hitchon et al., 1971). These investigators estimated that 80% of all the sedimentary strata in Alberta were deposited under marine conditions. This led to the conclusion that one could assume with negligible error that all sedimentary strata originally contained seawater.

Further, the study indicated that evaporites form an important volumetric part of several of the stratigraphic units. Some of the stratigraphic units possibly contain bitterns subsequent to halite precipitation, but preceding precipitation of potassium salts. They calculated an average formation water salinity of about 46,000 mg/l total dissolved solids, which indicated a net gain of dissolved salts. Of the major and some minor components, all showed a net gain with the exception of Mg^{2+} and SO_4^{2-}.

Factor analysis was used in interpretation of the analyses, and the following factors were considered to be major controls: composition of the original seawater, dilution by fresh-water recharge, membrane filtration, solution of halite, dolomitization, bacterial reduction of sulfate, formation of chlorite, cation exchange on clays, contribution from organic matter, and solubility relations.

It was concluded that the formation waters of western Canada are ancient seawaters in which the deuterium concentration was changed because of mixing with infiltrating fresh water, that ^{18}O was exchanged with carbonates in the rocks, and that the dissolved salts are in equilibrium with the rock matrix subsequent to their redistribution by membrane filtration and/or dilution by fresh-water recharge. Equilibrium was attained by such processes as dissolution of evaporites, new mineral formation, cation exchange on clays, desorption of ions from clays and organic matter, and mineral dissolution.

The majority of the published research studies seem to agree that all of the above controls and/or reactions are involved in establishing the composition of oilfield brines. Further, most investigators agree with the assumption that marine water usually is a part of the original material from which formation waters evolve. Opinions, however, are not unanimous with respect to how they evolved. The major disagreements are related to the membrane filtration theory and to other modes of concentrating the dissolved solids such as seawater evaporation.

It is possible to reconstruct the evolution of some oilfield brines in sedimentary basins if one reasons that they are genetically related to evaporites. For example, geochemical and geological studies of some very concentrated brines indicate that in deep quiescent bodies of water, strong bitterns can persist for long periods of time under a layer of near normal seawater. As

TABLE 6-2

Concentration ratios and excess factor ratios for some constituents in Smackover brines

Constituent	Average composition (mg/l)		Concentration ratio[a]	Excess factor[b]	Number of Smackover samples
	seawater	Smackover brines			
Lithium	0.2	174	870	18.1	71
Sodium	10,600	67,000	6	0.1	283
Potassium	380	2800	7	0.2	82
Calcium	400	35,000	88	1.8	284
Magnesium	1300	3500	3	0.1	280
Strontium	8	1900	238	5	85
Barium	0.03	23	767	16	73
Boron	4.8	130	27	0.6	71
Copper	0.003	1.1	359	7.5	64
Iron	0.01	41	4049	84.2	90
Manganese	0.002	30	14,957	311	69
Chloride	19,000	172,000	9	0.2	284
Bromide	65	3100	48	1	74
Iodide	0.05	25	501	10.4	73
Sulfate	2690	450	0.2	0.003	271
Mg′ [c]	1543	24,729	16	0.3	284

[a] Amount in brine/amount in seawater.
[b] Concentration ratio of a given constituent/concentration ratio of bromide.
[c] $Mg' = (24.31/40.08) \times mg/l\ Ca + mg/l\ Mg$.

a result, carbonates can precipitate from the less saline water and fall through the bittern at the bottom, and as compaction proceeds, the pore spaces remain filled with bitterns. Some fossil brines once trapped have not moved very far or very fast.

A geochemical model can be built to represent the origin and evolution of this type of brine, using the relatively simple operations and processes of (1) evaporation, (2) precipitation, (3) sulfate reduction, (4) mineral formation and diagenesis, (5) ion exchange, (6) leaching, and (7) expulsion of interstitial fluids from evaporites during compaction.

Experimental work indicates that about 14% of the bromide in seawater precipitates with the halite as seawater evaporates. The average concentration of bromide in the Smackover brines is 3100 mg/l (Table 6-2). The average degree of concentration of these brines compared to seawater, therefore, is $3100 \div 65 = 48$. Assuming that seawater is concentrated 50-fold, the approximate composition of a brine can be calculated.

Most of the Smackover brines are deficient in magnesium relative to an evaporite-formed brine, as illustrated in Table 6-2. Table 6-3 illustrates the approximate amounts of calcium, magnesium, bromide, and sulfate that could exist in a water before and after precipitation of gypsum.

Assuming that the sulfate (residual = 1644 mg/l) was removed by the dolomitization reaction as shown in equations (6-1), (6-2), and (6-3):

$$MgCl_2 + 2CaCO_3 \rightarrow CaCl_2 + CaMg(CO_3)_2 \tag{6-1}$$

$$CaCl_2 + MgSO_4 \rightarrow CaSO_4 + MgCl_2 \tag{6-2}$$

$$MgSO_4 + 2CaCO_3 \rightarrow CaSO_4 + CaMg(CO_3)_2 \tag{6-3}$$

then the Mg/Br ratio would be about $883 \div 65 = 13.6$, as illustrated by the data in Table 6-4. These reactions explain the limestone, anhydrite, and dolomite beds and interbeds commonly found in the Jurassic age strata in the Gulf Coast area. The reactions also explain where much of the sulfate originally in the brine has gone — namely, to calcium sulfate, which is

TABLE 6-3

Approximate seawater composition before and after gypsum precipitation (mg/l)

Ion	Before precipitation	After precipitation
Calcium	390	0
Magnesium	1300	1300
Bromide	65	65
Sulfate	2580	1644

genetically indistinguishable from the huge amounts of calcium sulfate precipitated before halite precipitation.

However, if the residual sulfate was removed by bacterial reduction (equation 6-4):

$$C_nH_m + Na_2SO_4 \rightarrow Na_2CO_3 + H_2S + CO_2 + H_2O \tag{6-4}$$

the Mg/Br ratio would be about $1300 \div 65 = 20$.

Magnesium will react with $CaCO_3$ (calcite) to form dolomite; thus, the concentration of calcium is increased in the brine, and the Smackover brines are enriched in calcium, as shown in Table 6-2. However, the total calcium plus magnesium in the brine should remain constant; this can be calculated as follows:

$$\frac{24.31}{40.08} \times \text{mg/l Ca} + \text{mg/l Mg} = \text{total equivalent Mg or Mg}' \tag{6-5}$$

The Mg'/Mg ratio will vary, depending upon the availability of calcite, and the ratio is indicative of the degree of dolomitization.

For example, brines in equilibrium with sandstones have a relatively low Mg'/Mg ratio; those in equilibrium with dolomite have higher ratios; and those in equilibrium with limestone have the highest ratios. The average Mg'/Mg ratio for the Smackover brines studied is seven, which indicates that the brines were in equilibrium with limestone and dolomite. Studies have shown that the average ratio in sandstones is about 2.5 and in limestones about 9.5.

The formation of chlorite from montmorillonite requires about 9.2 moles of MgO per mole of chlorite as follows:

$$1.7Al_2O_3 \cdot 0.9MgO \cdot 8SiO_2 \cdot 2H_2O + 9.2MgO + 6H_2O \rightarrow$$

$$10.1MgO \cdot 1.7Al_2O_3 \cdot 6.4SiO_2 \cdot 8H_2O + 1.6SiO_2 \tag{6-6}$$

Such a reaction can remove large amounts of magnesium from waters.

TABLE 6-4

Approximate seawater composition (mg/l) after dolomitization or bacterial reduction

Ion	After dolomitization	After bacterial reduction
Calcium	0	0
Magnesium	883	1300
Bromide	65	65
Sulfate	0	0

As evaporites form, bromide is accommodated in the halite crystal lattice and replaces chloride in solid solution. The weight percentage of bromide in solid solution in the halite lattice is related to its weight percentage in the parent brine as follows:

$$C = \frac{\text{wt.\% Br (in halite)}}{\text{wt.\% Br (in solution)}} \tag{6-7}$$

where C is the partition coefficient. In most natural environments for halite, $C = 0.14$ (Braitsch and Herrman, 1963).

The bromide relationships to chloride in saline deposition sequences are relatively constant. The relative weight proportion of bromide in halite, sylvite, carnallite, and bischofite of paragenetic crystallization at $25°C$ is: 1 (halite):10 ± 1 (sylvite):7 ± 1 (carnallite):9 ± 1 (bischofite), respectively. The bromide concentration within a given halite sequence theoretically should increase from the bottom to the top.

Bromide does not form its own minerals when seawater evaporates. Some of it is lost from solution because it forms an isomorphous admixture with chloride in the halite precipitate. However, more bromide is left in solution than is entrained in the precipitate. Relative to chloride, therefore, the bromide concentration in the brine increases exponentially. Because of this, the bromide concentration in the brine is a good indicator of the degree of seawater concentration, assuming that appreciable quantities of biogenic bromide have not been introduced.

Data presented in Table 6-2 were obtained by comparing the average composition of the Smackover brines to that of seawater. The concentration ratio was calculated by taking the mean average for a given constituent in the Smackover brines and dividing it by the amount of the constituent found in normal seawater. The excess factor was determined by dividing the concentration ratio of a given constituent by that of bromide. The calculation for Mg', or total equivalent magnesium, was previously explained, and the number of Smackover samples indicates how many samples were used in the calculation. For example, 71 Smackover brines were analyzed for lithium, whereas 283 were analyzed for sodium.

The concentration ratios indicate that, with the exception of sulfate, all of the determined constituents in the Smackover brines were enriched with respect to seawater. The excess factor ratios, however, indicate that sodium, potassium, magnesium, boron, chloride, sulfate, and total equivalent magnesium generally were depleted in the Smackover brines, whereas lithium, calcium, strontium, barium, copper, iron, manganese, and iodide were enriched. Furthermore, these ratios indicate that the Smackover brines have been altered considerably, assuming that they originally were seawater.

The concentration ratio of 48 for bromide is high. For example, bromide concentration ratios of 1.2, 4.4, 8.8, and 7.2 were found for brines from

Tertiary, Cretaceous, Pennsylvanian, and Mississippian age rocks, respectively, in other studies.

Strontium is also enriched in the Smackover brines relative to seawater, as shown in Table 6-2, and reactions that can aid in this are:

$$CaSr\,(CO_3)_2 + CaCl_2 \rightarrow 2CaCO_3 + SrCl_2 \tag{6-8}$$

$$2SrCO_3 + MgCl_2 \rightarrow SrMg\,(CO_3)_2 + SrCl_2 \tag{6-9}$$

and:

$$SrMg\,(CO_3)_2 + MgCl_2 \rightarrow 2MgCO_3 + SrCl_2 \tag{6-10}$$

The concentrations of potassium and sodium are depleted relative to seawater in some of the Smackover brines as indicated by the K/Br and Na/Br ratios and in Table 6-2. Some reactions of brines to form silicates that can account for the depletion of alkali metals are:

$$3Al_2Si_2O_5\,(OH)_4 + 2K^+ \rightarrow 2KAl_3Si_3O_{10}\,(OH)_2 + 3H_2O + 2H^+ \tag{6-11}$$

$$KAl_3Si_3O_{10}\,(OH)_2 + 6SiO_2 + 2K^+ \rightarrow 3KAlSi_3O_8 + 2H^+ \tag{6-12}$$

and:

$$Al_2Si_2O_5\,(OH)_4 + 4SiO_2 + 2Na^+ \rightarrow 2NaAlSi_3O_8 + H_2O + 2H^+ \tag{6-13}$$

These reactions account not only for the depletion of potassium or sodium, but also for a decrease in pH because of the release of hydrogen ions. The released hydrogen ions could react with constituents in the rocks. The most probable reaction would be with carbonate minerals:

$$2H^+ + CaCO_3 \rightarrow Ca^{2+} + CO_2 + H_2O \tag{6-14}$$

This reaction would increase the calcium concentration in the brine. The hydrogen ion also might dissolve other metallic minerals, convert bicarbonate to carbon dioxide, or convert bisulfide to hydrogen sulfide. Many Smackover brines contain hydrogen sulfide.

Table 6-2 indicates that calcium is enriched in all of the Smackover waters. In addition to the reactions (equations (6-1) and (6-12), cation exchange reactions with clays can account for some of this enrichment:

$$2Na^+\,(solution) + Ca\,(clay) \rightarrow Ca^{+2}\,(solution) + 2Na\,(clay) \tag{6-15}$$

This type of reaction also can explain additional sodium depletion as indicated by the excess factor in Table 6-2.

Seawater trapped in the deposited sediments has an index of base exchange (IBE) > 0.129 and a Cl/Na > 1.17 (Schoeller, 1955). Intruding meteoric water in sedimentary marine rock has an IBE < 0.129 and Cl/Na < 1.17. The IBE for all of the Smackover brines studied was positive, indicating that alkali metals in the waters exchanged for alkaline earth metals on the clays. If the IBE were negative, the indication would be exchange of alkaline earth metals in solution for alkali metals on the clays. In addition, the IBE for all of the samples exceeded 0.129, indicating that the brines were derived from ancient seawater trapped in deposited sediments.

Classification of the brines indicated that most of them are the chloride—calcium type according to Sulin's (1974) classification. All of the brines were very highly concentrated in chloride, according to Schoeller's (1955) classification. The sulfate concentrations were generally less than 6 equivalents per million (epm), indicating normal sulfation, and the ratio $SO_4 \times 100/Cl$ was less than 1, which is characteristic of brines associated with hydrocarbon accumulations. Many of the brines were saturated with calcium sulfate using an arbitrary value of 70 for $\sqrt{SO_4 \times Ca}$. The elevated pressures and temperatures and high concentrations of chloride and other ions can account for this. Many of the brines were saturated with calcium carbonate because they had values greater than seven for $\sqrt[3]{(HCO_3 + CO_3)^2 (Ca)}$.

If strontium in seawater precipitates with calcium carbonate in the same ratio that calcium and strontium occur in seawater, the ratio would be 400/8 or 50. The strontium content of the freshly precipitated calcium carbonate (limestone), therefore, would be 40/50 or 0.8%. Aragonite oolites from the Bahamas contain about 1.0% strontium. Analyses of the rocks from Jurassic age strata indicated that the strontium content ranged from less than 0.01% to about 0.5%. One rock sample from a depth of 12,000 ft in Choctaw County, Alabama, contained an anomalously high concentration of 5% strontium.

Finally, precipitation of tachyhydrite will remove both magnesium and calcium from solution. A Jurassic brine from Clarke County, Alabama, contained about 15% calcium chloride and 23% magnesium chloride and had a specific gravity of 1.387. At ambient temperature and pressure, crystals containing 40% tachyhydrite and 60% bischofite precipitated.

Fig. 6-1 is a plot of Na$'$ versus Br for the Jurassic brines, an altered relict bittern, evaporating seawater, and evaporite salt dissolved in water. Na$'$ = mg/l Na + $(2 \times 23/40)$ mg/l Ca because, in an exchange reaction, 2 moles of sodium are exchanged for 1 mole of calcium. The assumption is that the sodium in the brine exchanges for calcium in the clay, thus increasing the calcium concentration in the brine. Na$'$ is plotted versus Br because the bromide concentration should be proportional to the amount of salt redissolved. The Smackover brines contain some iodide, which indicates bioconcentration; therefore, some biogenic bromide is present, and this accounts for some of the point scatter.

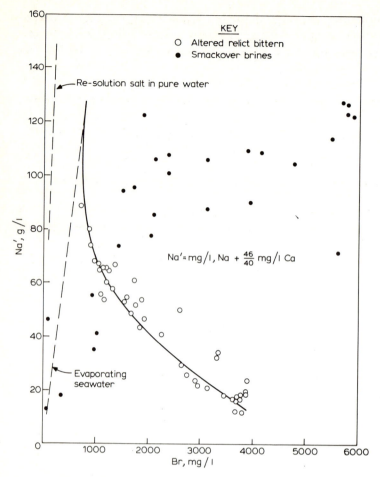

Fig. 6-1. Comparison of Na' and Br concentrations in the Smackover brines to an altered relict bittern, evaporating seawater, and re-dissolved salt (from Collins, 1975a, p. 361, fig. 11-9).

The plots in Fig. 6-1 indicate that the Smackover brine is altered relative to evaporated seawater; however, it appears to be an alteration that differs from the altered relict bittern. The high bromide content in the Jurassic brines indicates that the Jurassic brines reached the bittern state at one or more times, and it is further postulated that these brines are related to the Louann Salt. For example, the brines could have originated as the interstitial fluids in the Louann Salt, and with compaction resulting from sedimentation, they were expelled upward. The inert character of bromide makes it very useful in determining the chemical evolution of a brine derived from seawater. Knowledge of the bromide ion was used to differentiate some oilfield brines from Mississippi and Alabama (Collins, 1967). Ritten-house (1967) used the bromide ion to determine the origin and to classify oilfield brines. More recently, Carpenter (1978) used the bromide ion to study the origin and chemical evolution of brines in sedimentary basins. Knowledge of the bromide concentration is particularly useful in such studies, because as a water containing bromide evaporates, more bromide remains in solution than the amount of bromide entrained in the precipitates. When seawater is evaporated, most of the bromide, potassium, lithium and rubidium remain in solution until potassium salts precipitate. After that most of the bromide and lithium still remain in solution. The bromide ion is the only ion that is essentially immune to diagenetic reactions with the rock matrix.

Graphic plots can be made of the ions starting in seawater and progressing through the changes in ion concentration as seawater is evaporated (see Table 6-5). Collins (1975a) used such plots to study the concentrations of lithium, sodium, potassium, calcium, magnesium, strontium, boron, bromide, iodide, and chloride. The main inference that can be drawn from these plots is to determine if a given ion in an oilfield water is higher or lower in concentration than in a normally evaporated seawater.

TABLE 6-5

Five relative concentration changes of some dissolved ions during evaporation of seawater and brine[a]

Constituents	Concentrations (mg/l)					
	seawater	$CaSO_4$	NaCl	$MgSO_4$	KCl	$MgCl_2$
Lithium	0.2	2	11	12	27	34
Sodium	11,000	98,000	140,000	70,000	13,000	12,000
Potassium	350	3600	23,000	37,000	26,000	1200
Rubidium	0.1	1	6	8	14	10
Magnesium	1300	13,000	74,000	80,000	130,000	153,000
Calcium	400	1700	100	10	0	0
Strontium	7	60	10	1	0	0
Boron	5	40	300	310	750	850
Chloride	19,000	178,000	275,000	277,000	360,000	425,000
Bromide	65	600	4000	4300	8600	10,000
Iodide	0.05	2	5	7	8	8

[a] Approximate mg/l. Columns headed seawater, $CaSO_4$, etc., represent stages in seawater evaporation. For example, seawater contains 0.2 mg/l of lithium; after calcium sulfate has precipitated, the residual brine contains about 2 mg/l of lithium; after sodium chloride has precipitated, the residual brine contains about 11 mg/l of lithium. The residual brine contains about 12 mg/l of lithium after magnesium sulfate precipitates, 27 mg/l of lithium after potassium chloride precipitates, and 34 mg/l of lithium after magnesium chloride precipitates.

Bromide can be classified as a geochemical marker constituent, which is useful in studying the genesis and diagenesis of most oilfield brines. Most of the historical analyses of oilfield brines did not include the bromide ion. Therefore, the numerous U.S. oilfield brine analyses now available on computer tape do not include many bromide analyses (Collins, 1979). Carpenter (1978) has shown explicitly the importance of the bromide ion in determining the origin and chemical evolution of brines in sedimentary basins. His study is based on a starting material of Black Sea water with a salinity of about 20,000 mg/l, whereas some other studies are based on a starting material of seawater with a salinity of about 35,000 mg/l. The ion ratios, however, are not greatly different.

Carpenter (1978) used the Mg/K ratio to estimate the bromide concentrations in brines. This technique is useful when studying historical oilfield brine analyses.

Shale compaction and membrane filtration

Some investigators believe that the salinity variations found in oilfield waters are the result of filtration of water through shale (Clayton et al., 1966). The membrane filtration theory was first suggested by De Sitter (1947). Laboratory experiments were performed which indicated that natural shales can act as semipermeable membranes (Kharaka and Berry, 1973; McKelvey and Milne, 1962). This system working in nature should cause a water to be more saline on one side of the shale membrane and almost fresh or less saline on the other side. This follows because the shale membrane should filter out the dissolved ions on the up-flow side, causing the water on the down-flow side to contain few or no dissolved ions.

The greater the ion exclusion on the up-flow side, the more ideal the membrane. If the membrane is truly ideal, no ions will pass through with the water. According to various investigators, the shale membranes, if not ideal, will allow partial fractionation of the dissolved ions. Namely, some of the dissolved ions may pass through the shale with the water. Of course, even with fractionation the water on the up-flow side will continue to become more and more saline. Theoretically, however, if an electrolyte is forced to pass through the shale membrane, a high-salinity gradient (osmotic pressure) should develop between the solutions on both sides of the membrane. High hydraulic pressures are required to overcome the osmotic pressure and force solutions through some membranes. In fact, the required pressures could range up to hundreds of kilograms per square centimeter. Other investigators have evaluated difficulties of producing brines in natural geologic environments and have concluded that the pressure requirements for appreciable salt filtration are not satisfied by any known situations (Manheim and Horn, 1968). Bond (1972) determined that, prior to pumpage, the vertical head differences in the deep aquifers of the Illinois Basin were not sufficiently large to permit filtering through a shale membrane.

The stable isotopes have been used in several research studies to determine the origin of oilfield brines. With proper interpretation, these analyses have provided an excellent method to help differentiate an endogenetic water from an exogenetic water.

The most useful stable isotope data for these studies are δD and $\delta^{18}O$. Knowledge of these two isotopes is compared to the isotope values found in standard mean ocean water (SMOW) (Craig, 1961). The stable isotope value for $\delta^{13}C$ is used by determining the value found in the dissolved bicarbonate. Interpretations are made to determine if the $\delta^{13}C$ values result from inorganic or organic carbon sources.

CHEMICAL AND PHYSICAL PROPERTIES OF OILFIELD WATERS

Oilfield waters are analyzed for various chemical and physical properties. Most oilfield waters contain a variety of dissolved inorganic and organic compounds. However, the oil producer usually is interested in only a few of the macro properties. This is understandable, because the oil producer wishes to spend the least amount of money possible. It is necessary, therefore, to examine only the properties that are needed to evaluate any treatment for re-injection to recover more oil or to dispose of the oilfield waters.

Inorganic constituents

The major inorganic constituents dissolved in an oilfield brine quite often are measured by the petroleum company or the petroleum service company. The analytical data have several uses, including water identification, log evaluation, water treatment, environmental impact, geochemical exploration, and recovery of valuable minerals (Collins, 1975a).

Water identification often is done by constructing diagrams of the analytical data. These diagrams are useful in easy identification of some waters from some formations. The analytical data also are useful in evaluating the downhole logs, such as the resistivity log. Some constituents, such as boron and some of the organic acid salts, affect the log reading and its interpretation.

Environmental impact studies are made by using the analytical data to determine what environmental damage might occur under certain conditions. Geochemical exploration for petroleum and other minerals is done by using certain types of analytical data. Some oilfield brines contain economic concentrations of valuable elements, such as bromine, iodine, lithium, and magnesium. Oilfield brines can contain numerous dissolved and/or suspended constituents. The major elements usually present are sodium, calcium, magnesium, chloride, bicarbonate, and sulfate.

Rittenhouse et al. (1969) concluded that ions in oilfield waters commonly are present in the following concentrations:

%	Na^+, Cl^-
% or ppm	Ca^{2+}, Mg^{2+}, SO_4^{2-}
> 100 ppm	K^+, Sr^{2+}
1—100 ppm	Al^{3+}, B, Ba, Fe, Li
ppb (most oilfield waters)	Cr, Cu, Mn, Ni, Sn, Ti, Zr
ppb (some oilfield waters)	Be, Co, Ga, Ge, Pb, V, W, Zn

Cations

The major cation in most oilfield brines is sodium. Once sodium is in solution it has a tendency to stay in solution. It is not easily precipitated with an anion, and it is not as easily adsorbed by clays as are cesium, rubidium, potassium, lithium, barium, and magnesium.

The concentration of calcium in oilfield brines is a function of the origin of the water, the ages of the water and the enclosing rocks, type of rock, and type of clays in the rock. Reactions, such as ion exchange, formation of dolomite, formation of chlorite, etc. affect the concentrations of calcium in oilfield brines. The calcium concentration can range from less than 1000 mg/l to more than 30,000 mg/l. In general, waters associated with limestone, dolomite, gypsum, or gypsiferous shale contain more calcium than waters associated with sandstones.

Magnesium concentrations in oilfield brines, like calcium concentrations, are dependent upon origin of the water, rock, etc. Perhaps the prime reaction that affects the magnesium concentration in oilfield brines is the formation of dolomite. In general, this reaction depletes the concentration of magnesium and increases the concentration of calcium. Oilfield waters and brines contain from less than 0.1 g/l to more than 30 g/l magnesium.

The other cations which often are present in oilfield brines in concentrations greater than 10 mg/l are potassium, strontium, lithium, and barium. These alkali and alkali earth metals react similarly in mineral formation and/or ion exchange to their sister elements, i.e., sodium and calcium. Potassium in relation to sodium generally is depleted in oilfield waters, and this probably is caused by mineral formation. Concentrations ranging from 10 mg/l to more than 1 g/l are common. Strontium, like calcium, is enriched in many oilfield waters. The concentrations of strontium in oilfield waters range from a trace to more than 3 g/l.

Barium is found in some oilfield brines in concentrations greater than 100 mg/l. It also is a trace constituent, however, with less than 1 mg/l in many oilfield waters. The chemical properties of barium are similar to those of calcium and strontium. They precipitate as carbonates through loss of carbon dioxide from a solution containing bicarbonate, or as sulfates by the action of sulfuric acid, sulfides, or sulfates.

Some oilfield brines contain other inorganic cation constituents in concentrations greater than 10 mg/l. Examples of these constituents are aluminum, ammonium, iron, lead, manganese, and zinc. Aluminum is present in many clays, and the pH of the water is a prime control of the amount of dissolved aluminum. A pH of less than 4.0 would allow the dissolution of 1.0% or more of aluminum. Some oilfield brines possess a pH of about four in their in situ environment.

The ammonium ion is present in many oilfield waters, and often its concentration exceeds 10 mg/l. It is believed by some investigators to be an indicator of the presence of petroleum. Ammonia, NH_3, forms during the anaerobic decay of organic nitrogenous material. In the petroleum environment, ammonia transforms to ammonium, which contains nitrogen in the N^{3-} oxidation state, a reduced form. This reduced form is characteristic of a petroleum environment where the redox potential is low. If the redox potential were high, the ammonia could be oxidized to nitrate.

The solubility of iron in oilfield brines is a function of the iron compound involved, the amounts and types of other ions in solution, the pH, and the redox potential (Eh). Formation rocks contain carbonates, hydroxides, oxides, and sulfides of iron. When the equilibria is correct, therefore, iron could be dissolved. Oilfield brines contain from traces to over 1000 mg/l of iron. Knowledge of the amounts of ferrous and ferric iron dissolved in the oilfield brine can be used to calculate the Eh.

Lead occurs in several oilfield brines throughout the world. Its solubility is limited primarily by the solubility restrictions of its sulfide and sulfate in reducing and oxidizing systems. Lead is soluble in acetic acids and other acids, and it can be transported as the bicarbonate, which is a more soluble stable form than the carbonate. Lead concentrations as high as 100 mg/l are found in some Jurassic age oilfield brines in Mississippi (Carpenter et al., 1974). Metallic lead scale forms on some of the subsurface oil production equipment, causing severe oil production problems in the area. Most oilfield brines contain only traces of lead.

Manganese is present in many oilfield brines because it is readily dissolved by waters containing carbon dioxide and sulfate. In a reducing environment with a pH less than 7.0, such as that of most oilfield brines, the manganous ionic form of manganese predominates. Oilfield brines contain from trace amounts to about 10 mg/l of manganese.

The solubility of silica in water is a function of temperature, pressure, pH, and other ions in solution. Most silica in natural water probably is in the form of monomolecular silicic acid — H_4SiO_4 or $Si(OH)_4$. The solubility increases with increasing concentrations of sodium chloride, increasing pressure, and increasing temperature (up to about 125°C). The solubility increases with increasing pressure and with increasing temperature between

30° and about 100°C. Between about 125° and 200°C, the solubility decreases with increasing temperature. The solubility of silica in the presence of calcium chloride solutions decreases with increasing concentration of calcium chloride and with increasing temperature above about 100°C. Subsurface oilfield brines usually contain less than 30 mg/l of dissolved silica. Rittenhouse et al. (1969) reported that silica solubility ranges from about 1 to 500 ppm (as silicon), and that some low-salinity waters contain a higher median content of silica than more saline waters in other areas.

Acidity results from the solvated hydrogen ion H_3O^+, which is found in nature. Volcanic emanations produce HF, HCl, and H_2SO_4, probably formed by reactions between water and constituents associated with the magma. Waters associated with peat may contain organic acids, rain waters may contain carbonic acid, and waters associated with reducing conditions and anaerobic bacteria may contain H_2S.

Acidity, in contrast to alkalinity, is the capacity of a solution to neutralize a base. Most oilfield brines normally are not acid. New wells or reworked wells often are acidified or "acidized" with a strong mineral acid or a combination of mineral and organic acids. This treatment causes the produced water to contain a certain amount of acids until they are neutralized or diluted. Because of the large quantities of acids used in some treatments, it may take 6 months or longer for the water produced from a treated well to return to normal. The concentration of organic acids and organic acid salts, which are commonly found in oilfield waters, ranges from trace amounts to more than 3000 mg/l.

Krejci-Graf (1978) reviewed the geochemistry of oilfield brines, showing the concentrations of trace constituents found in oilfield waters in numerous areas of the world. Analyses were given which show the amounts of constituents, such as antimony, gallium, gold, mercury, silver, titanium, vanadium, uranium, etc. Numerous anions present at trace levels also are included in this review.

Anions

The major anion in most oilfield brines is chloride, which is very mobile in the total hydrosphere. It is also the predominant anion in seawater. The chloride ion does not form low-solubility salts, is not easily adsorbed on clays or other mineral surfaces, and does not play a significant role in most of the oxidation and reduction reactions in subsurface oilfield brines. It also does not form important solute complexes. The concentration of chloride in oilfield brines ranges from less than 19 g/l to more than 200 g/l.

Bromide ranks as the next most predominant anion in many oilfield brines. As a general rule, oilfield water samples are more likely to be analyzed for bicarbonate or sulfate than for bromide anion. The concentration of bromide in many oilfield waters, therefore, is less commonly known than the concentrations of bicarbonate and sulfate anions. Knowledge of the bromide concentration is important in determining the origin of an oilfield brine. Bromide anion is one of the better, if not the best, geochemical marker constituents of an oilfield brine. The concentration of bromide in oilfield brines ranges from less than 50 mg/l to more than 6 g/l.

Borates or boric acid are quite common in oilfield brines. The boron compounds in oilfield brines are treated here as anions. They form an important buffering system second only to the carbonate system. Boron is reported as HBO_2, H_3BO_3, BO_3 or B. In general, together with bromine and iodine, it is always associated with petroleum brines. Boron affects electric log deflections when its concentration approaches or exceeds 100 mg/l. It is present in oilfield brines in concentrations ranging from a trace to more than 200 mg/l, when calculated as B.

Bicarbonates usually are present in oilfield brines. The concentration reported for many oilfield brines is low because appropriate precautions often are not taken in determining the bicarbonate content or, as it is often called, the alkalinity. Alkalinity is defined as the capacity of a solution to neutralize an acid to a pH of 4.5. A solution with a neutral pH of 7.0 may have a considerable amount of alkalinity; therefore, alkalinity is a capacity function, in contrast to pH, which is an intensity function.

Alkalinity generally is caused by the presence of bicarbonate, carbonate, or hydroxyl ions in a water; however, weak acids, such as boric, phosphoric, and silicic, can contribute titratable alkalinity species. Carbon dioxide, dissolved in circulating waters as bicarbonate or carbonate, as a result of the carbon cycle, is the primary source of alkalinity in shallow ground waters. Additional carbon dioxide probably is dissolved in deep subsurface oilfield brines because of diagenesis of inorganic and organic compounds.

In general, oilfield brines do not contain hydroxyl ions, and most of them contain no carbonate ions; however, they do contain bicarbonate ions. Some oilfield waters in the Rocky Mountain area are alkaline and contain both primary and secondary alkalinity. Primary alkalinity is that associated with the alkali metals, whereas secondary alkalinity is that associated with the alkaline earth metals. For example, the Green River Formation waters that are in or near trona beds may contain more than 20,000 mg/l of carbonate and 5000 mg/l of bicarbonate. Most oilfield waters from other areas contain from about 100 to 2000 mg/l of bicarbonate.

It is important to note that organic acid anions can contribute a sizeable error to the alkalinity determination. The presence of these organic acid anions can be detected by the rancid odor similar to that of butyric acid when the brine sample is acidified. Also, the endpoint of the alkalinity titration is about pH 3.5 rather than pH 4.5, which is characteristic of bicarbonate.

Seawater contains 900 mg/l of sulfur as sulfate, and subsurface oilfield brines contain from none up to several thousand milligrams per liter. The amount of sulfate in the brine is influenced by bacterial activity and by the amounts of calcium, strontium, and barium that are present. If these three cations are present in relatively high concentrations, the amount of sulfate present will be low. Some brines containing high concentrations of magnesium and low concentrations of the other alkaline earth metals, however, may contain high concentrations of sulfate anion.

Hydrogen sulfide, often found in oilfield waters, is formed by anaerobic bacteria. One such species of bacteria is the *Desulfovibrio desulfuricans*, which obtains its oxygen from sulfate ions, causing them to be reduced to hydrogen sulfide.

Sulfur in surface water usually occurs in the S^{6+} form, complexed with oxygen as the sulfate anion SO_4^{2-}. As previously mentioned, the conversion of oxidized sulfur to a reduced form commonly involves a biogenic process, and such a reduction may not occur unless these bacteria are present. The Eh of subsurface oilfield brines usually is somewhat reducing, and the sulfur species in such environments can include hydrogen sulfide (H_2S), sulfite (SO_3^{2-}), and thionates ($S_4O_6^{2-}$). Detailed studies of the sulfur species in subsurface brines have not been made, and it is likely that other forms of sulfur are present in some brines. The temperature, pressure, Eh, pH, and various constituents in solution all influence the types of dissolved sulfur that occur in oilfield brines.

The iodide concentration of some oilfield waters is dependent on the proximity of argillaceous deposits containing organic matter, rather than on dissolved components of minerals. Gas may play an important part in the accumulation of iodide in subsurface waters. Iodide content is depleted at a distance from gas structures, some of which are bounded by iodide-rich waters.

Seawater contains about 0.05 mg/l of iodide, and most subsurface petroleum-associated brines contain less than 10 mg/l; however, some have been found to contain up to 1400 mg/l of iodide. Theoretically, only iodate is thermodynamically stable in seawater. The exact form of iodine in oilfield brines has not been investigated; probably, it will vary with the salinity, Eh, and other factors.

The solubility of calcium fluoride (fluorite) in water at 25°C is about 8.7 ppm of fluoride; this solubility is affected by other dissolved constituents. The solubility of sodium fluoride is very high and magnesium fluoride is more soluble than calcium fluoride. An oilfield brine that is deficient in calcium and has been in contact with rocks containing fluoride minerals can contain appreciable quantities of fluoride. Not many oilfield brines have been analyzed for fluoride, but a few are known to contain from a trace amount to 10 mg/l.

The phosphorus species present in most natural waters probably are the phosphate anions, but they are usually reported as an equivalent amount of the orthophosphate ion ((PO_4)$^{3-}$), which is the final dissociation product of phosphoric acid (H_3PO_4). This dissociation occurs in four steps, giving four possible phosphate forms: H_3PO_4, $H_2PO_4^-$, HPO_4^{2-}, and PO_4^{3-}. In the alkalinity titration, any HPO_4^{2-} is converted to $H_2PO_4^-$ and appears as bicarbonate. Seawater contains about 0.07 mg/l of phosphate. A detailed study of the content of phosphate in oilfield brines has not been made, but of the few that have been analyzed, most have contained less than 5 mg/l.

In an acidic environment, AsO_4^{3-} (the oxidized ion) is mobile, and mineral arsenates tend to be solubilized. The arsenates usually are formed in oxidation zones in contact with atmosphere and free oxygen, and arsenic will precipitate with ferric iron hydroxide. Oilfield brines may contain arsenic as $HAsO_2^-$ or H_2AsO_4, depending upon the Eh and pH. A low Eh may favor the $HAsO_2^-$ form. Seawater contains about 0.003 mg/l of $HAsO_2^-$ and subsurface oilfield brines contain from 0 to 10 mg/l. Compounds containing arsenic sometimes are used in corrosion inhibitors; therefore, information concerning well treatments should be obtained before assuming that any arsenic found occurs naturally.

Trace amounts of numerous other anions are present in some oilfield brines. The interested reader should consult the review by Krejci-Graf (1978).

Physical properties

The pH of oilfield waters usually is controlled by the carbon dioxide—bicarbonate system. Because the solubility of carbon dioxide is directly proportional to temperature and pressure, the pH measurement should be made in the field if a close-to-natural-conditions value is desired. The pH of the water is not used for water identification or correlation purposes, but it will indicate possible scale-forming or corrosion tendencies of a water. The pH also may indicate the presence of drilling-fluid (mud) filtrate or well-treatment chemicals.

Virtually no environment exists on or near the earth's surface where the pH/Eh conditions are incompatible with organic life. Because CO_2 is the main byproduct of organic oxidation and the building material of plant and much bacterial life, it plays a dominant role. It dissolves in H_2O, producing the bicarbonate ion and a free-hydrogen ion. The concentration of the hydrogen ion is 1×10^{-7} moles per liter (pH = 7) at 25°C in pure water, but when saturated with CO_2, it rises to 1×10^{-5} moles per liter (pH = 5). In an ocean water system in equilibrium with carbon dioxide, the pH at each equilibrium step is approximately:

$$H_2O + CO_2 \rightleftharpoons H_2CO_3 \ (pH = 5)$$

$$H_2CO_3 \rightleftharpoons HCO_3^- + H^+ \ (pH = 6.3)$$

$$HCO_3^- + H^+ \rightleftharpoons 2H^+ + CO_3^{2-} \ (pH = 10.3)$$

This reaction moves to the right with increasing temperature in a closed system. In the presence of organic constituents, the equilibria are modified, and the pH range can extend from 2 to 12.

The pH of concentrated brines, which is usually less than 7.0, will rise during laboratory storage, indicating that the pH of the water in the reservoir probably is appreciably lower than many published values. Addition of the carbonate ion to sodium chloride solutions will raise the pH. If calcium is present, calcium carbonate will precipitate. The reason the pH of most oilfield waters rises during storage in the laboratory is because of the formation of carbonate ions as a result of bicarbonate ion decomposition.

The redox potential, which is often abbreviated as Eh, may also be referred to as oxidation potential, oxidation—reduction potential, or pE. At equilibrium it is related to the proportions of oxidized and reduced species present and is expressed in volts.

Attempts to obtain useful results from Eh measurements in natural media involve numerous difficulties. In a natural medium, such as oilfield brines, there are many variables, none of which is controlled, which individually or collectively may have little or great influence on Eh measurements made on the water. Many chemical substances, such as ferric or ferrous ions, various organic oxidation—reduction systems, sulfides, and sulfates, may be present in the water in large or small amounts. Even controlled systems in the laboratory often produce unaccounted-for variances. In the field, the lack of knowledge of actual participating species may seriously impair proper interpretation of Eh readings. The Eh measurements made on poorly poised media (media with poor Eh stability), such as some oilfield waters, involve additional uncertainties. Response of electrodes in such solutions is sluggish, and electrodes are easily contaminated with trace amounts of substances which will produce invalid readings.

Knowledge of the redox potential is useful in studies of how compounds such as uranium, iron, sulfur, and other minerals are transported in aqueous systems. The solubility of some elements and compounds is dependent upon the redox potential and the pH of their environment.

Some water associated with petroleum is "connate" water, which commonly has a negative Eh; this has been proven in various field studies. Knowledge of the Eh is useful in determining how to treat a water before it is reinjected into a subsurface formation. For example, the Eh of the water will be oxidizing if the water is open to the atmosphere, but if kept in a closed system in an oil-production operation, the Eh should not change appreciably as it is brought to the surface and then reinjected. In such a situation, the Eh value is useful in determining how much iron will stay in solution and not deposit in the wellbore.

Organisms that consume oxygen cause a lowering of the redox potential. In buried sediments aerobic bacteria remove the free oxygen from the interstitial water. Sediments laid down in a shoreline environment will differ in degree of oxidation, as compared to those laid down in a deep-sea environment. For example, the Eh of the shoreline sediments may range from —50 to 0 mV, but the Eh of deep-sea sediments may range from —150 to —100 mV.

The aerobic bacteria die when the free oxygen is totally consumed. The anaerobic bacteria attack the sulfate ion, which is the second most important anion in the seawater. During this attack, the sulfate is first reduced to sulfite and then to sulfide, the Eh drops to —600 mV; H_2S is liberated, and $CaCO_3$ is precipitated as the pH rises above 8.5.

Dissolved gases

Large quantities of dissolved gases are contained in oilfield brines. Most of these gases are hydrocarbons; however, other gases such as carbon dioxide, nitrogen, and hydrogen sulfide often are present. The solubility of the gases generally decreases with increased water salinity, and increases with pressure.

Sokolov (1956) published solubility coefficients for some gases dissolved in distilled water at 20°C and atmospheric pressure (see Table 6-6).

As illustrated in Table 6-6, ethane is more soluble in water than any of the other four hydrocarbons, whereas hydrogen sulfide is the most soluble of the inorganic gases. Carbon dioxide is more than 18 times as soluble as ethane.

Hundreds of drill-stem samples of brine from water-bearing subsurface formations in the Gulf Coast area of the United States were analyzed to determine the amounts and kinds of hydrocarbons present (Buckley et al., 1958). The chief constituent of the dissolved gases usually was methane, with measurable amounts of ethane, propane, and butane present. The concentration of the dissolved hydrocarbons generally increased with depth in a given formation and also increased basinward with regional and local variations. In close proximity to some oilfields, the waters were enriched

TABLE 6-6

Solubility coefficients[a] of gases in distilled water at 20°C and atmospheric pressure

Gas	Solubility coefficient	Gas	Solubility coefficient
Methane	0.033	Carbon dioxide	0.87
Ethane	0.047	Nitrogen	0.016
Propane	0.037	Oxygen	0.031
Butane	0.036	Hydrogen sulfide	2.58
Isobutane	0.025		

[a] Volume of gas, corrected to 0°C and 760 mm Hg, absorbed by one volume of water when the pressure of the gas itself, without the aqueous tension, is equal to 760 mm Hg.

in dissolved hydrocarbons, and up to 14 standard cubic feet of dissolved gas per barrel of water were observed in some locations.

Experimental data indicates that 1 cu.ft of sedimentary rock, having 20% porosity, buried 300 m deep, and saturated with a brine containing 50,000 ppm of NaCl, can accommodate 0.3 mole of methane in solution. Many oil-field waters contain methane; however, in Japan, there is a type of natural-gas deposit called "suiyōsei-tein'nengasu", a dry gas, which is found dissolved in subsurface brines (Collins, 1975). The major reservoirs in which methane is found occur in marine or lagoonal sedimentary basins with thick sediments and of wide areal extent. Some of the associated brines contain more than 80 mg/l of iodide, which is the only commercial source of iodine in Japan. Some of these brines also contain dissolved ethane, propane, isobutane, butane, isopentane, and pentane.

Large Soviet deposits of natural gas in a solid state hydrate totaling about 15 trillion cubic meters were reported to the U.S.S.R. Committee for Inventions and Discoveries. According to Soviet investigators, molecules of groundwater attract molecules of natural gas and convert them to a hydrate, which resembles silvery-grey ice, where the pressure is 250 atm and the temperature is 25°C or less. A single cubic meter of the hydrate contains up to 200 m^3 of natural gas. These solid hydrate deposits are found in permafrost zones at depths down to 2500 m. Because of the high electrical resistance, they are discoverable by geophysical methods. The hydrate can be converted to gas by drilling a well and reducing the pressure and/or pumping a catalyst such as methyl alcohol into it.

Molecular hydrogen was found in oilfield waters in the Lower Volga region. Up to 43% of the dissolved gas in these waters was hydrogen. Other gases dissolved in the waters were methane, ethane, butane, pentane, carbon dioxide, nitrogen, helium, and argon. The pH of these waters was as low as 3.4, and the iron content was as high as 1100 ppm.

Research currently is being conducted to determine the amount of recoverable methane from geopressured geothermal areas in the United States. For example, estimates ranging from 3.54×10^{12} m^3 to 32.5×10^{12} m^3 have been made of potentially recoverable methane from geopressured waters in the northern Gulf of Mexico basin. Experimental solubility data for methane in geopressured brines are being investigated (Blount et al., 1979). Adequate solubility data are needed for waters containing up to 250,000 mg/l dissolved solids, at temperatures up to 360°C and at pressures up to about 1410 kg/cm².

Stable isotopes

The isotopic compositions of some oilfield brines have been determined and used to make interpretations concerning the origin of the brines. These stable isotopes include deuterium, ^{18}O, ^{13}C and the $^{87}Sr/^{86}Sr$ ratio. The deuterium concentrations usually are reported relative to SMOW in units of δ D. Oxygen isotopic concentrations are also given relative to SMOW, but as parts per thousand $\delta^{18}O$ ($^0/_{00}$) (Bailey et al., 1973). Carbon isotopic concentrations are reported as $\delta^{13}C$ ($^0/_{00}$) and usually are relative to a standard prepared from a Cretaceous belemnite, *Belemnitella americana*, from the Peedee Formation of South Carolina (Carothers, 1976).

The $^{87}Sr/^{86}Sr$ ratios are determined by measuring both the ^{87}Sr and ^{86}Sr with a mass spectrometer. The ratios then are calculated and compared to appropriate reservoir rock analyses. For example, it has been shown that layered ultramafic rocks contain relatively high $^{87}Sr/^{86}Sr$ ratios, whereas intermediate and silicic rocks show an apparent differentiation. The differences in the values, therefore, are commonly attributed to selective migration of radiogenic strontium (Sunwell and Pushkar, 1979). Further it is believed that subsurface fluids play a role in the selective migration.

Organic constituents

In addition to the simple hydrocarbons, a large number of organic constituents in colloidal, ionic, and molecular forms occur in oilfield brines. In recent years, some of these organic constituents have been quantitatively measured. However, there are numerous organic constituents present, which have not been determined in some oilfield brines primarily because the analytical methods are difficult and very time consuming.

Knowledge of the dissolved organic constituents is important, because these constituents are related to the origin and/or migration of oil, as well as to the disintegration or degradation of an oil accumulation. The concentrations of organic constituents in oilfield brines vary widely. In general, the more alkaline the water, the more likely that it will contain higher concentrations of organic constituents. The bulk of the organic matter consists of anions and salts of organic acids; however, other compounds also are present.

Knowledge of the concentrations of benzene, toluene, and other components in oilfield brines is used in exploration. The solubilities of some of these compounds in water at ambient conditions and in saline waters at elevated temperatures and pressures have been determined (McAuliffe, 1966; Price, 1976). For example, hexane, cyclohexane, and benzene, each with six carbon atoms have solubilities of 9.5, 60, and 1750 mg/l, respectively, at ambient conditions, whereas heptane, methylcyclohexane, and toluene, each with seven carbons, have solubilities of 2.5, 15, and 530 mg/l. For a given carbon number, the aromatics are much more soluble than the alkanes, e.g., benzene is 185 times more soluble than hexane, and toluene is 210 times more soluble than heptane.

The actual concentrations of these and other organic constituents in subsurface oilfield brines, however, is another matter. It has been shown experimentally that the solubilities of some organic compounds found in crude oil increase with temperature and pressure if pressure is maintained on the system. The increased solubilities become significant above 150°C. The solubilities decrease with increasing water salinity. Waters associated with paraffinic oils are likely to contain fatty acids, whereas those associated with asphaltic oils more likely contain naphthenic acids.

It has been shown that oilfield brines contain many amino acids in amounts of up to 50—70 ppm for each amino acid. Generally the bulk of these acids consist of aspartic acid, glycine, serine, alanine-B, and threonine, which are amino acids with up to four carbon atoms (Degens et al., 1964). Fatty acids containing one to eight carbon atoms (C_1—C_8) are soluble in water, whereas those with nine or more carbon atoms essentially are insoluble. If present in water as dissociated molecules, they would contribute to the alkalinity. Cooper (1962) found some fatty acids in oilfield brines in concentrations up to 600 ppb. The bulk of these fatty acids consisted of C_{14}—C_{30}. He found none with less than eight carbon atoms (C_8).

Carothers and Kharaka (1978) determined the aliphatic acid anion concentrations of some oilfield brines. They found that the concentrations ranged from below detection limits up to about 5000 mg/l. According to them, the aliphatic acid anions contribute more than 50% of the acidity in some oilfield brines. They further contend that these anions occur primarily because of thermocatalytic degradation of kerogen and that subsequent decarboxylation of the acid anions is extensive and capable of producing huge quantities of natural gas. The short-chain aliphatic acid anions that they found included acetate, propionate, butyrate, and valerate.

Quantitative recovery of organic constituents from oilfield brines is difficult. Temperature and pressure changes, bacterial activity, absorption, and the high inorganic/organic constituents ratio in most oilfield brines are some reasons why quantitative recovery is difficult.

SAMPLING

A representative analysis cannot be made without a representative sample. Many, if not all, oilfield water samples are contaminated with some material extraneous to the subsurface environment from which they are taken. This is true because even the well tubing through which the sample passes from its subsurface environment to the surface will add to or take some material (ions, etc.) from the water sample. If the well tubing is ruptured, cracked, corroded, etc., at any point other than in the producing zone, there can be contamination from another fluid-bearing horizon.

The fact that the sample is contaminated does not necessarily mean that it should not be analyzed. It is the extent and type of contamination that is important. Quite likely the well tubing will not change the concentrations of the major constituents. Extraneous water from another subsurface or surface zone will contaminate a sample. To resolve such a problem, it is necessary to analyze the sample and compare the results with a sample that is known to be representative.

A recently drilled and completed well probably will not immediately produce a representative sample. Important contributing factors are the presence of: drilling fluids, cement, acids, fracturing fluids, emulsion-breaking chemicals, corrosion inhibition chemicals, bactericides, and scale inhibition chemicals. The most likely contaminants in a recently drilled and completed well are the first four materials mentioned above.

Numerous types of drilling fluids are used (both oil-base and water-base), and each type of drilling fluid (or mud) contains different kinds and amounts of chemicals (Collins, 1975b). New cement used in completing a well will introduce the hydroxide ion which results in a high pH. Acids are used in well stimulation operations to increase the permeability of the reservoir rocks. The volume may vary from 500 to several thousand gallons. Hydrochloric acid or a mixture of hydrochloric and hydrofluoric acids are the more common; however, other acids, including nitric, sulfuric, formic and acetic, also are used. Fracturing fluids are used to fracture the reservoir rock and thus increase the permeability. These fluids contain mixtures of water, oil, chemicals, and sand. In some operations, large quantities of the fracturing fluid are forced under pressure into the well.

Most of the other chemicals, such as the emulsion breakers, corrosion inhibitors, etc., are added after fracturing or acidizing operations. Many of these chemicals are used on a regular routine schedule.

Knowledge of the above factors is important in sampling the formation water from a well. Drill-stem tests are taken from some wells almost immediately after drilling and completion operations. Correct sampling and interpretation for some constituents are possible even with a drill-stem test. These constituents will be only the macro-constituents as a general rule.

A representative formation water sample can be obtained after the well has been on production long enough to permit complete ejection of the drilling fluids, soluble cement additives, acids, and fracturing fluids. Chemicals which are added routinely will be present. Their addition can be stopped or arrangements can be made to obtain a sample when their concentrations are low. Knowledge of the chemicals that are added is necessary if the sample is to be analyzed for trace inorganic or organic constituents.

Most oil wells produce an emulsion consisting of a mixture of the oil and formation water. Some of these emulsions are quite stable, and heat and/or chemicals must be applied to break them. Thus, it is necessary to either obtain a sample of water effluent from the heater-treater tank or to obtain a sample of the emulsion and then separate the oil and water. In either case, some of the unstable constituents will change with respect to their original composition within the formation. Some of these constituents are dissolved gases, bicarbonate anion (HCO_3^-), and elements susceptible to a valence change as a result of oxidation. The Eh of most formation waters is reducing; therefore, constituents such as iron and manganese are present in the formation water in a reduced form.

The value of the sample is directly proportional to the facts known about its source; therefore, sites should be selected for which the greater source knowledge is available.

For surveillance purposes, samples can be collected from the same site at sufficiently frequent intervals in order to detect any important change in quality that could occur between sampling times. Change in composition may result from changes in rate of water movement, pumpage rates, or infiltration of other water. Changes that can occur in petroleum-associated water are illustrated in Table 6-7. Well 1 shows the sort of change that commonly occurs. The water from well 2 did not change between 1947 and 1957, within the accuracy of the analytical determination. Water from well 3 changed drastically, suggesting the intrusion of water from a different source.

There is a tendency for some oilfield waters to become more dilute (fresher) as the oil reservoir is produced. Such dilution may result from the movement of fresher water from adjacent overpressured shales into the petroleum reservoirs as pore pressure declines in sands with the continued removal of oil and brine (Wallace, 1969).

The composition of oilfield water varies with the position within the geologic structure from which it is obtained. For example, if the water table is tilted, the more dilute water probably will be on the structurally high side. In some cases, the salinity increases upstructure (with decreasing depth) to a maximum at the point of oil—water contact. Usually this is caused by infiltrating meteoric waters.

TABLE 6-7

Example of changes in the composition of oilfield waters (mg/l) with time

Constituent	Well 1		Well 2		Well 3	
	1947	1957	1947	1957	1956	1959
Sodium and potassium	29,100	25,000	46,000	45,900	1500	860
Magnesium	1100	1200	2100	2200	30	2
Calcium	5900	5500	14,200	14,400	60	10
Bicarbonate	30	10	25	10	600	1800
Sulfate	15	50	3	50	200	0
Chloride	58,500	51,800	102,100	102,800	2000	300
Total dissolved solids	94,600	83,600	164,400	165,400	4400	3000

Drill-stem test

The drill-stem test, if properly made, can provide a reliable formation water sample. Mud filtrate will be the first fluid to enter the drill-stem test tool, and it will be found at the top of the fluid column immediately below the oil. At some point down the column a representative formation-water sample can be found. The point is variable and will be influenced by rock characteristics, mud pressure, type of mud, and duration of the test. It is best to sample the water after each stand of pipe is removed. Normally, the total dissolved solids content will increase downwards and become constant when pure formation water is obtained. If the concentration continues to increase all the way to the bottom, no representative sample can be obtained. A test that flows water will give even higher assurance of an uncontaminated sample. If only one drill-stem test water sample is taken for analysis, it should be taken just above the tool, as this is the last water to enter the tool and is least likely to show contamination.

The drill-stem test can provide pressure head and head decline and buildup data useful in permeability calculation (Bredehoeft, 1965) and other information for the determination of additional reservoir conditions, such as the gas/oil ratio and reservoir depletion (McAlister et al., 1965). A stratigraphic interval of interest is isolated in the drilled hole by use of packers attached to the drillstring. Opening the tester valve in the test string allows the formation fluid to enter the drillpipe. Pressures are recorded by gages in the bottom of the test tool.

To insure that a representative sample is obtained, the pH, resistivity, and chloride content of samples taken at various intervals down the drillpipe can be determined. Usually a transition zone will be found below which apparently uncontaminated formation water is located. The pH, resistivity, and chloride content will vary above the transition zone, and they will become constant below it. The sample taken for analysis in the laboratory can yield positive evidence of contamination, if present. The two most indicative tests are pH and the color of a filtered sample. If the filtered sample remains tan or brown and the color cannot be removed even with pressure filtration, it probably is contaminated with drilling-mud filtrate. A sample can be placed on a white-spot plate for color evaluation. For positive identification of the presence of mud filtrate, a sample of the drilling mud used in drilling the well can be obtained and allowed to react with distilled water. Then this water is analyzed to determine the mud-contributed ions, and the suspected contaminated sample is analyzed to determine if it contains these ions.

Analyses of water obtained from a drill-stem test of Smackover Limestone water in Rains County, Texas, show how errors can be caused by improper sampling of drill-stem test water. Analyses of top, middle and bottom samples taken from a 15-m fluid recovery are shown in Table 6-8. These data show an increase in salinity with depth in the drillpipe, indicating that the first water sample was contaminated by mud filtrate (Noad, 1962). The middle sample is approximately half mud filtrate and half formation water. The bottom sample is the most representative of Smackover water.

No single procedure is universally applicable for obtaining a sample of oilfield water. For example, information may be desired concerning the dissolved gases or hydrocarbons in the water, or the reduced species present — such as ferrous or manganous compounds. Sampling procedures applicable to the desired information must be used.

Dissolved hydrocarbons

Knowledge of certain dissolved hydrocarbon gases is used in exploration and production. Methane is quite soluble in water, but samples must be collected in a sampler that keeps the subsurface pressure on the sample until it is opened in the laboratory. The testing tool is kept open until the head of water in the drillpipe is equalized with the formation pressure or until water flows at the surface. The pressure equalization may require four or more hours. A surface recording subsurface pressure gage, however, can be lowered into the drillpipe to determine when the pressure has equalized. After equalization of pressure, formation-water samples can be obtained by lowering a subsurface sampler into the drill-pipe (Buckley et al., 1958). Zarrella et al. (1967) determined the concentration of dissolved benzene. In such a case, it is not necessary to use a subsurface sampler. Instead, the samples are caught in buckets on opening the pipe string, and immediately transferred from the buckets to new narrow-necked glass or metal containers.

A method of obtaining a sample for subsequent gas analysis is to catch the aqueous sample in a metal container of about one-quart capacity. This sample is immediately transferred to another metal sample container. The second container should be filled completely to the top. Then the sides of

TABLE 6-8

Drill-stem test recovery of Smackover Limestone water

Constituent ion	Concentration (mg/l)		
	top	middle	bottom
Sodium	29,600	43,500	71,800
Calcium	8100	13,100	22,400
Magnesium	600	900	1400
Bicarbonate	500	500	400
Sulfate	2000	1300	500
Chloride	59,900	91,800	154,000
Total dissolved solids	101,000	151,000	251,000

the can are lightly squeezed to allow for fluid expansion, and the lid is sealed tightly. A foil-lined (not plastic) lid should be used. If possible, the sample should be analyzed immediately. If this is not possible, the sample should be cooled or frozen.

Sampling at the flowline

Another method of obtaining a sample for analyses of dissolved gases is to place a sampling device in a flowline. Fig. 6-2 illustrates such a device. The device is connected to the flowline, and water is allowed to flow into and through the container, which is held above the flowline, until ten or more volumes of water have flowed through. Then the lower valve on the sample

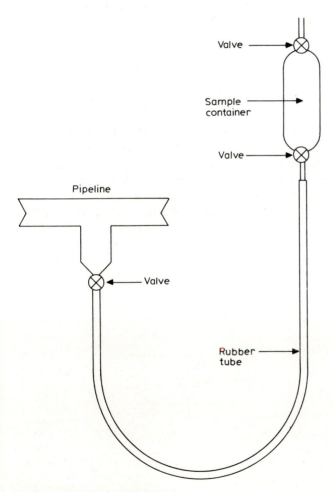

Fig. 6-2. Flowline sampler.

container is closed and the container removed. If any bubbles are present in the sample, the sample is discarded and a new one is obtained.

Sampling at the wellhead

It is common practice in the oil industry to obtain a sample of formation water from a sampling valve at the wellhead. A plastic or rubber tube can be used to transfer the sample from the sample valve into the container (usually plastic). The source and sample container should be flushed to remove any foreign material before a sample is taken. After flushing the system, the end of the tube is placed in the bottom of the container, and several volumes of fluid are displaced before the tube is slowly removed from the container and the container is sealed. Fig. 6-3 illustrates a method of obtaining a sample at the wellhead. An extension of this method is to (1) place the sample container in a larger container, (2) insert the tube to the bottom of the sample container, (3) allow the brine to overflow both containers, (4) withdraw the tube, and (5) cap the sample under the fluid.

At pumping wellheads, the brine surges out in heads and mixes with oil. In such situations, a larger container equipped with a valve at the bottom can be used as a surge tank, an oil-water separator, or both. To use this device: (1) the sample tube is placed in the bottom of the large container, (2) the wellhead valve is opened, (3) the large container with the well fluid is rinsed, (4) the large container is allowed to fill, and (5) a sample is withdrawn through the valve at the bottom of the large container. This method serves to obtain samples that are relatively oil-free.

In some studies it is necessary to obtain a field filtered sample. The filter-

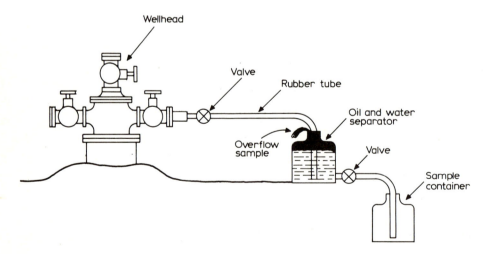

Fig. 6-3. Example of method of obtaining a sample at the wellhead.

ing system shown in Fig. 6-4 was designed and has proven successful for various applications.

This filtering system is simple and economical. It consists of a 50-ml disposable syringe, two check valves, and an inline-disc-filter holder. The filter holder accommodates filters 47 mm in diameter and 0.45 μm in pore size, with the option of a prefilter and a depth prefilter.

After the oilfield brine is separated from the oil, it is drawn from the separator into the syringe. With the syringe, the brine is forced through the filter into the collection bottle. The check valves allow the syringe to be used as a pump for filling the collection bottle. If the filter becomes clogged, it can be replaced in a few minutes. Approximately two minutes are required to collect a 250-ml sample. Usually two samples are taken, with one being acidified to pH of 3 or less with concentrated HCl or HNO_3. The system can easily be cleaned or flushed with brine to prevent contamination.

Sample for determining unstable properties or species

A mobile analyzer was designed which measures pH, Eh, O_2, resistivity, S^{2-}, HCO_3^-, CO_3^{2-} and CO_2 in oilfield water at the wellhead. When oilfield brine samples are collected in the field and transported to the laboratory for analysis, concentrations of many of the unstable constituents change. The amount of change depends on the sampling method, sample storage, ambient conditions and the amounts of the constituents in the original sample. An analysis of the brine at the wellhead, therefore, is necessary to obtain reliable data (Hoke and Collins, 1981).

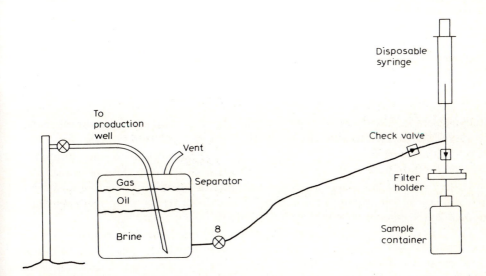

Fig. 6-4. Example of field filtering equipment.

The mobile analyzer consists of a flow-through system using ion-selective electrodes as the main analytical tools. The design and construction of the analyzer was done by Orion Research Inc., Cambridge, Mass., under U.S. Department of Energy Contract No. EW-78-C-19-0021. The analyzer, having inside dimensions of 7 ft × 12 ft × 74 inches, is mounted on the side of the wall of a mobile truck van.

Sample for stable-isotope analysis

A sample that is to be analyzed for stable isotopes should be collected with care. If possible, such a sample should be taken at reservoir temperatures and pressures to minimize any isotope fractionation. Since this usually is impossible, however, caution should be exercised to insure that a representative sample is collected at the prevailing wellhead temperature and pressure.

The sample should be collected at the wellhead. If this proves impossible, it may be feasible to collect the sample from a nonheated separator or heater. Samples should not be taken of water that has been heated or treated with any chemicals. Glass sample bottles (about 100 ml usually is sufficient) should be used, and the sample should overflow the bottle. The bottle should be closed with a cap equipped with a plastic insert, and the top should be sealed with wax to minimize exchange reactions with air.

Sample containers

Various factors influence the type of sample container that is selected. Containers that are used include polyethylene, other plastics, hard rubber, metal cans, and borosilicate glass. Glass adsorbs various ions such as iron and manganese, and may contribute boron or silica to the aqueous sample. Plastic and hard rubber containers are not suitable if the sample is to be analyzed to determine its organic content. A metal container is used by some laboratories if the sample is to be analyzed for dissolved hydrocarbons such as benzene.

The type of container selected is dependent upon the planned use of the analytical data. Probably the more satisfactory container, if the sample is to be stored for some time before analysis, is the polyethylene bottle. All polyethylenes are not satisfactory because some contain relatively high amounts of metal contributed by catalysts in their manufacture. The approximate metal content of the plastic can be determined using a qualitative emission spectrographic technique. If the sample is transported during freezing temperatures, the plastic container is less likely to break than glass.

The practice of obtaining two samples and acidifying one sample so that the heavy metals will stay in solution works better if the plastic container is used. Some of the heavy metals are adsorbed by glass even if the sample is acidified.

Tabulation of sample description

The sample is of little value if detailed information concerning it is not available. Information such as that in Table 6-9 should be obtained for each sample of oilfield water. For certain types of studies, additional information may be needed.

TABLE 6-9

Description needed for each oilfield water sample

Sample number _____ Field _____
Farm or lease _____ Well No. _____ in the _____
of Section _____ Township _____ Range _____
County _____ State _____ Operator _____
Operator's address (main office) _____
Sample obtained by _____ Date _____
Address _____ Representing _____
Sample obtained from (lead line, separatory flow tank, etc.) _____

Completion date of well _____ Evaluation of well _____

Name of productive zone from which sample is produced _____

Sand _____ Shale _____ Lime _____ Other _____

Name of productive   Names of formations
formation _____ well passes through _____

Depths: Top of formation _____ Bottom of formation _____

      Top of producing zone _____ Bottom of producing zone _____

      Top of depth drilled _____ Present depth _____

Bottomhole pressure and date of pressure _____

Bottomhole temperature _____

Date of last workover _____ Are any chemicals   If yes,
                        added to treat well? _____ what? _____

Well production	Initial	Present	Casing service record
Oil, barrels/day	_____	_____	_____
Water, barrels/day	_____	_____	_____
Gas, cubic feet/day	_____	_____	_____

Method of production (primary, secondary, or tertiary)

_____

Remarks: (such as casing leaks, communication or other pays in same well, lease or field)

_____

METHODS OF ANALYSIS OF OILFIELD WATERS

Analytical methods for analyzing oilfield waters are improving with respect to speed of analysis and in precision and accuracy. There have been at least two groups trying to standardize methods of oilfield water analyses during the past 20 years. They are the American Petroleum Institute (API) and the American Society for Testing and Materials (ASTM). The API (1968) published a "Recommended Practice for Analysis of Oil-Field Waters". The API has sponsored no additional work since 1968 in this area.

The ASTM has an active group in Committee D-19 who still are trying to standardize methods of analyzing oilfield brines. Several methods have been standardized by rigorous round-robin testing by several laboratories and sub-

TABLE 6-10

Geochemical water analyses

Property or constituent	Produced water	Injection water	Steam generation water	Disposal water
pH	X	X	X	X
Eh	O	X		O
Specific resistivity	X			
Specific gravity	X	X	X	X
Bacteria	O	X		O
Barium	X	X		X
Bicarbonate	X	X	X	X
Boron	O			
Bromide	O			
Calcium	X	X	X	X
Carbonate	X	X	X	X
Carbon dioxide	O	X	X	O
Chloride	X	X	X	X
Hydrogen sulfide	O	X		O
Iodide	O			
Iron	X	X	X	O
Magnesium	X	X	X	X
Manganese	O	O	O	O
Oxygen	O	X	O	O
Potassium	O			
Residual hydrocarbons		X		O
Sodium	X	O	O	O
Silica	O	X	X	O
Strontium	O	X	O	O
Sulfate	X	X	X	X
Suspended solids		X		X
Total dissolved solids	X	X	X	X

x = usually requested; O = sometimes requested.

sequent ASTM committee balloting procedures. These methods are found in the Annual Book of ASTM (1982), "Standards, Part 31, Water: Section VII — Saline and Brackish Waters, Sea Waters and Brines".

Additional methods appear in the "Gray Pages" of the same publication. The methods in the "Gray Pages" are methods that have not yet undergone rigorous round-robin testing, but are recognized as the currently best available methods.

Table 6-10 illustrates the analyses for various properties or constituents of a water that are usually made for (1) a produced oilfield water, (2) water used in injection for pressure maintenance, or for secondary recovery, (3) water used to generate steam for steam injection, and (4) water that is injected into a disposal well. Methods to determine most of these properties or constituents can be found in publications by API (1968), ASTM (1981), Collins (1975a), and Hoke and Collins (1981). Bacteriological methods can be found in Beestecher (1954), Davis (1967), Patton (1974), and Postgate (1979). As noted previously, the ASTM approved methods are subjected to round-robin testing by cooperating laboratories.

ENHANCED OIL RECOVERY OPERATIONS

EOR operations can be separated into three major types, namely: (1) chemical, (2) miscible displacement, and (3) thermal. The objective of each is to recover as much oil as possible. The optimum application of each type is dependent upon reservoir characteristics including type of oil.

Table 6-11 illustrates the screening parameters that are useful in preliminary evaluations of reservoirs for the various EOR technologies. One of the most convenient methods of applying these screening criteria is to use a digital computer. However, to do this one needs a computer data base which contains the necessary engineering and geological data for petroleum reservoirs. The Bartlesville Energy Technology Center has developed such a data base.

After the preliminary screening of the data base for candidate reservoirs, it is necessary to do a detailed reservoir analysis of the best reservoirs selected. This involves both an engineering and geologic in-depth study. It can involve drilling wells for the purpose of obtaining more accurate residual oil and water saturations, permeability, porosity, etc.

Surfactant and polymer floods

Surfactant and polymer floods are chemical techniques. Surfactants are micellar or surface-active agents including soaps and soap-like substances. To be useful in enhanced oil recovery, they must reduce the interfacial tension between water and oil. They have an amphiphilic molecule that is

TABLE 6-11

Screening parameters for reservoir characterization for enhanced oil recovery by steam injection, in-situ combustion, surfactant—polymer flooding, polymer flood, caustic waterflood, and hydrocarbon miscible flooding

Screening parameters	Steam injection	In-situ combustion	CO_2 flood	Surfactant—polymer	Polymer	Alkaline waterflood	Hydrocarbon miscible
1. Oil viscosity, cp	n.c.	n.c.	< 12	< 20	< 200	< 200	< 5
2. Oil gravity, °API	10–25	10–45	> 30 (Calif. > 26)	> 25	> 18	15–35	> 30
3. Depth, ft	200–5000	> 500	> 2300	< 8500	< 8500	n.c.	n.c.
4. Reservoir temperature, °F	n.c.	n.c.	n.c.	< 250	< 200	< 200	n.c.
5. Initial reservoir pressure, psig	n.c.	n.c.	> 1200	n.c.	n.c.	n.c.	n.c.
6. Net pay, ft	> 20	> 10	n.c.	> 20	> 20	> 50	n.c.
7. Permeability, md	n.c.	n.c.	n.c.	> 25	> 50	> 25	n.c.
8. Residual oil saturation, %	50	50	25	n.c.	n.c.	n.c.	> 25
9. (kh/μ) transmissibility, md ft/cp	> 100	> 20	n.c.	n.c.	n.c.	n.c.	n.c.
10. Porosity, %	> 10	> 10	n.c.	n.c.	n.c.	n.c.	n.c.
11. Salinity (TDS), ppm	n.c.	n.c.	n.c.	< 50,000	n.c.	< 2500	n.c.
12. Hardness (Ca and Mg), ppm	< 2500	n.c.	n.c.	< 1000	n.c.	n.c.	n.c.
13. Operating pressure, psi	> 500	n.c.	> 1100	n.c.	n.c.	n.c.	> 1300
14. Target oil, bbl/acre-ft	n.c.	> 400	n.c.	n.c.	n.c.	n.c.	n.c.
15. Lithology		n.c.	n.c.	sandstone	n.c.	sandstone preferred	n.c.
16. Gas cap	none to minor	none to minor	none to minor	none to minor	none to minor	none to minor	n.c.
17. Natural water drive	none to weak	none to weak	none to weak	none to minor	none to minor	none to minor	
18. Fractures	none to minor	none to minor	none to minor	none to minor	none to minor	none to minor	
19. Well spacing	n.c. unless extreme ⩽ 10	⩽ 20	n.c.	n.c.	n.c.	n.c.	n.c.
20. Comments	porosity and thickness (high); economic fresh water available; economic fuel available; high net to gross pay; low clay content	high dip preferred; porosity and thickness (high); low vertical permeability preferred; temperature > 150°F preferred; high net to gross pay	thin pay preferred; high dip preferred; homogeneous formation preferred; porosity and thickness (low); natural CO_2 available; low vertical permeability	homogeneous formation preferred; low clay content; $\phi \times h$ (high); waterflood sweep > 50%	use with or prior to waterflood; low-calcium clay content; $\phi \times h$ (high)		

n.c. = not critical.

attracted, at one end, to water (the hydrophilic or water-loving end) and at the other to oil (the oleophilic or oil-loving end).

Alcohol improves the quality of some micellar solutions and, when used, is a cosurfactant. The cosurfactant also aids the micelle in solubilizing oil or water, stabilizes the solution, and reduces adsorption.

The micellar costs are dependent upon the cost of oil since many of these chemicals are sulfonated crude oil fractions, e.g., gas oil. Therefore, if the cost of the crude oil is \$35/bbl, the sulfonate cost will be the original cost of the crude plus the cost of fractionation and sulfonation. Obviously, this can be an expensive chemical.

The water-soluble polymers used in EOR consist of high-molecular-weight chainlike molecules with molecular weights up to or exceeding 20 million. Polymers such as polyacrylamides and polysaccharides often are used as mobility-control buffers for permeability reduction and/or increased viscosity. Polysaccharides sometimes are called biopolymers. Polymers increase the viscosity of the water and prevent it from running ahead of the oil. Increased resistance to flow, particularly in high-permeability zones, improves the volumetric reservoir sweep efficiency resulting in increased oil recovery.

Water-soluble synthetic polyacrylamides consist of high-molecular-weight chainlike molecules with $CONH_2$, $COOH$, and $COONa$ groups attached to every other carbon atom on a carbon chain. Naturally occurring polysaccharides consist of cyclic carbohydrate monomers alternating in the polymer structure. These additives aid oil recovery by decreasing the flood-water's mobility.

Table 6-12 indicates a relative comparison of the potential advantages of polyacrylamides and polysaccharides. The polyacrylamides, for example, are most susceptible to breakdown because of mechanical shear degradation and are more likely to adsorb on clay or silicate surfaces than the polysaccharides. The fact that the polysaccharides react with polyvalent cations (low concentrations) and bacteria, however, and in general plug filters or well sand faces because of numerous reactions, gives polyacrylamides a wider acceptance in oil recovery operations.

The following description is an enhanced recovery operation that utilizes

TABLE 6-12

Relative comparison of polymers

	Mechanical degradation	Salinity	Bacteria	Adsorption	Polyvalent ions	Oxygen
Polyacrylamides	−	−	+	−	+	−
Polysaccharides	+	+	−	+	−	+

a preflush to remove connate or saline interstitial water from the reservoir followed by surfactant, polymer, and floodwater. As noted previously, the amount of water used in an EOR process to recover a barrel of oil can be significant. Additionally, some or even much of this water may be classified as a fresh water. For example, Table 6-1 contains estimates of 10—15 barrels of water required per barrel of oil recovered.

In many of the surfactant—polymer EOR operations a preflush is utilized. This preflush often consists of a fresh water to which sodium chloride is added. More specifically, it probably will consist of fresh water, sodium chloride, a bactericide, and a corrosion inhibitor. The purpose of the preflush is to remove the connate brine from the reservoir or at least in the area of the reservoir where the operator wants to form an oil bank in front of the surfactant that he plans to inject. The preflush can be large, up to 0.8 pore volume ($0.8V_p$), or smaller — $0.05V_p$. If possible an optimum value should be selected for a specific reservoir. Pang et al. (1981) have developed a method to design a preflush for a commercial flood. Holm and Josendal (1982) recommended a preflush containing a high-pH sodium silicate. This reduces the adsorption of the surfactants and polymers.

After completion of the preflush, the sulfonate solution is injected. The preflush theoretically removes most of the divalent cations (calcium and

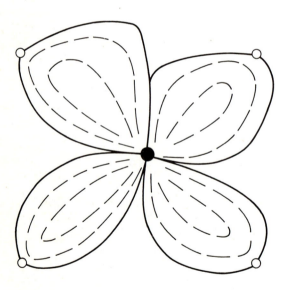

●   Production well

○   Injection well

Fig. 6-5. Five-spot well pattern illustrating enhanced recovery flood fronts.

magnesium) that were in the connate brine. These divalent cations react with many sulfonates causing them to precipitate or become inactive (or useless) in the entrainment or entrapment of the oil phase.

Other constituents in this surfactant or micelle phase may be sodium hydroxide, sodium chloride, polymer, crude oil and, of course, fresh water. The polymer is added to increase the viscosity of the solution. Sodium hydroxide, if used, may aid in forming a multiphase microemulsion system. The microemulsion has at least three components (oil, water and surfactant) according to Healy et al. (1975).

Sufficient micelle solution is injected to fill a specified V_p. In the field it often is called a slug size. Several published designs specify an injection of about $0.10V_p$. Next, the micelle bank is displaced with a polymer-thickened water. This polymer slug may be $0.45V_p$ or even $1.05V_p$ depending upon the gradient desired. It usually is made by mixing polymer with fresh water. The polymer slug is followed by a drive water. The first drive water usually consists of a fresh water. It often is followed by produced water.

Fig. 6-5 illustrates a five-spot well pattern showing a schematic of EOR flood fronts. Not all EOR operations utilize the five-spot pattern.

Fig. 6-6 illustrates a surfactant—polymer oil recovery operation. Oil and water are produced at the production well on the right-hand side. The figure does not illustrate a preflush. If a preflush were used, it, of course, would be injected before the micellar solution.

Brine-tolerant chemical systems are being developed. Widmyer and Pindell (1981) described a surfactant—polymer flood, in which field brine was utilized in the surfactant solution. A preflush with fresh water was not used.

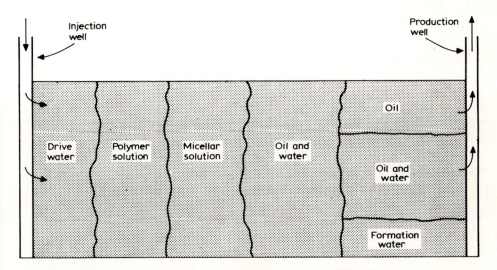

Fig. 6-6. Example of a micellar—polymer enhanced recovery operation.

A water buffer, however, was injected after the surfactant slug. Also, the polymer slug consisted of a low-salinity water, as did the final displacement water slug.

Polymer operation

A polymer operation is similar to a surfactant—polymer operation. The notable exception is that the surfactant phase is not injected. The polymer phase only is used and, therefore, it might be called a thickened or polymer augmented waterflood. The polymer increases the mobility ratio of the flood and tends to move more oil without allowing the flood to finger through the oil. Table 6-1 indicates that a polymer flood may require 16—50 barrels of water per barrel of oil recovered.

A preflush usually is used. Fresh water is used in many of the preflushes, in the polymer phase, and in the first drive water phase. Brine-tolerant polymers will decrease the necessity of using fresh water. Many polymers react with divalent cations such as calcium and magnesium.

Alkaline flood operation

In general, an alkaline (caustic) flood is used only in a sandstone reservoir because of the abundance of calcium in a carbonate reservoir brine. Although the most common chemical used in caustic flooding is sodium hydroxide, sodium orthosilicate and sodium carbonate also are used. Other chemicals that have been used include ammonium hydroxide, potassium hydroxide, sodium silicate, trisodium phosphate, and polyethylenimine. The cost is important, e.g., a chemical such as sodium hydroxide costs less than potassium hydroxide. Ammonium hydroxide has been used in a carbonate reservoir according to Ban et al. (1977).

Divalent cations such as calcium and magnesium in the connate water can deplete a caustic slug by precipitation of hydroxides. Also, if anhydrite or gypsum are present in the rock, calcium will react with the slug to precipitate calcium hydroxide. In addition, clays having high ion exchange capacity will exchange hydrogen for sodium, rendering the caustic slug ineffective by producing water and tying up the sodium.

In general, caustic reacts too slowly with silica in sandstone to cause problems. Also, most dolomites and limestones will not react with the caustic to cause deleterious effects. As noted in Table 6-1, the amount of water in barrels required to produce a barrel of oil with caustic flooding ranges from 22 to 33.

Carbon dioxide flooding

The data in Table 6-1 indicates that one to three barrels of water are required to produce one barrel of oil using a carbon dioxide (CO_2) flood. It is necessary to reach a miscibility pressure of about 1200 psi in a CO_2 flood. This is true only if the API° gravity of the oil is 30 or greater and the temperature is 120°F or less. If the oil is 27°API, the miscibility pressure is about 4000 psi. Temperatures greater than 120° F also must be considered.

The depth of the reservoir, therefore, must be 2500 ft or more. If it is not, the overlying rock may be fractured. In addition, if the pressure in the reservoir containing an oil of 30°API gravity or greater has been depleted to less than 1200 psi, the pressure must be built up by injection of water before the CO_2 injection starts. This is the cheapest method. Of course, CO_2 could be injected to build up the pressure; however, this would be very expensive at the current prices for CO_2.

Most CO_2 floods utilize a water injection phase as a preflush and as water alternating with gas injection (WAG). For example, the preflush may be a fresh water to which salt is added or it may be a softened salt water. In some areas softened seawater is used.

There are at least four methods of carbon dioxide and water injection that have been studied or used. They are: (1) continuous injection of carbon dioxide during the life of the flood; (2) injection of carbon dioxide followed by water, (3) injection of alternate slugs of carbon dioxide and water, and (4) simultaneous injection of carbon dioxide and water. The water in some field designs consists of polymer-thickened water. Also, some projects utilize a preflush. Carbon dioxide floods are useful in both carbonate and sandstone reservoirs.

Miscible hydrocarbon flooding

As the name of the miscible hydrocarbons flood implies, the injected gas or liquid hydrocarbons becomes miscible with the hydrocarbons in the reservoir. This miscibility usually is accomplished at elevated temperatures and pressures. Depth of the reservoir, therefore, is important because of the need to maintain a high pressure.

Three different techniques are commonly used: (1) Miscible Slug Process, whereby a slug of liquid hydrocarbon about $0.05 V_p$ is injected followed by gas and water as the drivers; (2) Enriched Gas Process, whereby a slug of enriched gas is injected followed by lean gas and water as the driver; and (3) High-Pressure Lean Gas Process, whereby lean gas is injected at high pressure to cause evaporation of the crude oil and formation of a miscible phase.

Inert gas injection

Increased costs of natural gas and even carbon dioxide have prompted operators to look at other methods to maintain the pressure in petroleum reservoirs. With natural gas, miscibility could be achieved in some reservoirs.

The miscibility state allows almost 100% displacement efficiency in the swept zone; however, this was not always the goal. Often pressure maintenance was the goal.

Inert gases are not miscible with many oils at low pressures. Also, the API gravity of the oil should be 35° or higher (Peterson, 1978).

In-situ combustion

There are two fundamental processes of in-situ combustion, namely: (1) forward combustion and (2) reverse combustion. Water is used in variations of the forward combustion process. When water is injected with the air, it forms superheated steam near the injection well. At the combustion front it mixes with nitrogen, carbon monoxide, carbon dioxide and other gases. This hot gas mixture displaces the oil. Heat reduces the viscosity of the oil, allowing the oil to flow toward the production well.

Table 6-1 indicates that a wet in-situ combustion technique requires 0.5—1 barrel of water to produce a barrel of oil. The benefit of the wet method is that it results in a threefold reduction in air volume required to produce a barrel of oil.

Steam operations

According to Table 6-1, a steamflood requires about 2—5 barrels of water per barrel of oil recovered. Water used in a steamflood usually is a high-quality water. Not only is it a high-quality water, but the water usually is softened before it enters the steam generator. This is done to prevent scale problems.

Steam and hot-water flooding account for most of the oil recovered by all EOR operations. There are two steam recovery techniques: (1) steam stimulation sometimes called cyclic steam injection, steam soak, or huff and puff, and (2) steamflooding which is a process similar to waterflooding. Heat from the steam reduces the viscosity of heavy or highly viscous oils, allowing the oil to flow more freely to the production well.

In any event, the steamflooding technique accounts for more oil recovered than by all other EOR technologies. Steamflooding recovers worldwide about 410 million barrels of oil per day (mbpd), steam stimulation plus in-situ combustion about 57 mbpd, and all other EOR techniques about 220 mbpd (van Poolen, 1980).

INJECTION WATER

Items that should be considered in detail prior to implementation of any EOR project involving any type of injection water include: (1) formation type; (2) formation quality such as sand quality and clay content; (3) formation porosity and permeability; (4) depth of formation; (5) fracture pressure of formation; (6) fracture pressures of overlying and underlying formations; (7) compatibility of injection solutions with fluids already in the formation as well as with the formation rocks.

Petroleum reservoir rocks are filters and are susceptible to plugging by any type of solid material which may be suspended in or precipitated from an injection fluid. Even materials such as oil and grease from the pumps, corrosion inhibitors and bactericides can cause plugging problems.

Water sources

There are three major types of water that are used as injection water: (1) formation water, (2) seawater, and (3) fresh water. Formation water, as here defined, is subsurface brackish or brine water produced from a petroleum productive formation or from a non-petroleum productive formation. Under seawater one may also include water from a salty (non-potable) lake. Fresh water primarily is water that can be made potable by the usual treatment procedures such as flocculation, filtration, and chlorination. It contains less than 2000 ppm dissolved solids (DS).

An estimate of the amounts of water that are present in various reservoirs was made for the State of Oklahoma. The estimate indicated that Oklahoma has about 3.4 trillion gal of surface water possessing a quality of 100—1000 ppm DS; about 5.0 trillion gal of ground water with a quality of 280—4000 ppm DS; about 23.6 trillion gal of formation water down to a depth of 5500 ft with a quality of 15,000—110,000 ppm DS; and 35.8 trillion gal of formation water from a depth of 5500 ft to 8500 ft with a quality of 15,000 to 110,000 ppm DS. Further, it was determined that the state of Oklahoma has no exact information on the quantity or quality of water injected or produced in petroleum operations involving primary recovery, secondary recovery, and EOR. Related information for other states was not determined.

An analysis was made of the approximate amount of water produced with crude oil in 14 states. These states and their percent of total U.S. crude oil production are: Alabama, 0.3%; Alaska, 19.9%; California, 11.7%; Colorado, 1.0%; Florida, 1.4%; Louisiana, 13.4%; Montana, 1.0%; Mississippi, 1.2%; Nebraska, 0.2%; New Mexico, 2.3%; North Dakota, 1.4%; Texas, 31.2%; Utah, 0.8%; and Wyoming, 4.2%.

Fig. 6-7 indicates the crude oil and water production from wells in these 14 states. The figure indicates that about 4.3 barrels of water are produced per barrel of oil. Fig. 6-8 is a similiar graph for 13 states excluding Alaska. This figure indicates that about 5.2 barrels of water are produced per barrel of oil. Further it can be shown that oil wells produce more water as cumulative oil production increases. In other words, the older the well, the higher the water to oil ratio.

In any EOR project, a first consideration must be given to the water source. In some projects where a fresh water preflush is necessary, it is obvious what the water source must be. Usually sodium chloride is added to the fresh water to inhibit clay swelling. Some of the more salt resistant EOR chemicals can tolerate a more salty water. In such cases, formation water, a mixture of formation waters, a mixture of formation water and fresh water, or even seawater might be used. The major problems with

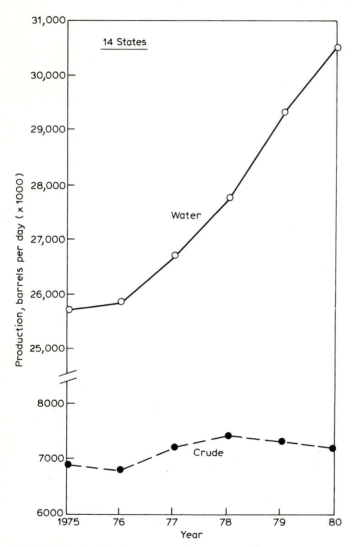

Fig. 6-7. Crude oil and water produced (× 1000 bbl/day) from wells in 14 states including Alaska. (Source: Petroleum Information Corporation.)

these waters and EOR chemicals such as surfactants, polymers, and caustics are caused by the multivalent cations. The two major-problem cations are calcium and magnesium, primarily because they are so highly concentrated in some waters.

Fresh water

The first step in determining the suitability of any water for injection is to

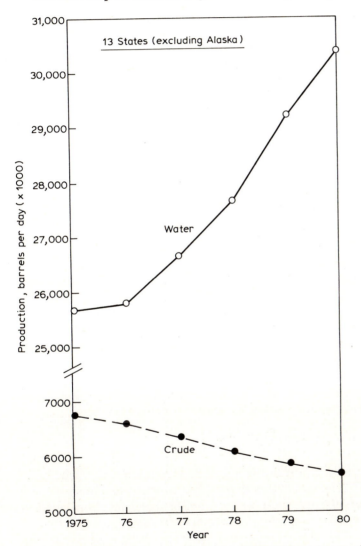

Fig. 6-8. Crude oil and water produced (× 1000 bbl/day) from wells in 13 states excluding Alaska. (Source: Petroleum Information Corporation.)

analyze the water for chemical, physical and biological constituents. Next, the composition of the formation into which it is to be injected should be determined. Clays such as smectites, kaolinites, chlorites and illites are sensitive to fresh water. Permeability reduction may occur because of clay dispersion and clay swelling (Mungan, 1965). Increasing the salinity of the water usually minimizes this effect.

Smectites and illites are the more common clays which are sensitive to fresh water. They can absorb water on their edges and surfaces. Fresh water can penetrate between the layers of a smectite to cause the plates to separate and swell. Formation damage by fresh water, therefore, usually is most severe in a formation that contains smectite.

Seawater

Several companies utilize seawater for water injection as a pressure maintenance technique or for secondary recovery in some giant oil reservoirs (Davis, 1974; Kennedy, 1977; *Oil and Gas Journal*, 1977; *World Oil*, 1977; Mitchell, 1978; Carlberg, 1979). It is injected into both sandstone and carbonate reservoirs. Some of the disadvantages of seawater injection are elaborated by Ogletree and Overly (1977).

Eventually, seawater will be utilized as an injection fluid in EOR technology. The use of seawater presents the same problems associated with any open system, i.e., where air—water contact exists. In addition, seawater presents some additional problems. One of the most notable of these is the biomass, e.g., organisms such as copepods, diatoms, and dinoflagellates.

Mitchell and Finch (1978) in their paper outline some of the necessary water quality tests including: membrane filter test; examination of the filtered particulates with light and scanning electron microscopy; onsite core injectivity tests; particle size distribution in the injection water with respect to the pore size distribution in the reservoir; amount and type of biomass (other than bacteria) in the raw seawater, and bacterial levels, aerobic and anaerobic. They found that cores are superficially plugged by lipids derived from copepods plus inorganic debris. They also emphasized the plugging of cores by bacterial debris which was documented by Fekete (1959).

Seawater vs. formation water as source water

The cost of seawater (if available) is less than formation water usually by a factor of about 10. The water treatment costs of seawater, however, are at least 10 times that of formation water. The installed injection facilities will cost about the same for seawater and formation water.

A major problem that can be encountered when using seawater is clay swelling, which can cause severe injectivity losses. Seawater is an acceptable injection water if it is adequately treated, filtered and if the formation is not sensitive to the seawater salinity. Another major problem often encountered when using seawater vs. formation water is corrosion. The run time for centrifugal-injection pumps for formation water is greater than for seawater probably by a factor of about 3. The extended run time for formation water can be attributed to the absence of chlorine and oxygen in the injection water. Repair and/or replacement of injection pumps is a major cost consideration.

WATER QUALITY TESTS

Water quality tests described in this section are tests that are necessary to determine such variables as the amount and composition of suspended solids, compatibility of two or more waters, compatibility of the injection solution with the reservoir rock, clay sensitivities, degree of corrosion, and bacteria.

Membrane filter test

Injection waters must be free of all particles in suspension. These waters are purified by mixing the waters, allowing them to settle, and then filtering.

It is important to know the quantity and composition of suspended solids in injection waters. This information can be obtained by filtering of an aliquot of the water through a 0.45-μm membrane filter, determining the weight of the solids, and analyzing the solids by use of appropriate techniques. The National Association of Corrosion Engineers has a standard method of determining the quantity of suspended solids (NACE, 1976a). Patton (1974) described quality evaluation specifications using the results of the test. Cerini et al. (1946) and Doscher and Weber (1957) also used the test.

Volumetric rate of flow versus cumulative volume (after 130 minutes, for example) is a common method of evaluating the membrane filter test (Cerini, 1953; NACE, 1976a). No direct relationship, however, exists between the slope and injectivity. Detrimental changes usually occur in the membrane filter cake before injectivity loss occurs. Examination of the filter cake enables one to pinpoint the source of trouble. For example, bacterial growth on the filter may be observed under the microscope.

The slope (rate vs. cumulative volume filtered) ratings show warning signals of future trouble. They are not a direct measure of the condition of the water. Slopes must be equated with the type of chemical treatment and the amount and nature of solids.

There are two flow stages during a rate versus cumulative volume determination:

(1) The initial portion of the filtration takes place at high rate until the

entire surface of the membrane filter is covered with suspended solids. This portion of the curve may be very short or very long depending upon the amount, the size, and the stickiness of the suspended solids. This first phase may be attributed to the plugging or filter cake formation phase.

(2) During the steady flow stage, the flow occurs through a gradually thickening filter cake of suspended solids. The flow rate is governed by the permeability of the filter cake of suspended solids. Flow gradually decreases due to the continual thickening and plugging of the filter cake by deposition of solids, with a consequent increase in resistance to flow.

The final flow rate can be combined with the determination of total suspended solids to determine the permeability of the filter cake, using Darcy's Law:

$$k = 1000 \, q\mu L/A\Delta p$$

where k = permeability, md; q = flow rate, cm^3/s; μ = viscosity of water, cp; L = cake thickness, cm; A = filtration area, cm^2; and Δp = filtration pressure, atm. All of the necessary data are either determined or readily available except for cake thickness. The cake thickness may be reasonably approximated by assuming a density for suspended solids and determining approximate composition and weight of suspended solids. Table 6-13 gives useful values for this purpose.

The rationale behind the rating of permeability of the filter cake comes from an examination of the classification of waters proposed by Stormont

TABLE 6-13

Probable cake thickness

Proximate suspended solids composition	Probable density (g/cm^3)	Probable cake thickness per mg of solids (cm)
A — Hydrocarbons, algae, bacterial growth, or hydrous compounds	1	0.000075
$\frac{2}{3}$ of A + $\frac{1}{3}$ B	1.5	0.000050
$\frac{1}{3}$ of A + $\frac{2}{3}$ B	2	0.000035
B — Silt, sand, or calcium carbonate scale	2.5	0.000030
$\frac{1}{4}$ C + $\frac{3}{4}$ B or $\frac{1}{2}$ A + $\frac{1}{2}$ C	3	0.000025
$\frac{1}{2}$ B + $\frac{1}{2}$ C or $\frac{1}{4}$ A + $\frac{3}{4}$ C	3.5	0.000021
$\frac{1}{4}$ B + $\frac{3}{4}$ C or $\frac{1}{8}$ A + $\frac{7}{8}$ C	4	0.000019
C — Iron sulfide or iron oxides	4.5	0.000017

(1958). Stormont had three classifications: perfect, minimum quality standard, and substandard. His ratings were based on injectivity experience in one particular California oilfield. Stormont's (1958) perfect water rating is set at a rating of excellent as shown below and in Table 6-14. The upper limit of minimum quality standard is acceptable, and the upper limit of substandard water quality is fair.

Permeability (k) of filter cake (md)	*Rating*
$k > 4$	excellent
$4 < k > 0.4$	very good
$0.4 < k > 0.04$	acceptable
$0.04 < k > 0.004$	moderate
$0.004 < k > 0.0004$	fair
$0.0004 < k$	excessive

There are problems as well as benefits in looking at the permeability of the filter cake. The problems concern the correlation of filter cake permeability with injectivity.

The pressure differential across the membrane filter is 20 psi, whereas the pressure differential across the formation face in the injection well can range from low to several thousand psi in the case of "skin" formation.

Depending upon the nature of the suspended solids, this pressure differential in the well could prevent the formation of a skin or filter cake. Certain solids such as hydrous solids are so soft that they are sheared by the pressure differential and thus are injected into the formation and dispersed radially. These same hydrous solids form a stable low-permeability filter cake at $p \leqslant$ 20 psi. The presence of significant percentages of strong solids with the hydrous solids will prevent this shearing, provided they are large enough to bridge on the formation pores.

The benefits gained by examining the permeability of the filter cake are:

(1) An indication of the presence of bonding substances which lead to the formation of deposits.

(2) An understanding of the danger of permitting skin formation in the injection well.

(3) The permeability of the filter cake is sensitive to changes in the water and/or system, such as bacterial growth, incompatible chemicals, and oxygen entry into a closed system.

Table 6-14 is a rating chart for several variables of an injection water. Suspended solids are rated from 1 to 20 where 1 indicates an excellent water, whereas 20 indicates a very poor water. Calculation can be made as follows:

$$\text{Suspended solids increase, mg/sq. ft day} = \frac{\text{solids increase, mg/l} \times (\text{vol. water, bbl/day}) \times 607}{(\text{diameter, in.}) \times (\text{length, ft})}$$

TABLE 6-14

Injection water rating chart

	Rating					
	1	2	3	4	10	20
Total sulfide increases, equivalent deposit thickness in inches	$S<0.007$ none	$0.007<S<0.014$ very low	$0.014<S<0.035$ low	$0.035<S<0.072$ moderate	$0.072<S<0.14$ large	$S>0.14$ excessive
Iron count increases, equivalent mil/year	$I<0.01$ none	$0.01<I<0.1$ very low	$0.1<I<1.0$ low	$1.0<I<5$ moderate	$5<I<10$ large	$I>10$ excessive
Increase in sulfate-reducing bacteria, colonies/ml	none	sporadic random appearance; questionable	absent more times than present; acceptable	present more times than absent; moderate	always present, 1—9 colonies/ml; large	greater than 9 colonies/ml; excessive
Increase in total bacteria count, colonies/sq. ft day	$B<10^7$ negligible	$10^7<B<10^8$ very low	$10^8<B<10^9$ low	$10^9<B<10^{10}$ moderate	$10^{10}<B<10^{11}$ large	$B>10^{11}$ excessive
Corrosion rate (30 days, insulated coupon), mil/year	$C<0.01$ none	$0.01<C<0.1$ very low	$0.1<C<1$ low	$1<C<5$ moderate	$5<C<10$ high	$C>10$ excessive
Pit depth (30 days, insulated coupon), mil	0 none	1 shallow	2—3 minor	4—5 moderate	6—10 deep	≥10 excessive
Pit frequency (30 days, insulated coupon), pits/sq. in.	0 none	1 very low	2 low	3 moderate	4 high	5 excessive

TABLE 6-14 (continued)

	Rating					
	1	2	3	4	10	20
Permeability of filter cake, md	$k>4$ excellent	$4>k>0.4$ very good	$0.4>k>0.04$ acceptable	$0.04>k>0.004$ moderate	$0.004>k>0.0004$ fair	$k<0.0004$ poor
Suspended solids increase, mg/sq. ft day	$SS<170$ excellent	$170<SS<350$ very good	$350<SS<520$ acceptable	$520<SS<870$ moderate	$980<SS<1300$ fair	$SS>1300$ excessive

where solids increase, mg/l = mg/l of solids found in the water using membrane filter test; diameter = diameter of injection pipe system, and length = length of injection pipe system. This calculation is useful in determining conditions existing in the injection system.

Table 6-15 describes some of the factors that were used in preparing the rating data shown in Table 6-14.

Water compatibility

Waters used for the recovery of oil by waterflooding usually contain a number of inorganic salts and sometimes organic salts in solution. It is common practice to test the compatibility of the injection water and water in the formation before starting a waterflood operation. Often this test is performed by mixing the injection water with the formation water in various

TABLE 6-15

Descriptors used in preparing rating chart

Rating value	Rating	Philosophy
1	Excellent, negligible, or none	System in best possible condition with regard to this variable — the ideal
2	Very good, very low, or shallow	System is in very good condition with regard to this variable — less than ideal, but substantially better than a system in normal trouble-free operation.
3	Acceptable, low, or minor	System in good condition — normal condition for trouble-free operation
5	Moderate	Systems in reasonable condition, However, condition is not as good as normal condition for trouble-free operation. System could be drifting towards trouble; hence, the extra increase in the rating number
10	Fair, large, high, or deep	System in fair condition. System will be in serious trouble if these conditions continue to prevail; hence, the heavy weighting of the rating number
20	Excessive	System in trouble. These conditions will cause serious loss of injectivity or serious corrosion or both if continued; hence, the extra heavy weighting of the rating number

proportions in a glass container and determining the amount of precipitate. The precipitate or scale can then be analyzed to determine its composition.

Waters are compatible if they can be mixed without producing chemical reactions between the dissolved solids in the waters and, thus, precipitating insoluble compounds. The precipitated insoluble compounds are undesirable because they can reduce the permeability of a porous petroleum-producing formation, plug injection wells in waterflood systems, and cause scale formation in water pumps, flow lines, and face of the formation.

Some of the more common ions that frequently occur in oilfield waters and cause precipitation if formation and injection waters are incompatible are: Ca^{2+}, Sr^{2+}, Ba^{2+}, Fe^{2+}, Fe^{3+}, HCO_3^- and CO_3^{2-}. Common reactions are:

$$CaCl_2 + Na_2SO_4 \rightarrow 2NaCl + CaSO_4 \downarrow$$
$$CaCl_2 + MgSO_4 \rightarrow MgCl_2 + CaSO_4 \downarrow$$
$$Ca(HCO_3)_2 \rightarrow CO_2 + H_2O + CaCO_3 \downarrow$$
$$CaCl_2 + 2NaHCO_3 \rightarrow 2NaCl + CO_2 + H_2O + CaCO_3 \downarrow$$
$$SrCl_2 + Na_2SO_4 \rightarrow 2NaCl + SrSO_4 \downarrow$$
$$SrCl_2 + MgSO_4 \rightarrow MgCl_2 + SrSO_4 \downarrow$$
$$BaCl_2 + Na_2SO_4 \rightarrow 2NaCl + BaSO_4 \downarrow$$
$$BaCl_2 + MgSO_4 \rightarrow MgCl_2 + BaSO_4 \downarrow$$
$$Fe + H_2S \rightarrow H_2 + FeS \downarrow$$
$$Fe_2O_3 + 6H_2S \rightarrow 6H_2O + 2Fe_2S_3 \downarrow$$

A relatively insoluble compound CA, where C is the cation and A is the anion, will precipitate from an aqueous solution if:

$$a_C \cdot a_A > S_{CA}$$

Where a_C = the cation activity, a_A = the anion activity in the solution, and S_{CA} = the solubility product of the compound CA. When two salts with a common cation (CA_1 and CA_2) are in equilibrium in a solution, the following will hold:

$$S_{CA_1} = a_C \cdot a_{A_1}$$

$$S_{CA_2} = a_C \cdot a_{A_2}$$

and:

$$a_{A_1}/a_{A_2} = S_{CA_1}/S_{CA_2}$$

If $a_{A_1}/a_{A_2} > S_{CA_1}/S_{CA_2}$, CA_1 will precipitate, whereas CA_2 will precipate if $a_{A_1}/a_{A_2} < S_{CA_1}/S_{CA_2}$.

Deposition of scale in water tanks, lines, equipment, injection wells, production wells and formations can be very costly. The scale not only restricts production but also causes inefficiency and equipment failure.

Important variables to scaling are:

(1) Temperature of the formation in relation to solubility of the possible scale former in the fluids passing through it. $CaSO_4$ and $SrSO_4$ become less soluble with increasing temperature (Blount and Dickson, 1969; Fletcher et al., 1981), whereas $BaSO_4$ becomes more soluble (Templeton, 1960).

(2) Subsurface pressures change for any system, with the highest pressure found while the fluid flows through the formation. The greatest pressure change is at the sand face of the producing well (Vetter and Phillips, 1970), which causes this area to be where solubility changes are the greatest. Deposition of scale at this point is the most damaging to oil production and the most difficult to discover or to remedy. Limited data are available on pressure—solubility relations of most scale-forming compounds, but $CaSO_4$ has been shown to decrease in solubility with decrease in pressure at NaCl concentrations up to 10% (Fulford, 1968).

(3) Brine concentration, exclusive of precipitating compounds, also influences scale formation. Most electrolytes in ionic form cause an increase in the solubility of compounds which form scales. The solubility normally increases with increasing electrolyte concentration unless some other solubility equilibrium is reached. This can occur, for example, when $BaSO_4$ saturation level is reduced because of increasing amounts of Ca^{2+} ion in the solution (Davis and Collins, 1971). Other properties of brine known to influence the solubility levels of scale formers are gases in solution, hydrogen ion concentration, ion pairs, and dissolved organic chelates (Weintritt and Cowan, 1967).

Mitchell et al. (1980) studied scaling problems in the North Sea Forties field. They encountered injectivity inhibitions resulting from barium and strontium sulfates. The injection water consisted of commingled formation water and seawater. Seawater contains about 2500 mg/l of sulfate, whereas the formation water contained about 250 and 660 mg/l of barium and strontium, respectively. According to them, the worst condition exists during an initial period of injection into a well, when there are relatively large amounts of formation water and small amounts of injection water. For example, this situation might exist during the initial injection and while the water migrates through the formation and finally breaks through at the producing well. Also according to them, plugging could occur if seawater inadvertently enters the formation and kills a wet production well.

They developed a scaling test rig which utilizes radial flow rather than linear flow, because apparently barium sulfate is more efficiently precipitated during radial flow. They were able to block (plug) a core completely in about eight minutes utilizing this test rig.

They found that polymeric scale inhibitors (Types A and B) were more effective than organic phosphonates (Types C and E) for barium sulfate inhibition. They also noted that a scale inhibitor only delays scale formation. To prevent scale buildup in the formation around the production well, squeeze treatments must be used. Timing of these treatments is vital and requires careful monitoring.

Anhydrite and gypsum

Both anhydrite, $CaSO_4$, and gypsum, $CaSO_4 \cdot 2H_2O$, are common in evaporites. A typical zonation from the base upwards is anhydrite, gypsum, and limestone. These same minerals often are associated with salt-dome oilfields and comprise what is called "cap rock". In an evaporite sequence, limestone is deposited first, followed by calcium sulfate.

Anhydrite can form even when water is present if the salinity of the water is high enough. The gypsum—anhydrite reaction is reversible and gypsum will form when the salinity of water is low enough. This reaction can be detrimental or perhaps "fatal" to a petroleum reservoir. For example, when a volume of anhydrite is hydrated, the volume of gypsum is 1 to $1\frac{1}{2}$ times greater.

Anhydrite is present in some reservoirs including some sandstones. It may serve as a cementing agent between sand grains or even oolite grains. Attention should be paid to the mineralogical composition of the reservoir and chemical composition of interstitial fluids, because injection of water less saline or even of a different composition may trigger the reversal of the anhydrite—gypsum reaction. The injection water chemistry and injection pressures must, therefore, be carefully monitored.

Water compatibility test

If the planned EOR project involves using chemicals such as surfactants, polymers, caustics, etc., the compatibility tests become even more complex. For example, various studies indicate that sulfonates and polymers react with the multivalent cations in formation waters (see Meister et al., 1979, 1980; French and Collins, 1980; French et al., 1981).

The tolerance of petroleum sulfonates to the multivalent cations depends upon the average equivalent weight (AEW) of the sulfonate. In general, the amount of cations tolerated increases as the AEW of the sulfonate decreases.

Ostroff (1979) presented two methods of determining water compatibilities and information on how to predict scale formation. Collins (1975a) presented some information on brine stabilization and methods for calculating over- and undersaturation of some relatively insoluble compounds. A method tentatively approved by the American Society for Testing and Materials (ASTM) Subcommittee D-19.09 appeared in the gray pages of Part 31 of the ASTM Annual Book of Standards in 1982. Essentially two

ASTM methods were approved, both of which involve filtering the solutions through 0.45-μm filters prior to mixing at various ratios. The mixtures are agitated continuously for 24 hours at specific temperatures and pressures. Subsequently, these solutions are filtered through 0.45-μm filters. If any suspended solids are filtered, the mixture is incompatible and the filtered solids must be analyzed to determine the nature of incompatibility.

Clay sensitivity

Many reservoirs contain as much as 10—15% clay. Such reservoirs can be damaged with respect to porosity and permeability on using an injection water that is not compatible to the clays. The clays in some sedimentary formations were deposited with the sand grains, i.e., each sand grain lies in a bed of clay. In such a formation, the permeability is about zero. In some other formations (some of the better petroleum reservoirs), the clays either formed in place or were deposited from infiltrating waters after the sand deposition (Dickey, 1979).

About seven different clay mineral groups exist in reservoir rocks; kaolinite, illite, smectite (formerly called montmorillonite), chlorite, vermiculite, mixed-layer, and lath form. The mixed-layer group contains a variety of intermediate forms of clays that actually are in a state of evolution from one mineral group to another. The lath-form group includes sepiolite and attapulgite (Millot, 1979).

The clay minerals are silicate minerals and montmorillonite (a species of the smectite group) consists of two layers of silica in tetrahedral coordination and one layer of magnesium in octahedral coordination, with oxygen and hydroxyl in the anion positions. All members of the smectite group swell in water because of introduction of water in the interlayer positions perpendicular to the C-axis. Also, they readily exchange their cations with cations in solution. Because of these two factors, swelling and ion exchange, montmorillonite can have a very important influence upon the porosity and permeability of a petroleum reservoir rock.

Ion exchange reactions on clay minerals are reversible and follow the law of mass action. The number of exchange sites governs the reaction. Other important factors include temperature, pressure, solution concentrations, and bonding strength of exchangeable ions. Ion exchange between clay minerals and a brine will stop when equilibrium is attained.

As the waters move in their subsurface environment, their dissolved ions have a tendency to exchange with those in the rocks. There are two extreme types of adsorption in addition to intermediate types of adsorption: (1) a physical adsorption or van der Waals' adsorption with weak bonding between the adsorbent and the constituent adsorbed, and (2) a chemical adsorption with strong valence bonds.

Cations can be fixed at the surface and in the interior of minerals. These fixed cations can exchange with cations in the water. Under the proper physical conditions of the adsorbent, similar exchange can occur with the anions. Some of the constituents in formations that are capable of exchange and adsorption are argillaceous minerals, zeolites, ferric hydroxide, and certain organic compounds.

Particle size influences the exchange rates and capacities if the solids are clays such as montmorillonite, illite, and kaolinite. The rate increases with decreasing particle size. However, if a larger mineral has a lattice, the exchange can easily occur on the plates. The concentrations of exchangeable ions in the adsorbent and in the water are important. More exchange usually occurs when the solution is highly concentrated with dissolved solids.

According to Grim (1952), the replacing power of some ions in clays is:

(1) In NH_4-kaolinite: $Cs > Rb > K > Ba > Sr > Ca > Mg > H > Na > Li$

(2) In NH_4-smectite: $Cs > Rb > K > H > Sr > Ba > Mg > Ca > Na > Li$

The replacing order indicates that lithium and sodium are more likely to be left in solution, whereas cesium and rubidium are more likely to be removed from solution.

Lithium has a small radius, a low atomic number, and a larger hydrated radius and polarization than sodium. Consequently, its replacing power in the lattices of clay minerals is low (Kelley, 1948). Other ions such as barium, strontium, calcium, magnesium, cesium, rubidium, and potassium will preferentially replace lithium and sodium in clay minerals, thus releasing lithium and sodium to solutions.

As sediments are buried deeper, smectite is altered to a mixed-layer illite—chlorite—smectite, and the regularity of this mixed-layer phase tends to increase with depth (Weaver and Beck, 1969). Smectite retains its swelling ability up to temperatures of about 210—230°C. Even after being dehydrated at temperatures of 400—500°C, it will rehydrate quite readily. When it is chemically altered, for example to chlorite, however, the resulting chlorite is not expandable.

Fig. 6-9 taken from Cassan et al. (1981), illustrates the general distribution of clay minerals and associated minerals as a function of depth.

Sandstone reservoir heterogeneity is an important geological constraint on micellar flooding success. The factors which have the greatest effect on sandstone reservoir heterogeneity are:

(1) Clay mineralogy and percentage of various clays.

(2) "Primary" vs. "diagenetic" clays: (a) diagenetic clays not in pore system; (b) "depositional" clays not in pore system.

(3) Environment of deposition of the sandstone.

(4) Lateral and vertical continuity (communication) of the sandstone reservoir; exclusive of faults.

These four factors are interdependent and interrelated to each other.

Somerton and Radke (1979) determined that the ion exchange capabilities of clay can have the following effects:

210

(1) A release of divalent ions, which can cause surfactant precipitation, loss of ultra-low interfacial tension, and polymer degradation.

(2) Ion exchange reactions occurring with surface charge effects, which can cause structural damage in the formation.

Additional problems include: (a) expanding (swelling) clays, (b) mechanical migration of the clays, and (c) chemical precipitates.

These five problems can be generated in a clay-rich sandstone reservoir by the introduction of almost any fluid.

"Primary" clays, as defined here, are clays (1) which were deposited at the

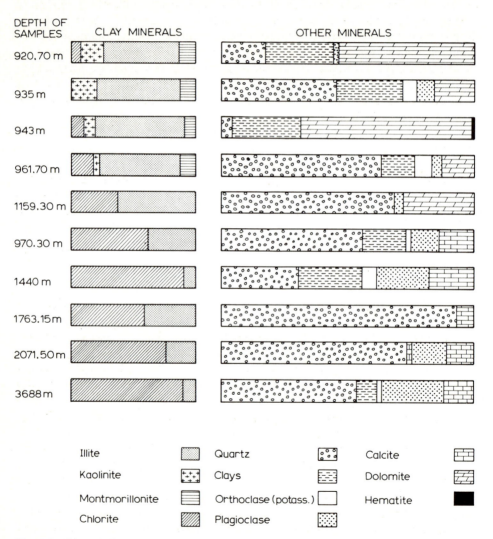

Fig. 6-9. General distribution of minerals as a function of depth (from Cassan et al., 1981).

time of deposition (2) were introduced into a sandstone by burrowing organisms, or (3) resulted from the reworking of the soft unlithified sediment by waves or currents. These clays, most commonly extremely fine-grained, are found in all sandstones as an essential constitutent of the matrix, e.g., the fine-grained debris between the larger detrital clasts of quartz, feldspar, and rock fragments. Under normal hydrologic conditions, the clays will settle into discrete layers and, unless reworked by waves, water currents or burrowing organisms, will become shales or claystones and, thus, are responsible (generally) for non-reservoirs. When reworking of the clay minerals occurs, then the clays become an integral part of the sandstone.

With additional burial, compaction, and increased temperature, the kaolinites, smectites, and mixed-layer groups decrease in concentration, whereas the illites and chlorites increase in abundance. With increasing compaction and temperature, however, metamorphism and recrystallization transforms the clay minerals into micas, feldspars, and other primary minerals. In this environment, petroleum hydrocarbons would be converted to graphite and some form of hydrogen.

Core flow test

Core flow testing is the only good method of determining the effects of the proposed injection fluid upon the permeability of the formation reservoir. McCune (1977) described a core test equipment that can be used to make a flow test.

Corrosion

Ostroff (1979) lucidly defined corrosion and the forms of corrosion found in oilfield operations. As he pointed out, electrochemical corrosion of steel is the usual type of corrosion in the oilfield. Further, he noted: "it is necessary to have an (1) anode, (2) cathode, (3) electrolyte, and (4) external connection. Remove any one of these and corrosion will cease." Inasmuch as the electrolyte is brine, it is impossible to remove it from an oilfield water system. Also, usually it is impossible to remove the anode, cathode, or the external connection in most oilfield systems. Complete coating of the steel lines and vessels or use of non-conducting lines and vessels (cathodic protection) would solve the problem, but this is not yet feasible for all systems.

The gases that often are found in injection waters and which are deleterious because of potential corrosion problems are O_2, H_2S, and CO_2. The presence of these gases in a salt water presents severe corrosion problems because of the following: (a) salt water is an electrical conductor and is corrosive, and (b) the corrosivity increases as the water becomes saltier and as the concentration of O_2, H_2S, or CO_2 increases. These dissolved gases drastically increase the corrosivity of salt water (Patton, 1974). Fewer corrosion problems exist if they are removed and excluded and if the injection water is maintained at a neutral (or slightly higher) pH. From the standpoint of the formation rock itself, rather than corrosion or fluid compatibility, it might be undesirable to have the pH above 7, because of the effect of high pH on clay swelling. Of course, there are some high-pH floods in operation which are attempting to alter the wettability of reservoir rocks.

Among gases, oxygen presents the worst corrosion problem. Concentrations of less than a part per million of oxygen in an injection water can cause severe corrosion in an injection water system. Oxygen solubility in water is a function of temperature, pressure, and concentrations of dissolved solids. It is less soluble in salt water than in fresh water.

The CO_2 dissolved in water forms carbonic acid which lowers the pH and increases the corrosivity of the water. Waters that contain bicarbonate alkalinity (and most oilfield waters do) present a greater corrosion problem because of the carbon dioxide—bicarbonate—carbonate equilibria. Corrosivity increases as the pH decreases because bicarbonate converts to carbonic acid and additional CO_2 is formed. If the pH is raised, the result is scale formation or precipitation of carbonates.

The solubility of CO_2 is influenced by the partial pressure of CO_2 in the atmosphere above the water. The temperature, pressure, and dissolved solids also influence the solubility. Considerable research is needed to determine more accurately the solubility of CO_2, O_2, and H_2S in saline waters. .

H_2S is present in many oilfield waters. Its origin sometimes is attributed to sulfate-reducing bacteria. This gas is very soluble in water, and its solubility also is a function of temperature, pressure, and dissolved solids. The corrosivity of a water increases with increasing concentrations of dissolved H_2S.

Table 6-14 shows four criteria that can be used to evaluate corrosion: (1) iron count increases, equivalent mil/year; (2) corrosion rate (30 days, insulated coupon), mil/year; (3) pit depth (30 days, insulated coupon), mil; and (4) pit frequency (30 days, insulated coupon), pits/sq. in.

Iron count increase is the increase in the concentration of iron in the water between two points in the system. For example, if point 1 is the water supply intake and point 2 the injection well, it can be calculated as follows:

$$\text{iron count increases, equivalent mil/year} = \frac{(\text{iron count increase, mg/l}) \times (\text{water, bbl/day}) \times 11.95}{(\text{diameter of line, in.}) \times (\text{length of line, ft})}$$

where, iron count increase, mg/l = amount of iron in mg/l found in a sample taken from point 2 minus amount of iron in mg/l found in sample taken from point 1.

The other tests involving the corrosion coupons are described by Wright (1963), Ostroff (1979), and NACE (1976a). The NACE (1976b) publication covers standard materials for pumps, valves, equipment, and filters used in oilfield systems. Ostroff (1979) also presented considerable information on oilfield systems and corrosion.

Bacteria

Injection waters must be free of bacteria because they can cause corrosion, as well as plugging of equipment and the injection wellbore. Bacteria can reproduce rapidly in extremely diverse conditions such as low and high pH, temperature, pressure. They even grow in the absence of oxygen. Patton (1974) presented the following classification of bacteria found in oilfield water systems:

(1) Aerobic bacteria — growth dependent upon oxygen.

(2) Anaerobic bacteria — grow best in the absence of oxygen.

(3) Facultative bacteria — growth independent of oxygen.

The primary bacterial problems found in oilfield injection waters are caused by sulfate reducers, iron bacteria and slime formers. The sulfate reducer is an anaerobic bacteria *(Desulfovibrio desulfuricans)* and is a primary problem in injection waters. They reduce sulfate ions to sulfide ions, resulting in H_2S which causes corrosion. The end product, iron sulfide, causes plugging problems.

Iron bacteria, such as *Crenothrix*, *Gallionella* and *Sphaerotilus*, sheath themselves in iron hydroxide, utilizing soluble iron ions in water. They cause both corrosion and plugging problems. These bacteria are an indirect cause of corrosion in that they shield part of the surface and thus create oxygen concentration cells.

The slime formers include *Aerobacter, Bacillus, Escherichia, Flavobacterium* and *Pseudomonas*, and are aerobic. They cause plugging and corrosion problems, and contribute to corrosion in a manner similar to iron bacteria. An API (1975) report presents methods to determine bacteria in oilfield waters. Ostroff (1979) and Patton (1974) described techniques to test bacteria and control them in oilfield injection systems.

The sulfate reducers use sulfates as an oxidant in the presence of various organic compounds. Organic material plus oxidation of molecular hydrogen provide the energy for the sulfate reducers. *Desulfovibrio desulfuricans* and other sulfate-reducing bacteria contain the enzyme hydrogenase, which also is implicated in corrosion processes (Zajic, 1969).

Sulfate-reducing bacteria (SRB) cause corrosion of oilfield pumps, lines, tanks, etc. Also, SRB can cause plugging problems. Two books (Beestecher, 1954; Davis, 1967) have been written about SRB activity related to petroleum. A symposium (Anderson, 1957) was held for the purpose of improving information about SRB and how to control SRB. Postgate (1979) provided additional information about SRB and listed various Quaternary amines and other inhibitors that have been developed as bactericides to control them. Also, coating metal pipes on both the inside and outside to prevent the bacteria from accessing the metal is a successful corrosion prevention technique.

Postgate (1979) pointed out that mere bacteria counts are not adequate, because they do not distinguish between active and dormant bacteria. According to him. Eh is the only adequate guide. An Eh below +100 mV indicates severe corrosion, whereas a reading above 400 mV indicates absence of corrosion.

In a simple small system, a net change in sulfides is adequate for determining bacterial activity. Length of the system, however, should be considered in determining the dangerous level. The factors which influence the total sulfide increase are: (a) growth of sulfide-producing bacteria, (6) rate of flow of the water, (c) length of the line, and (d) diameter of the line.

The rating for sulfate-reducing bacterial infection as measured by sulfide increase is:

Equivalent deposit thickness (inches)	Rating
0	none
0.0072—0.013	very low
0.014 —0.035	low
0.036 —0.071	moderate
0.072 —0.13	large
0.14 to thicker	excessive

where equivalent deposit thickness is calculated as follows:

$$\text{Equivalent deposit thickness, inches} = \frac{(\text{sulfide increase, mg/l}) \times (\text{water, bbl/day}) \times 0.0719}{(\text{diameter of line, in.}) \times (\text{length of line, ft})}$$

The rationale behind expressing sulfide increase as probable deposit thickness is that high sulfide-producing bacterial growth requires significant amounts of deposits to support the populations. Presence of significant sulfide increase indicates that the system is dirty, with pitting corrosion probably occurring. Remedial measures can be any means that remove the deposits (and bacterial growth), as well as bactericidal treatment.

The experience factors, which led to the development of this rating scale for sulfide increases, were developed in systems having water temperatures around 70—75°F. Higher water temperatures would require decreased equivalent deposit thicknesses due to increased bacterial growth rates. Temperatures above about 110°F would result in decreased growth rates and require increased equivalent deposit thicknesses. Temperatures below 70°F would likewise decrease growth rates and require an increase in equivalent deposit thicknesses.

Sulfate-reducing bacteria count

The presence of sulfate-reducing bacteria (SRB) in a system does not necessarily mean trouble. However, when continued checks show they are multiplying across the system, the bactericide is not effective. Growth may take place in deposits, under scale, or under a film of general bacterial growth hidden from the moving stream of water. Bacterial counts of the water, therefore, must be supplemented by bacterial counts on scrapings of deposits on the interior walls of the system whenever possible.

When changing from an ineffectual bactericide to an effective one, sometimes bacterial counts temporarily increase. This is particularly true when using the second chemical at threshold levels at which its action is bacteriostatic instead of bactericidal. This increase in bacterial count is due to erosion of earlier thriving growths.

The rationale for developing the rating for SRB is not like that used to develop the rating for general bacteria. The detection of growth of SRB in a system is accomplished only after there are enough colonies growing to enable detectable numbers to be eroded off into the moving stream of water. Thus, routine detection of growth means a well established infection. The following rating is commonly used:

Degree of bacterial growth	*Rating*
No growth detected	none
Sporadic detection of growth	very low
Growth detected less than half the time	acceptable
Growth detected more than half the time	moderate
Growth present in numbers below 10 colonies/ml	large
Growth present in numbers greater than 10 colonies/ml	excessive

Table 6-16 shows the differences of these ratings in some typical systems.

Total bacterial count

The total bacterial count is made under aerobic conditions. Counts should be made on both the supply water and at the injection well to determine if growth is occurring across the system.

The interest in aerobic plate counts lies in the ability of most bacteria to grow under either aerobic or anaerobic conditions. It is much simpler to make an aerobic count than an anaerobic count. The presence of significant numbers of general bacteria usually makes an environment more favorable for sulfate-reducing bacteria. Large numbers of general bacteria mean that a loss of injectivity probably is occurring in tight sands. Normally, it is necessary to detect changes in numbers of general bacteria across the system; however, large numbers of general bacteria are not desirable even if no growth is occurring in the system.

TABLE 6-16

SRB ratings in some typical oilfield systems

System	A	B	C
Line diameter, inches	6	24	60
Line length, miles	1	10	75
Flow rate, ft/s	3	3	2
Increase of 1 colony/ ml is equivalent to:	7×10^5 colonies/sq. ft per day	7×10^4 colonies/sq. ft per day	2×10^4 colonies/sq. ft per day
Alternatively, increase of 1 colony/sq. ft is equivalent to:	1 colony/ 2×10^5 ml or 200 liters	1 colony/ 7×10^4 ml or 70 liters	1 colony/ 2×10^4 ml or 20 liters

As in the case of SRB, large general bacterial populations may be present in deposits within the system, hidden from the moving stream of water. The problem is that bacteria do not grow primarily in water, but on water-wet surfaces. Thus, the bacteria culture (or count) in water samples are those which have been torn off the walls of the pipe by turbulence or are traveling through the system. The limiting factor on bacterial growth is the interior surface of the pipe.

To obtain insight into the meaning of a bacteria count, it is necessary to translate the number of colonies per milliliter into a universally usable number. Increase in the number of colonies per square foot of interior pipe surface appears useful:

$$\text{Increase in colonies/sq. ft day} = \frac{\left(\dfrac{\text{increase in bacteria}}{\text{colonies/ml}}\right) \times (\text{water, bbl/day}) \times 607{,}200}{(\text{diameter of line, in.}) \times (\text{length of line, ft})}$$

The increase in bacteria expressed as the number of colonies per square foot of surface has the same limitation as iron counts and sulfide increases, i.e., assuming uniform conditions between the two points sampled. Localized hot spot conditions cannot be detected except by sampling at closer distances along the system.

The finding of high numbers of general bacteria colonies per square foot of surface has the same connotation as a thick average deposit thickness as determined by sulfide increases. Simply stated, it means that the system is dirty. Remedial measures can be the same as for high sulfide increases.

An industry yardstick is that in a specific waterflood there is no loss of injectivity when the general bacteria counts are 10,000 colonies per milliliter or lower. When the general bacteria counts become 10,000 colonies per milliliter or higher there is loss of injectivity. This observation has become the rule of thumb in the oilfield for many field personnel.

On using this break-point and a 6-inch-diameter injection line one mile long with a flow rate of 3 ft per second, one can develop numbers for the rating scale. The assumptions are reasonable since the end result is in the right order of magnitude and any errors in the assumptions would be substantially less than one order of magnitude. The break point between 10,000 and 100,000 colonies per milliliter was placed between the large and excessive position on the rating chart. The rest of the numbers then fall in place.

INJECTIVITY LOSSES

Filtration of small particles in the injection water by the formation near the wellbore region will cause injection rates to decrease. Eventually the permeability will become lower further back in the formation. For example, it is not unusual for water injection rates to decline by 50% in 12 months. The only way to circumvent this is to inject water containing no suspended solids and which is compatible with the formation water and the formation minerals, especially the clays. Workovers can improve the injection rates after a decline, but they are expensive and time consuming. Even a workover will have little or no effect if the injection well and formation contain deposits such as barium sulfate.

CONCLUSIONS

Water is important in EOR operations. Adequate considerations should be given to the type, quality, and quantity of water available. Necessary tests should be made to insure that the water used is compatible with the EOR technology planned, the reservoir rock, and associated indigenous fluids. After the EOR operation is started, necessary tests should be conducted on a routine basis to insure that the system is maintained at optimum conditions.

REFERENCES

Anderson, K.E., 1957. *Sulfate Reducing Bacteria, Their Relation to the Secondary Recovery of Oil.* St. Bonaventure University, New York, N.Y., 110 pp.

API, 1968. Recommended practice for analysis of oilfield waters. *Am. Pet. Inst.*, RP-45: 49 pp.

API, 1975. Recommended practice for biological analysis of flood injection waters. *Am. Pet. Inst.*, RP-38: 7 pp.

ASTM, 1982. *Annual Book of ASTM Standards, Part 31. Water. Section VII — Saline and Brackish Waters, Sea Waters and Brines.* American Society for Testing and Materials, Philadelphia, Penn., pp. 1075—1107.

Bailey, N.J.L., Krouse, H.R., Evans, C.R. and Rogers, M.A., 1973. Alteration of crude oils by water and bacteria — evidence from geochemical and isotope studies. *Bull. Am. Assoc. Pet. Geol.*, 57 (7): 1276—1290.

Ban, A., Balint, V. and Pach, F., 1977. Ammonia enhanced production from Hungarian field. *Oil Gas J.*, 75 (Jul. 11): 149—152.

Beestecher, E., 1954. *Petroleum Microbiology — An Introduction to Microbiological Engineering.* Elsevier, New York, N.Y., 375 pp.

Blount, C.W. and Dickson, F.W., 1969. The solubility of anhydrite ($CaSO_4$) in $NaCl-H_2O$ from 100 to 450°C and 1 to 1000 bars. *Geochim. Cosmochim. Acta*, 33: 227—245.

Blount, C.W., Price, L.C., Wenger, L.M. and Tarullo, M., 1979. Methane solubility in aqueous NaCl solutions at elevated temperatures and pressures. In: *Proceedings, Fourth United States Gulf Coast Geopressured—Geothermal Energy Conference. Res./Dev.*, 3: 1225—1262.

Bond, D.C., 1972. Hydrodynamics in deep aquifers of the Illinois Basin. *Ill. State Geol. Surv. Circ.*, 470: 72 pp.

Buckley, S.E., Hocott, C.R. and Taggart, M.S., Jr., 1958. Distribution of dissolved hydrocarbons in subsurface waters. In: L.G. Weeks (Editor), *Habitat of Oil.* American Association of Petroleum Geologists, Tulsa, Okla., pp. 850—882.

Braitsch, O. and Herrman, A.G., 1963. Zur Geochemie des Broms in Salinaren Sedimenten, Teil 1. Experimentelle Bestimmung der Br-Verteilung in Naturlichen Salzsystem (Geochemistry of bromine in a primary saline sediment: experimental distribution of bromine in a salt system). *Geochim. Cosmochim. Acta*, 27: 361—391.

Bredehoeft, J.D., 1965. The drill-stem test: the petroleum industry's deep-well pumping test. *Ground Water*, 3: 15—23.

Carothers, W.W., 1976. *Aliphatic acid anions and stable carbon isotopes of oil field waters in the San Joaquin Valley, California.* Master's Thesis, San Jose State University, San Jose, Calif., 69 pp.

Carothers, W.W. and Kharaka, Y.K., 1978. Aliphatic acid anions in oil-field waters — implications for origin of natural gas. *Bull. Am. Assoc. Pet. Geol.*, 62: 2441—2453.

Carlberg, B.L., 1979. How to treat seawater for injection projects. *World Oil*, 189 (1): 67—71.

Carpenter, A.B., 1978. Origin and chemical evolution of brines in sedimentary basins. *Okla. Geol. Surv. Circ.*, 79: 60—79.

Carpenter, A.B., Trout, M.L. and Pickett, E.E., 1974. Preliminary report on the origin and chemical evolution of lead- and zinc-rich oilfield brines in central Mississippi. *Econ. Geol.*, 69: 1191—1206.

Cassan, J.P., Palacios, M.G., Fritz, B. and Tardy, Y., 1981. Gabon Basin diagenesis of sandstone reservoirs. *Bull. Cent. Rech. Explor.-Product., Elf-Aquitaine*, 5 (1): 113—135.

Cerini, W.F., 1953. How to test quality of injection waters. *World Oil*, 137 (3): 189—190.

Cerini, W.F., Battles, W.R. and Joes, P.H., 1946. Some factors influencing the plugging characteristics of an oil-well injection water. *Trans. AIME*, 165: 52—63.

Clayton, R.N., Friedman, I., Graf, D.L., Mayeda, T.K., Meents, W.F. and Shimp, N.F., 1966. The origin of saline formation waters, 1. Isotopic composition. *J. Geophys. Res.*, 71: 3869—3882.

Collins, A.G., 1967. Geochemistry of some Tertiary and Cretaceous age oil-bearing formation waters. *Environ. Sci. Technol.* 1: 725—730.

Collins, A.G., 1975a. *Geochemistry of Oilfield Waters*. Elsevier, Amsterdam, 496 pp.

Collins, A.G., 1975b. Chemical applications in oil- and gas-well-drilling and completion operations. *Conference Proceedings — Environmental Aspects of Chemical Use in Well-Drilling Operations*. EPA-560/1-75-004, pp. 231—260.

Collins, A.G., 1979. Oilfield brine data system. *Am. Chem. Soc./Chem. Soc. Jpn., Chem. Congr., Honolulu, Hawaii, April 1—6, 1979.*

Cooper, J.E., 1962. Fatty acids in recent and ancient sediments and petroleum reservoir waters. *Nature*, 193: 744—746.

Craig, H., 1961. Isotopic variations in meteoric waters. *Science*, 133: 1702—1703.

Davis, J.B., 1967. *Petroleum Microbiology*. Elsevier, New York, N.Y., 604 pp.

Davis, J., 1974. Big waterflood begins off Abu Dhabi. *Oil Gas J.*, 73 (33): 49—51.

Davis, J.W. and Collins, A.G., 1971. Solubility of barium and strontium sulfates in strong electrolyte solutions. *Environ. Sci. Technol.*, 5: 1039—1043.

Department of Energy (DOE), 1981. DOE updates enhanced recovery economics. *Oil Gas J.*, 79 (28): 43—44.

De Sitter, L.U., 1947. Diagenesis of oil-field brines. *Bull. Am. Assoc. Pet. Geol.*, 31: 2030—2040.

Degens, E.T., Hunt, J.M., Reuter, J.H. and Reed, W.E., 1964. Data on the distribution of aminoacids and oxygen isotopes in petroleum brine waters of various geologic ages. *Sedimentology*, 3: 199—225.

Dickey, P.A., 1979. *Petroleum Development Geology*. PPC Book, Tulsa, Okla., 398 pp.

Doscher, T.M. and Weber, L., 1957. The use of the membrane filter in determining quality of water for subsurface injection. *Drill. Prod. Pract. API*, pp. 169—179.

Fekete, T., 1959. *The plugging effect of bacteria in sandstone systems*. M.S. Thesis, University of Alberta, Edmonton, Alta.

Fletcher, G.E., French, T.R. and Collins, A.G., 1981. A method for calculating strontium sulfate solubility. *DOE/BETC/RI-80/10*, 17 pp.

French, T.R. and Collins, A.G., 1980. Precipitation of petroleum sulfonates by magnesium ions. *DOE/BETC/RI-80/3*, 13 pp.

French, T.R., Stacy, N.and Collins, A.G., 1981. Polyacrylamide polymer viscosity as a function of brine composition. *DOE/BETC/RI-80/12*, 13 pp.

Fulford, R.S., 1968. Effects of brine concentration and pressure drop on gypsum scaling in oil wells. *J. Pet. Technol.*, 20: 559—564.

Grim, R.E., 1952. *Clay Mineralogy*. McGraw-Hill, New York, N.Y., 396 pp.

Healy, R.N., Reed, R.L. and Carpenter, C.W., Jr., 1975. A laboratory study of micro-emulsion flooding. *Soc. Pet. Eng. J.*, 15 (1): 87—103.

Hitchon, B., Billings, G.K. and Klovan, J.E., 1971. Geochemistry and origin of formation waters in the Western Canada Sedimentary Basin, III. Factors controlling chemical composition. *Geochim. Cosmochim. Acta*, 35: 567—598.

Hoke, S.H. and Collins, A.G., 1981. Mobile wellhead analyzer for the determination of unstable constituents in oil-field waters. *Am. Soc. Test. Mater., Spec. Tech. Publ.*, 735: 34—48.

Holm, L.W. and Josendal, V.A., 1972. Reservoir brines influence soluble-oil flooding process. *Oil Gas J.*, 70 (46): 158—168.

Kelley, W.P., 1948. *Cation Exchange in Soils*. Reinhold, New York, N.Y., 144 pp.

Kennedy, J.L., 1977. Offshore water-injection system at Dubai is expanded. *Oil Gas J.*, 75 (22): 85—89.

Kharaka, Y.K. and Berry, F.A.F., 1973. Simultaneous flow of water and solutes through geological membranes, I. Experimental investigations. *Geochim. Cosmochim. Acta.*, 37: 2577—2603.

Krejci-Graf, K., 1978. Data on the geochemistry of oilfield waters. *Geol. Jahrb., Reihe D*, 21—25: 3—174.

Manheim, F.T. and Horn, M.K., 1968. Composition of deeper subsurface waters along the Atlantic coastal margin. *Southeast. Geol.*, 8: 215—236.

McAlister, J.A., Nutter, B.P. and Lebourg, M., 1965. A new system of tools for better control and interpretation of drill-stem tests. *J. Pet. Technol.*, 17: 207—214.

McAuliffe, C.D., 1966. Solubility in water of paraffin, cycloparaffin, olefin, acetylene, cyclo-olefin and aromatic hydrocarbons. *J. Phys. Chem.*, 70: 1267—1275.

McCune, C.C., 1977. On-site technology to define injection-water quality requirements. *J. Pet. Technol.*, 1: 17—24.

McKelvey, J.G. and Milne, J.H., 1962. The flow of salt solution through compacted clay. In: *Clays and Clay Minerals, 9th Natl. Conf. Clays and Clay Minerals*, pp. 248—259.

Meister, M.J., Wilson, C.A. and Collins, A.G., 1979. Tolerance of petroleum sulfonates to the presence of calcium ions. In: K.L. Mittal (Editor), *Solution Chemistry of Surfactants*. Plenum Press, New York, N.Y., pp. 927—940.

Meister, M.J., Wilson, C.A. and Collins, A.G., 1980. Effect of temperature and sulfonate concentration on calcium tolerance. *BETC/RI-79/3*: 13 pp.

Millot, G., 1979. Clay. *Sci. Am.*, 240 (4): 108—118.

Mitchell, R.W., 1978. The Forties Field sea-water injection system. *J. Pet. Technol.*, 30: 877—884.

Mitchell, R.W. and Finch, T.M., 1978. Water quality aspects of North Sea injection water. In: *SPE (UK) Europe Offshore Petroleum Conference, Proceedings*, 1: 263—276.

Mitchell, R.W., Grist, D.M. and Boyle, M.J. 1980. Chemical treatments associated with North Sea projects. *J. Pet. Technol.*, 32 (5): 904—912.

Mungan, N., 1965. Permeability reduction through changes in pH and salinity. *J. Pet. Technol.*, 12: 1449—1453.

NACE, 1976a. Methods for determining water quality for subsurface injection using membrane filters. *Natl. Assoc. Corros. Eng.*, TM-01-73: 4 pp.

NACE, 1976b. Recommended practice for selection of metallic materials to be used in all phases of water handling for injection into oil-bearing formations. *Natl. Assoc. Corros. Eng.*, RP-04-75: 10 pp.

Noad, D.F., 1962. Water analysis data, interpretation and applications. *J. Can. Pet. Technol.*, 1 (2): 82—89.

Ogletree, J.O. and Overly, R.J., 1977. Sea-water and subsurface-water injection in West Block 73 waterflood operation. *J. Pet. Technol.*, 25: 623—628.

Oil and Gas Journal, 1977. Emphasis on water injection still growing in the Middle East. *Oil Gas J.*, 75 (20): 105—106.

Ostroff, A.G., 1979. *Introduction to Oilfield Water Technology*. National Association of Corrosion Engineers, Houston, Texas, 394 pp.

Pang, H.W., Fleming, P.D. and Boneau, D.F., 1981. Design of a preflush for commercial scale polymerflood in the North Burbank Unit. In: *Proceedings Second Joint SPE/ DOE Symposium on Enhanced Oil Recovery, Tulsa, Okla., April 5—8, 1981*. Society of Petroleum Engineers, Tulsa, Okla., pp. 97—122.

Patton, C.C., 1974. *Oilfield Water Systems*. Campbell Petroleum Series, Norman, Okla. 65 pp.

Peterson, A.V., 1978. Optimal recovery experiments with N_2 and CO_2. *Pet. Eng.*, 50 (Nov.): 40—50.

Postgate, J.R., 1979. *The Sulfate Reducing Bacteria*. Cambridge University Press, New York, N.Y., 151 pp.

Price, L.C., 1976. Aqueous solubility of petroleum as applied to its origin and primary migration. *Bull. Am. Assoc. Pet. Geol.*, 60: 213—244.

Rittenhouse, G., 1967. Bromine in oilfield waters and its use in determining possibilities of origin of these waters. *Bull. Am. Assoc. Pet. Geol.*, 51: 2430—2440.

Rittenhouse, G., Fulton, R.B., Grabowski, R.J. and Bernard, J.L., 1969. Minor elements in oilfield waters. *Chem. Geol.*, 4: 189—209.

Schoeller, H., 1955. Géochemie des eaux souterraines (Geochemistry of subterranean waters). *Rev. Inst. Fr. Pét.*, 10: 181—213; 507—552; 823—874.

Sokolov, K.A., 1956. *Migratsiya Gaza i Nefti (Gas and Oil Migration)*. Izd. Akademii Nauk SSSR, Moscow, 45 pp.

Somerton, W.H. and Radke, C.J., 1979. Role of clays in enhanced oil and gas recovery. In: *Fifth DOE Symposium on Enhanced Oil and Gas Recovery and Improved Drilling Technology, 1*. Petroleum Publishing Co., pp. D-7/1—D-7/16.

Stormont, D.H., 1958. Filter test gives data on quality of flood waters. *Oil Gas J.*, 56 (44): 84.

Sulin, V.A., 1947. *Vody Neftyanykh Mestorozhdeniy v Sisteme Prirodnykh Vod (Waters of Petroleum Formations in the System of Natural Waters)*. Gostoptekhizdat, Moskow, 96 pp.

Sunwall, M.T. and Pushkar, P., 1979. The isotopic composition of strontium in brines from petroleum fields of southeastern Ohio. *Chem. Geol.*, 24: 189—197.

Templeton, C.C., 1960. Solubility of barium sulfate in sodium chloride solutions from $25°$ to $95°C$. *J. Chem. Eng. Data*, 5: 514—516.

van Poolen, K.H. and Associates, Inc., 1980. *Fundamentals of Enhanced Oil Recovery*. PennWell Books, Tulsa, Okla., 155 pp.

Vetter, O.J.G. and Phillips, R.C., 1970. Prediction of deposition of calcium sulfate scale under down-hole conditions. *J. Pet. Technol.*, 22: 1299—1308.

Wallace, W.E., 1969. Water production from abnormally pressured gas reservoirs in South Louisiana. *J. Pet. Technol.*, 21: 969—982.

Weaver, C.E. and Beck, K.C., 1969. Changes in the clay—water system with depth, temperature and time. *Ga. Inst. Technol., Completion Rep.*, (Off. Water Resour. Res. Proj. No. A-008-GA WRC-0769), 95 pp.

Weintritt, D.J. and Cowan, J.C., 1967. Unique characteristics of barium sulfate scale deposition. *J. Pet. Technol.*, 19: 1381—1394.

White, D.E., 1957. Magmatic, connate and metamorphic waters. *Geol. Soc. Am. Bull.*, 68: 1659—1682.

Widmyer, R.H. and Pindell, R.G., 1981. Marvel enhanced recovery pilot-performance evaluation. *Proc. Second Joint SPE/DOE Symp. on Enhanced Oil Recovery, Tulsa, Okla., April 5—8, 1981*, SPE/DOE 9793: 301—316.

World Oil, 1977. Worlds largest offshore waterflood goes on stream. *World Oil*, 184 (5): 89—90.

Wright, C.C., 1963. Rating water quality and corrosion control in waterfloods. *Oil Gas J.*, 61: 154—157.

Zajic, J.E., 1969. *Microbial Biogeochemistry*. Academic Press, New York, N.Y., 345 pp.

Zarrella, W.M., Mousseau, R.J., Coggeshall, N.E., Norris, M.S. and Schrayer, G.T., 1967. Analysis and significance of hydrocarbons in subsurface brines. *Geochim. Cosmochim. Acta*, 31: 1155—1166.

Chapter 7

SOME CHEMICAL AND PHYSICAL PROBLEMS IN ENHANCED OIL RECOVERY OPERATIONS

MUKUL M. SHARMA, T.F. YEN, G.V. CHILINGARIAN and E.C. DONALDSON

INTRODUCTION

Design and development of EOR projects requires consideration of a wide variety of chemical problems. The entire subject dealing with the physics and chemistry of surfactants, polymers and other EOR agents has been looked upon with a new perspective — potential use in EOR. This subject is treated in detail in Volume II of this book.

Some of the more important problems include (a) design of suitable EOR agents that will be effective under reservoir conditions, (b) minimizing the requirements for these expensive agents, and (c) predicting and reducing losses of these agents. There can be no doubt that such problems merit a great deal of attention. It is unfortunate, however, that certain aspects of these problems have not received as much attention as they should, perhaps because they are "chemical problems".

Some common chemical problems, which are encountered in EOR operations, are briefly presented in this chapter. This discussion will hopefully serve to bring into perspective some of the highlights of these problems.

PRECIPITATION AND DEPOSITION OF ASPHALTENES AND PARAFFINS

Many of the EOR projects underway today are aimed at recovering low-API crude oil with high asphaltene and wax content; therefore it is important to explore the potential problems and complications that such "heavy crudes" may present. In addition to the physical and chemical properties of the crude oil and the formation, such as very high oil viscosities and low permeabilities, which are beyond the direct control of the producer, there are certain other problems which the producer may be able to avoid. One of the most troublesome problems is that of asphaltene and wax deposition in and around the wellbore, in flowlines and surface equipment.

Almost every oil producer encounters paraffin problems to some degree and attempts to solve them. Although less widespread, asphaltene precipitation is a common problem with asphaltic crude oils such as those from Venezuela and California. These problems, which are prevalent in conventional primary and secondary operations, are greatly accentuated in miscible

224

displacement processes where the phase equilibrium of the crude oil is disturbed.

Multiple contact miscible gas flooding has the potential for recovering a considerable amount of residual oil over the conventional waterfloods. The injection of CO_2 has proven to be technically and economically successful in many reservoirs, and other mixtures of hydrocarbon gases also show considerable promise. Under suitable conditions of temperature and pressure these gaseous mixtures will miscibly displace the oil that they contact. Although prohibitively expensive, liquid hydrocarbons may also be used to achieve miscible displacement.

All these miscible processes have extremely complicated phase behaviors, which have been extensively studied (Rutherford, 1962; Jacoby and Yarborough, 1967; Shelton and Yarborough, 1977). As many as four phases may be in equilibrium at any given temperature and pressure, i.e., two liquids, a gas and a solid. A phase diagram of a recombined reservoir oil and CO_2 mixture as determined by Shelton and Yarborough (1977) is shown in Fig. 7-1. A similar phase diagram using a mixture of hydrocarbon gases instead of CO_2 at 105°F is shown in Fig. 7-2. These authors also proposed a method for predicting the amount of asphaltics and heavy liquid components that are left behind in such a flood. One of the reasons for the concern about phase behavior arises from the danger of a large amount of asphaltenes precipitating due to compositional disturbances caused by miscible

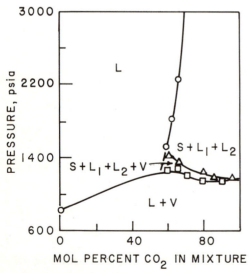

Fig. 7-1. Phase diagram of a mixture of a recombined reservoir oil and CO_2 at 94°F. (Modified after Shelton and Yarborough, 1977, fig. 7; courtesy of the SPE of AIME.)

displacement. If occurring deep in the formation, such precipitates would reduce the overall recovery efficiency of the process, whereas if occurring near or in the wellbore, they would cause serious plugging and clogging problems. The formation of solid precipitates in the case of many oils is more severe than indicated by experiments on mixtures of pure CO_2 and oil in a static equilibrium cell. As lighter hydrocarbons are extracted into the CO_2 gas phase, the ability of the maltene phase to keep the asphaltenes peptized decreases; the solubility of the paraffins also decreases. If the injected gas contains some hydrocarbon gases, the tendency of solid precipitates to form is greatly reduced (Huang and Tracht, 1974).

Sometimes water is injected alternately with gas to improve the mobility ratio of the flood, as described by Harvey et al. (1977) for a project in West Texas. Large reductions in water injectivity were attributed to the deposition of solid and heavy liquid fractions left behind by the miscible flood. This resulted in trapping of gas and reductions of water relative permeabilities (Schneider and Owens; 1976), which may be quite pronounced if gas and water are being injected alternately.

Although uncommon, miscible displacement using liquid hydrocarbons is susceptible to similar problems.

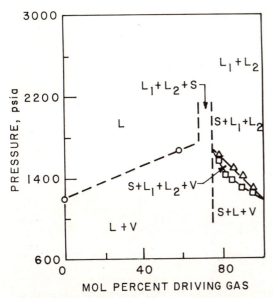

Fig. 7-2. Phase diagram of a mixture of recombined reservoir oil and a typical hydrocarbon drive gas consisting mainly of methane, ethane and propane at 105°F. The exact composition of the recombined oil and the drive gas is presented by Shelton and Yarborough (1977). S, L_1, L_2 and V are the solid (asphaltene), two liquid and the vapor phases, respectively. (Modified after Shelton and Yarborough, 1977, fig. 5; courtesy of the SPE of AIME.)

Chemistry of asphaltenes and paraffins

The accumulated experimental results of the last few decades have helped develop a hypothetical structure of the asphaltenes in petroleum. The complex nature of the average structure requires analysis in terms of both a microstructure and a macrostructure (Yen, 1974). The microstructure of the asphaltenes reveals a peri-condensed aromatic nucleus as determined by densimetric ring analysis (Yen et al., 1961), ring compaction (Yen and Dickie, 1968), and a combination of NMR and X-ray methods (Yen and Erdman, 1962). These Π systems are linked by a number of saturated systems, mainly short-chain alkanes and kata-condensed naphthenics. The aromaticity ranges from 0.25 to 0.5. The aromatic nuclei of asphaltenes are 50—70% substituted by other alkyls and naphthenics. The majority ($\approx$ 15%) of

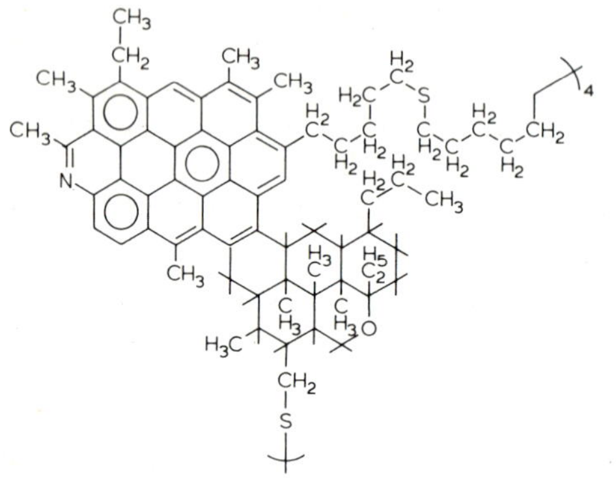

		Formula	Spectra
$H_A = 4$	$\dfrac{H_A}{H}$	0.046	0.055
$H_\alpha = H_{\alpha Me} + H_{\alpha CH_2}(P,N) + H_{\alpha CH}(P,N)$ $= 15 + 6 + 1 = 22$	$\dfrac{H_\alpha}{H}$	0.25	0.27
$H_N = H_C - H_{\alpha CH_2}(N) - H_{\alpha CH}(N) + H_{\beta CH_2}(P)$ $= 17 - 2 - 1 + 2 = 16$	$\dfrac{H_N}{H}$	0.18	0.18
$H_R = (H_P - H_{Me}) - H_{\alpha CH_2}(P) - H_{\beta CH_2}(P) + H_{\beta Me}(P)$ $= (66 - 36) - 4 - 2 + 3 = 27$	$\dfrac{H_R}{H}$	0.31	0.31
$H_{SMe} = H_{Me} - H_{\alpha Me}(P) - H_{\beta Me}(P)$ $= 36 - 15 - 3 = 18$	$\dfrac{H_{SMe}}{H}$	0.21	0.19

Fig. 7-3. Hypothetical molecular formula of an asphaltene from a Lagunillas oil in Venezuela. The positions of all substituents are quite arbitrary. In the lower portion of the figure, the five major hydrogen types from NMR are given together with their definitions. Number of hydrogens are indicated respectively as H_A (aromatic); H_α (α-substituted); H_N (naphthenic); H_R (methylene); and H_{SMe} (saturated methyl). Hydrogen-type ratio determined both from spectra and from the above formula is shown. (From Yen, 1974, fig. 1; courtesy of *Energy Sources*.)

the alkyl substituents are methyls. Polarography indicates that quite a few sulphur atoms are situated in aromatic systems as benzothiophene structures (T.F. Yen, unpublished results, 1964). Although carbonyl oxygen is detected by the infrared method, it is scarce and there is absolutely no quinone or semiquinone type of oxygen in the asphaltenes. Ether-type oxygen and quinoline-type nitrogen sites were detected.

The average microstructure for a Lagunillas oil asphaltene sample from Venezuela is shown in Fig. 7-3. Both the carbon types and the proton types are indicated in the figure. The empirical formula along with all the structure parameters are also shown. The complexity of the structure necessitates the use of a large number of structural parameters to characterize it.

The macrostructure of the asphaltene describes the association of the basic molecular species with each other to form clusters and micelles. One of the properties of the peri-condensed polynuclear aromatic systems is the attraction between the II electron systems and the resulting formation of layer structures through association. Association can take place either due to II-II interaction or due to coordination through heteroatoms in the sheets (Tynan and Yen, 1969). It is known that many of the aromatic portions of the sheets have defective centers (gaps and holes), which usually are the sites for the unpaired spins. For petroleum asphaltenes this stacking of sheets is about five layers thick and is shown schematically in Fig. 7-4. A number of these small particles can aggregate under appropriate conditions to form a large micelle. The asphaltene in petroleum then behaves like a true colloidal system.

The colloidal nature of asphaltenes has been long recognized and the preceding discussion on its structure supports this view. Historically, the macro-

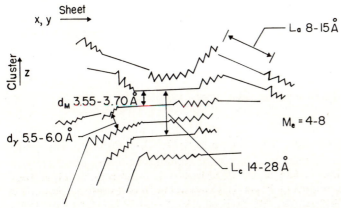

Fig. 7-4. Cross-sectional view of an asphaltene model based on X-ray diffraction. Zig-zag line represents the zig-zag configuration of a saturated carbon chain or loose net of naphthenic rings; the straight line represents the edge of flat sheets of condensed aromatic rings. (From Yen, 1974, fig. 2; courtesy of *Energy Sources*.)

structure of asphaltenes was postulated by Nellensteyn (1938); however, his theory of colloidal nature of asphalt, and the elementary carbon nucleus concept have long since been abandoned. Pfeiffer and Saal (1940) proposed the concept of a dispersed asphaltene phase peptized by oils and resins. Eilers (1949) showed that asphalts behave rheologically like suspensions. Eldib et al. (1960) found that upon centrifuging crude oils at 80,000 times gravity, the asphaltene colloidal particles would separate out. Attempts have also been made to study this complex colloidal system using electrophoresis techniques (Wright and Minesinger, 1963; Hau et al., 1982). These studies may ultimately explain the flocculation and precipitation behavior of the asphaltenes under different conditions. It is clear, however, that any change in composition of the maltene phase (caused by depletion of aromatics or light ends which peptize the asphaltenes) will cause the precipitation of asphaltenes.

The paraffins or waxes in the oil, which are chemically a much simpler group of compounds, are easier to describe. They consist of high-molecular-weight alkanes having a general formula $C_n H_{2n+2}$ and varying from 20 to 60 units in size, with melting points ranging from 98 to 215°F. These non-polar and inert paraffins are resistant to acids, bases, or oxidizing agents.

Unlike the asphaltenes, the paraffins and waxes are in true solution in crude oils under stable conditions. The primary reason for their separation from the crude oil is reduction in their solubility. The change in solubility could be due to changing temperature or pressure, loss of dissolved gases, or loss of lighter components from the crude oil. Pour point and cloud point measurements are generally used to refer to the ability of a crude oil to hold paraffins in solution. The cloud point is defined as the temperature at which paraffins begin to come out of solution. If the crude oil is cooled without agitation below its cloud point, the wax crystals form a stable network, causing the oil to lose its fluid properties and to behave more like an elastic solid. This temperature is called the pour point. The exact procedures for determining these properties are listed in ASTM D2500-66 (cloud point) and ASTM D97-66 (pour point).

Factors determining asphalt and paraffin deposition

The deposition of asphaltenes and paraffins is interrelated. Asphaltene micelles form the nucleating centers around which insoluble paraffin crystals precipitate. In the absence of suitable nucleating sites, paraffins can remain in supersaturated solution for a long time without precipitating. Rough edges on tubing walls or formation grains may sometimes form good nucleating sites, and deposition of paraffins may directly occur on these solid surfaces.

The most important factors controlling paraffin deposition are temperature and pressure. Cooling of the crude oil below the cloud point will cause crystallization of paraffins. Such cooling may take place due to: (1) a sudden release of pressure, causing expansion of the gas and a Joule Thompson cooling effect; (2) cooling produced by reduction in ambient temperatures as the oil moves up the production string; and (3) cooling caused by the release of gas from solution or the evaporation of lighter components from the oil. Any one of these potential cooling mechanisms may be operative in a miscible gas drive operation at the producing well. Unfortunately, very little can be done economically to prevent this from happening.

The effect of temperature on asphaltene flocculation is still not a very well understood phenomena. A recent study by Hirschberg et al. (1982) attempted to explain both the temperature and pressure effects on asphaltene precipitation by using the Flory-Hildebrand Theory. Much more research needs to be done, however, before a satisfactory understanding of the problem is attained. The solubility parameter of asphaltenes varies almost linearly with temperature as shown in Fig. 7-5. This behavior is very similar to that of the heavy aromatics such as naphthalene. Data were obtained for an oil having a high asphaltene content (oil 1) and oil with a low asphaltene content (oil 2; Fig. 7-5).

Although pressure itself has practically no effect on the solubility of the paraffin, it affects the position of the system on the phase diagram and, thus, controls the composition of the oil, gas, and solid phases. Large and sudden pressure drops are usually anticipated at the formation face and in the well-

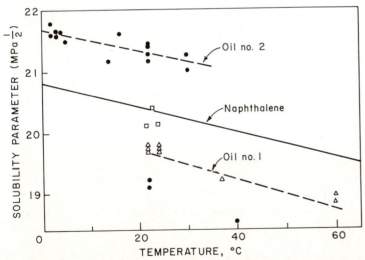

Fig. 7-5. The temperature dependence of the solubility parameter of asphalts and asphaltenes for two crude oils as compared to a model polynuclear aromatic compound. Oil 1 was a highly asphaltic oil, whereas oil 2 was low in asphaltene content. Oil 1: assuming asphaltene precipitation; oil 2: assuming asphalt precipitation. (Modified after Hirschberg et al., 1982, fig. 2; courtesy of the SPE of AIME.)

bore of a producing well, where paraffin deposition is most likely to occur. A similar conclusion was reached by Hirschberg et al. (1982) who studied the effect of pressure on the precipitation of asphaltenes. They found that asphaltene solubility attained a minimum value at the bubble point as shown in Fig. 7-6. Such data will provide some idea of the amount of asphaltenes that may precipitate in conventional oilfield operations. In miscible displacement processes, however, this phase behavior is much more complicated and will differ significantly.

Compositional changes in the crude oil during miscible flooding directly affects the precipitation and flocculation of asphaltenes and paraffins. This could be brought about by either temperature or pressure changes, and is determined by the phase diagram for the particular system at hand. Loss of lighter components of the crude oil reduces the quantity of paraffins that the oil can hold in solution at a specific temperature. The peptization ability of the maltene phase is also reduced if it is depleted in the aromatics and

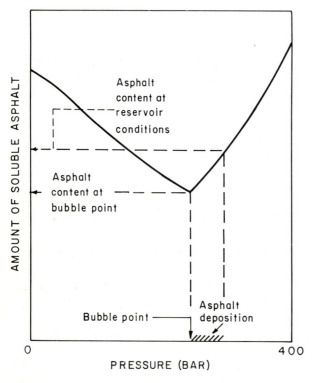

Fig. 7-6. Pressure dependence of asphalt solubility for a North Sea crude oil, showing the possibility of asphalt deposition in the well tubing. (Modified after Hirschberg et al., 1982, fig. 4; courtesy of the SPE of AIME.)

resin components, which act as peptizing agents for the asphaltenes. The injected CO_2 progressively strips the lighter components from the crude oil until attainment of multiple contact miscibility. As experimentally observed by Shelton and Yarborough (1977), the precipitation of both paraffins and asphaltenes occurs in the mixing zone, i.e., the zone of stripping and solubilization. This is a direct result of compositional changes in the oil. As noted before, the stripping of the lighter components results in cooling, which acts to accentuate the problem.

Frequently, paraffin problems become more acute with the age of the well, because during production the reservoir is progressively depleted of the lighter components and the saturation of paraffins and asphaltics is constantly increasing. Often EOR projects are initiated on such depleted reservoirs, necessitating additional caution.

Suspended particulate matter in the crude oil such as drilling fluid solids, formation fines, and corrosion debris provide additional nuclei for paraffin crystals to nucleate and aggregate. Sand and silt seem to cause separation and deposition of the paraffins to occur more readily. In addition to helping deposition, the solid particles increase plugging problems.

Removal of deposits

Although the EOR projects may pose significantly greater problems with organic deposits, the methods of removal used are the same as for conventional workover operations. Excellent reviews on the subject have been published by Shock et al. (1955), Knox et al. (1962), Jorda (1966), and Bilderback and McDougal (1969).

There are essentially three methods of removing organic deposits: (1) mechanical means (scrappers, etc.), (2) thermal methods (hot oil or other liquids that will melt the wax and dislodge it, and (3) chemical methods (solvents of varying compositions are used to dissolve the deposit).

In conclusion, the deposition of asphaltenes and paraffins can pose a greater problem in some EOR processes than in conventional recovery methods due to the following reasons:

(1) Stripping of light ends in the crude oil during miscible flooding processes results in a decrease in solubility of the paraffins.

(2) Depletion of aromatics and resins, which are peptizing agents for asphaltenes in the crude oils, cause the asphaltenes to flocculate and separate out.

(3) Large temperature drops in the wellbore region due to the injection gas coming out of solution as the pressure is lowered.

(4) A higher potential for fines to be present in the producing stream due to the multiphase flow in the reservoir and more extensive corrosion problems.

SCALING PROBLEMS

One of the problems that arises with injecting a wide range of chemicals down the wellbore is that it increases the potential for scale formation. Scale deposits are inorganic mineral deposits that form due to any changes in thermodynamic equilibrium, which results in the lowering of the solubility of a particular species in the water. Pressure drops, temperature fluctuations, and changes in pH or ionic strength are the most direct causes of scale formation. Precipitation of scale could occur in any part of the producing or injection system (perforations, flowlines, downhole pump, surface equipment) in the wellbore region, the formation matrix, vugs, fractures, and fissures. Severe scaling problems can cause drastic reductions in the productivity of a well and may cause formation damage that would require expensive well cleaning procedures. In many cases scaling problems have forced premature abandonment of wells or expensive well workover operations (James, 1970; Knowles, 1974).

The composition of the scale can vary greatly depending on the composition of the produced or injection water. The most commonly encountered deposits in conventional oilfield operations are calcium carbonate ($CaCO_3$), calcium sulphate ($CaSO_4$), barium sulfate ($BaSO_4$) and sodium chloride (NaCl). Frequently, salts of iron, such as iron carbonate ($FeCO_3$), iron oxide (Fe_2O_3) and iron sulphide (FeS), are found in the scale deposits. These, however, are more often products of corrosion rather than direct precipitation from the brine. Rarer precipitates include barium sulphate ($BaSO_4$), strontium sulphate ($SrSO_4$), strontium carbonate ($SrCO_3$), barium-strontium sulphate [$BaSr(SO_4)_2$] and other insoluble salts of the heavy metals. In general, an analysis of the reservoir brine and the injected brine will provide a fairly good indication as to what precipitates to expect in a scale deposit.

Any drastic changes in composition, temperature or pressure may cause scaling problems. The EOR processes such as CO_2 flooding, caustic flooding, steam flooding, and surfactant or micellar flooding, use design consideration in which scaling, unfortunately, is a relatively unimportant factor. Such injections could cause serious scaling problems. The potential for scale deposition in CO_2, steam, and caustic flooding is usually higher than in a conventional waterflood.

Another reason why EOR processes are more susceptible to scaling at the production wells is because the producing waters contain much higher percentages of dissolved solids than is the case in conventional waterfloods. As caustic or steam fronts move through the reservoir, they dissolve some of the minerals, changing the composition of the aqueous phase. If during production, producing brines of incompatible composition, from different zones, are brought into contact, severe scale deposition will occur.

It may be useful to understand and evaluate scaling potentials of any EOR process before its implementation. An overview of the causes and consequences of scaling, and some methods of prediction and of prevention during some typical EOR operations, is presented below.

Stability criterion for oilfield brines (CaCO₃ scale)

The most common scaling problems arise due to calcium carbonate precipitation. Tillman (1932) and Langelier (1946) studied this problem and their methods are widely accepted.

The precipitation of carbonate is governed by the bicarbonate—carbonate conversion equilibrium:

$$HCO_3^- \overset{k_1}{\rightleftharpoons} H^+ + CO_3^{2-} \tag{7-1}$$

$$k_1 = \frac{[CO_3^{2-}][H^+]}{[HCO_3^-]} = 5 \times 10^{-11} \tag{7-2}$$

where the concentrations are expressed in g-moles/liter. The pH is the main variable that determines which way the equilibrium will shift. High pH values, such as those encountered in caustic flooding, are favorable to the formation of carbonates. Consequently, it is of considerable concern in solving scaling problems. The metal salts may combine with the carbonate anion (CO_3^{2-}) and precipitate various carbonates:

$$Ca^{2+} + CO_3^{2-} \overset{k_s}{\rightleftharpoons} CaCO_3 \downarrow \tag{7-3}$$

$$k_s = [Ca^{2+}][CO_3^{2-}] = 5 \times 10^{-9} \tag{7-4}$$

$$\frac{[Ca^{2+}][HCO_3^-]}{[H^+]} = \frac{k_s}{k_1} = 100 \tag{7-5}$$

Therefore, $pH - pCa^{2+} - pHCO_3^- = 2$ at equilibrium, where pCa^{2+}, $pHCO_3^-$ and pH are the negative logarithms of the ionic concentrations. This equation, which is valid only at room temperature (25°C) and in distilled water was modified and generalized by Langelier (1946) and later Stiff and Davis (1952a) to account for other ionic species at any temperature and is known as the Langelier Stability Criterion. The Langelier Stability Index, *SI*, is defined as:

$$SI = pH - pCa - pHCO_3 - K \tag{7-6}$$

where K is a number whose value depends on the ionic strength and the temperature (Fig. 7-7). The stability index can be calculated from the ionic composition of a brine. A positive *SI* indicates a scaling condition, whereas a negative *SI* indicates a corrosive environment. For example, an *SI* of 0.3

234

indicates a situation where scaling may occur, whereas an *SI* of 2.0 indicates that scaling will occur.

Fig. 7-7 shows that *K* decreases with increasing temperature, which implies a more positive stability index and hence a greater tendency to form scale. Problems of scale deposition are accentuated in the case of high-temperature processes, such as hot-water drives, steam drives or steam soaks, and in-situ combustion operations, due to the high bottomhole temperatures which give rise to unstable brines. Rapid temperature changes, such as those encountered in cyclic steam injection or when the hot water zone ahead of a steam front or combustion zone hits a producing well, are potential causes for wellbore and formation damage. Rapid change in the temperature may result in the formation of large amounts of precipitates that would cause severe reductions in either the productivity or injectivity of a well. Surface equipment, i.e., heater treaters and steam generating and handling equipment, also suffer from similar problems due to the sudden temperature changes.

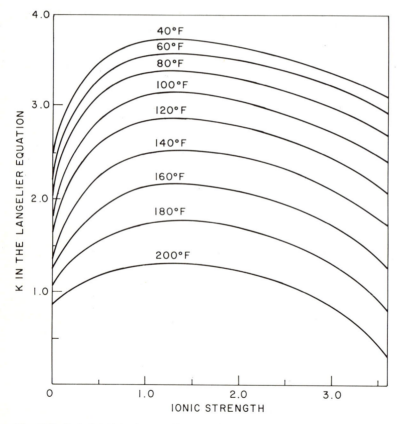

Fig. 7-7. Relationship between the value of *K* used in the Langelier equation and ionic strength. (Modified after Stiff and Davis, 1952b, fig. 1; courtesy of AIME.)

Although carbon dioxide flooding implies low pH, the producing wells in CO_2 floods may have scaling problems due to the sudden release of the CO_2 from solution as the pressure is reduced in the wellbore region. This escape of CO_2 results in a shift in equilibrium towards the formation of carbonates, i.e., high pH. Such problems may go undetected if the water samples are tested for stability after production, i.e., after depletion of CO_2. Such scale deposition can be prevented by the realization of this problem in advance and the suitable use of scale inhibitors or other means which are discussed later.

While $CaSO_4$ and $BaSO_4$ scaling tendencies are relatively easy to predict, the scales once formed are difficult to remove because they are insoluble in water or acid. The reverse is true for $CaCO_3$ scale. It is hard to predict $CaCO_3$ scale formation because of the difficulty in analysing for the HCO_3^- anion under bottomhole conditions. Water samples brought to the surface are depleted of CO_2 and are of little use for analysis. Further difficulty lies in the uncertainty of predicting the rate at which supersaturated solutions will crystallize out on the metal or rock surfaces. The rate of precipitation of solids depends on many factors which have not been fully studied.

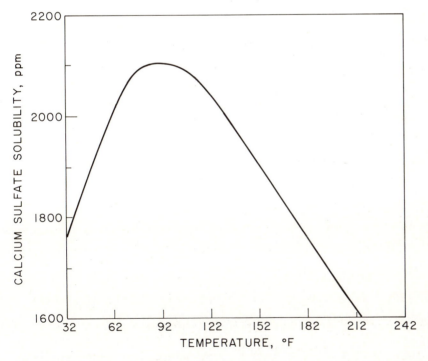

Fig. 7-8. Solubility of calcium sulfate in distilled water at various temperatures. (Modified after Stiff and Davis, 1952a, fig. 2; courtesy of the SPE of AIME.)

Calcium sulfate scale

Calcium sulfate precipitates as anhydride ($CaSO_4$) at temperatures greater than $130°F$, and as gypsum ($CaSO_4 \cdot 2H_2O$) at lower temperatures. Fig. 7-8 shows the solubility of $CaSO_4$ in distilled water as a function of temperature. The following equation enables determination of the overall solubility:

$$S = S_T \cdot F_1 \cdot F_2 \cdot F_3 \tag{7-7}$$

where S_T is the solubility from Fig. 7-8, F_1 is the correction factor for the excess Ca^{2+} or SO_4^{2-} ions present, F_2 is the correction factor due to the Na^+ ions, and F_3 is the correction factor due to the Mg^{2+} ions. The factors F_1, F_2, and F_3 may be determined from Figs. 7-9, 7-10, and 7-11, respectively.

The effects of the other ions that may be present in the brine may be significant. Indeed if the formation brine is rich in Ba^{2+} or Sr^{2+} ions, they may precipitate before the $CaSO_4$ does. Such cases are uncommon. The pH of the solution has a large effect on the solubility of the $CaSO_4$. The solubility increases considerably with a decrease in pH especially at high temperatures near the boiling point of water. Solubility also decreases with pressure and a quantitative relationship is:

$$S_p = S_1 \exp\left[\frac{(1-p)\,\Delta V}{RT}\right] \tag{7-8}$$

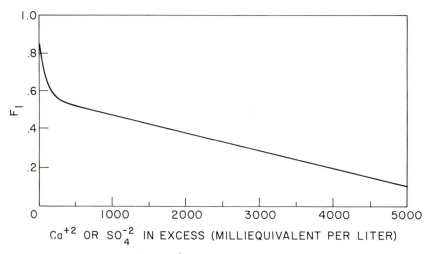

Fig. 7-9. The excess Ca^{2+} or SO_4^{2-} ion factor, F_1 (= solubility of $CaSO_4$ in the presence of excess Ca^{2+} or SO_4^{2-} ions/solubility of $CaSO_4$ in distilled water). (Modified after Stiff and Davis, 1952a, fig. 3; courtesey of the SPE of AIME.)

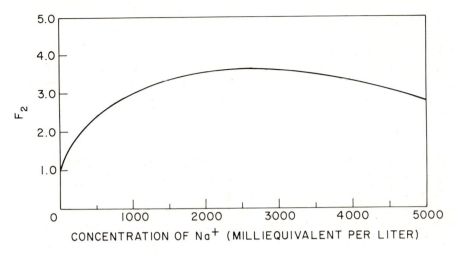

Fig. 7-10. The Na^+ ion factor, F_2 (= solubility of $CaSO_4$ in the presence of Na^+/solubility of $CaSO_4$ in distilled water). (Modified after Stiff and Davis, 1952a, fig. 4; courtesy of the SPE of AIME.)

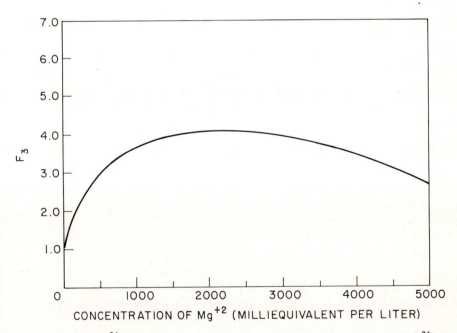

Fig. 7-11. The Mg^{2+} ion factor, F_3 (= solubility of $CaSO_4$ in the presence of Mg^{2+}/solubility of $CaSO_4$ in the distilled water). (Modified after Stiff and Davis, 1952a, fig. 5; courtesy of the SPE of AIME.)

where S_p = solubility at pressure p, S_1 = solubility at 1 atm, ΔV = change in molar volume upon solution in water, T = temperature, and R = universal gas constant.

For example, the solubility will approximately double for an 800 atm change in pressure, which is a relatively insignificant effect when compared to the other factors.

Although much less common than the carbonate deposits, sulfate scale deposits are usually caused by mixing of two incompatible waters. If an injection water is rich in sulfate ions and formation water has high concentrations of Sr^{2+}, Ca^{2+} and Mg^{2+} ions, severe formation plugging could result. Inasmuch as these ions are detrimental to many of the EOR processes, e.g., loss of Na^+ and OH^- due to cation exchange, precipitation of surfactants due to complexing with divalent ions, etc., they are frequently removed by preflushes and are present only in small concentrations. The $CaSO_4$ scaling in injectors in any EOR process is not greater than $CaSO_4$ scaling in an ordinary waterflood. It may cause considerable damage, however, and should be thoroughly evaluated prior to commencing a project. It should be kept in mind that producing waters in many EOR processes may solubilize a large variety of minerals. If two such producing brines from two different zones, one rich in Ba^{2+} and the other in SO_4^{2-}, were allowed to mix at the producer, severe scaling could occur as a result of $BaSO_4$ precipitation.

Miscellaneous scales

Barium sulfate is a very insoluble scale which may precipitate out of oilfield waters due to mixing of producing waters from two different zones. The precipitation is rapid and complete. This scale is hard and is extremely difficult to remove. The presence of barium in the formation water is a serious problem and scale inhibitors should be added, before producing waters from different zones are allowed to mix.

Iron sulfide scales, which are formed as a result of corrosion due to the presence of H_2S, may be present as hard brittle scales or as soft powdery accumulations. Iron sulfide scaling is much more pronounced in firefloods due to the large amounts of H_2S and SO_2 produced by the combustion process and the ensuing corrosion problems.

Scale control

Scale inhibitors modify both the solution and the colloidal properties of the scaling material, and the surface on which the scale develops. Hausler (1978) classified scale inhibitors into three classes on the basis of their physical and chemical action:

(1) *Chelating agents.* Chelating agents are chemicals, such as EDTA, which form complexes with the divalent cations in the solution so that

they can no longer react with the sulfate and carbonate anions and precipitate.

(2) *Threshold inhibitors.* Threshold inhibitors must be present in at least a minimum concentration to be effective. Inorganic polyphosphates act as inhibitors when present in concentrations of a few parts per million. By adsorbing on the nucleating crystals polyphosphates prevent their further growth. Polyphosphates also sequester Ca^{2+} and Mg^{2+} cations. Polyorganic acid LP-55, developed by Halliburton, also inhibits scaling through threshold treatment. A review of various organic phosphates and phosphonates, which have been developed over the years, is given by Featherston et al. (1959).

(3) *Crystal distortion inhibitors.* During nucleation and crystal growth, crystal distortion chemicals distort the shape of the crystals so that their size is limited and they do not agglomerate and form larger colloidal particles.

Scale removal

Scale removal techniques in EOR processes are the same as those in conventional well operations and are well reviewed in petroleum production books (e.g. Ostroff, 1965; Allan and Roberts, 1981). Although more important design considerations may determine the compostion of the injected fluids, it is important to evaluate the scaling problems, which could occur at both the injection and producing wells, and take necessary preventive measures in order to maximize the injectivity and producibility of the wells.

FORMATION DAMAGE DUE TO MIGRATION OF FINES

Reduction in permeability at or near the wellbore can have detrimental effects on the injectivity or productivity of a well in a radial flow system. One of the primary reasons for the reduction in absolute permeability both near the wellbore and deep in the formation is the swelling and migration of clays and other fines. This occurs when the formation is contacted by an external fluid that is incompatible with the formation. In conventional workover operations, such a fluid may be a drilling fluid filtrate, or some treating fluid, that comes into contact with the rock face. In waterflooding and particularly in EOR operations, this potential for formation damage due to contact with a foreign fluid is much higher. Many investigators have studied the swelling and migration of clays as a function of pH, salinity, and ionic contents of injection water.

The physical chemistry of clay migration and swelling

Clays constitute only a small fraction of a sandstone or a carbonate reservoir by weight, ranging from about 5% in clean formations to 20%

in clayey formations. Clays play a very important role in determining the electrochemical behavior of a reservoir, because of their large surface areas and easy accessibility.

Clays are composed of layers of silica tetrahedra and alumina octahedra stacked up parallel to each other in various fixed configurations. The most common clay minerals in reservoir rocks are illite (2 : 1), kaolinite (1 : 1), smectite (2 : 1), and some mixed-layer clays. An extensive discussion on clay minerals is presented by Weaver and Pollard (1973). Because of the layered structure of clays, clay platelets behave anisotropically; for example, the basal surface attains a net negative charge due to isomorphous subsitution of silicon or aluminum sites by lower valence elements, whereas the edge surfaces attain their surface charge due to broken bonds at the crystal edges and due to proton or hydroxyl ion exchange with the surrounding fluids. The basal surface charge is independent of the pH of the formation water, whereas the edge surface charge is pH dependent and will become more negative with increasing pH (Chilingarian and Vorabutr, 1981, p. 244).

Clay minerals (both authigenic and allothigenic) in reservoir rocks may occur (1) in a dispersed form in pores and pore throats, (2) lining detrital grains, (3) as laminations, or (4) as large discrete crystals or grains. Different types of clay mineral distribution affect effective porosity and permeability in a drastically different manner (Fertl, 1972).

Dispersed clays (leafy crystals, irregularly shaped plates, fibers, etc.) in the pore space occur as: (1) discrete particles, (2) coating of clastic grains by intergrown clay crystal linings, and (3) clay crystals bridging across the pore space (Fig. 7-12).

The detailed study of the rocks in the sedimentary mantle of the earth's crust shows that the clay minerals in them either are allothigenic (inherited from the parent substratum — the weathering crust) or were formed in the post-sedimentary diagenetic period (transformation, authigenesis).

Drastically different effects are exhibited on the permeability of reservoir rocks (Neasham, 1977). For a given porosity, pore lining allows reservoir permeabilities of about one to two orders of magnitude higher ($k = 1.0-200$ md). Discrete clay particles still allow high permeability — in excess of 100 md and into the darcy range. Thus, a porosity-dependent cut-off point for reservoir permeability also greatly depends on the type of clay distribution (Fig. 7-13).

Advanced analytical models have been developed by Ruhovets and Fertl (1981) that allow an estimate of the type of major clay distribution components from the combination of wireline nuclear and natural gamma ray spectral logging data.

Sarkisyan (1972) showed that the porosity and permeability of the reservoir rocks depend to a considerable degree not only on the type but on the nature of the clay material as well. The permeability of cap rocks, which consist of some clays such as mixed-layer formations of the montmorillonite—

hydromica, proved to be also related to their granulometric composition and microtextural and microstructural patterns. The influence of clay cement on the properties of reservoir rocks is clearly illustrated on studying the electron micrographs of clay cements. In the case of allothigenic clays, the haphazard arrangement of different particles and packets in the space among grains results in a considerable number of intercommunicating micropores. Thus, despite the high content of clays (up to 20%), the reservoir rock may remain

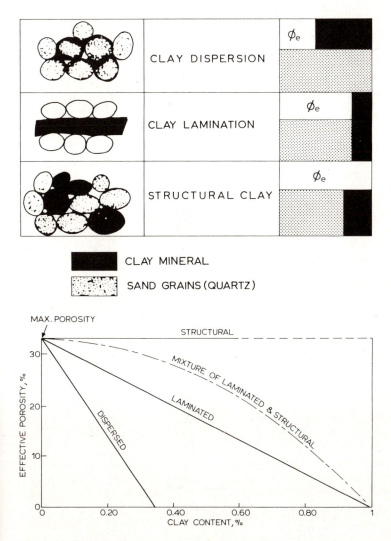

Fig. 7-12. Models of type of clay distribution in clastic rocks, and the effect of clay type and type of clay distribution on effective porosity. (After Fertl et al., 1982, fig. 1; courtesy of *Energy Sources.*)

242

productive. Diagenetic clays, on the other hand, are characterized by tight packing, as a result of which permeability is extremely low (see Sarkisyan, 1972; Fertl et al., 1982).

In sandstones, the detrital grains themselves are most often oxides of metals, primarily silica and alumina, with dolomite and calcite being present in smaller amounts. The surface charge properties of these oxides resemble those of metal oxides (Sharma and Yen, 1982) and are acquired by surface complexation of surface sites with protons on hydroxyl groups.

$$SO^- \overset{OH^-}{\rightleftharpoons} SOH \overset{H^+}{\rightleftharpoons} SOH_2^+ \qquad\qquad (7\text{-}9)$$

Such complexes may also be formed with other electrolytes present in solution:

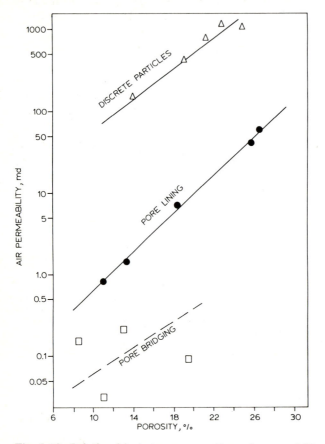

Fig. 7-13. Relationship between porosity and permeability to air (corrected for Klinkenberg effect) for selected sandstone samples having different types of clay distribution. (After Neasham, 1977; courtesy of the SPE of AIME.)

$$SOH_2^+ + Cl^- \rightleftharpoons SOH_2^+ Cl^- \quad \text{or} \quad SO^- + Na^+ \rightleftharpoons SO^- Na^+ \tag{7-10}$$

Sharma and Yen (1983a) have presented a more detailed analysis of this behavior.

Adsorption or detachment of clay particles from the surfaces of detrital grains is considered to be a physical adsorption or desorption process, where the free energy of escape of the clay particle is given by the free energy contribution due to electrostatic, dispersion (or Hamaker) and hydrodynamic forces:

$$G_{escape} = G_{elec.} + G_{disp.} + G_{hyd.} \tag{7-11}$$

A simple calculation (M.M. Sharma and T.F. Yen, unpublished results, 1982) shows that the hydrodynamic contribution to the free energy term is negligibly small even at flow rates much higher than those present in the reservoir. For highly charged surfaces such as clays and metal oxides, the electrostatic contribution to the total free energy is critical in determining whether attraction or repulsion occurs between the clay platelets and the oxide surfaces.

The free energy due to electrostatic interactions has been derived for sphere—plate and plate—plate interactions for a variety of boundary conditions (Hogg et al., 1966; Wiese and Healy, 1970; Kar et al., 1973). In the clay—oxide system, it would be most appropriate to regard the clay surface as one of constant charge and the oxide surface as one of constant surface potential. If a_1 and a_2 are the radii of the detrital grain and the clay particle, respectively, the free energy is given by (Kar et al., 1973):

$$G_{elec.} = \frac{\pi a_1 a_2}{(a_1 + a_2)} \left\{ \left(\frac{2\sigma_2 \psi_1}{\kappa} \right) \left(\frac{\pi}{2} - \tan^{-1} \sinh \kappa h \right) - \right.$$
$$\left. - \left[\frac{4\pi\sigma_2^2}{\epsilon \kappa^2} - \frac{\epsilon \psi_1^2}{4\pi} \right] \ln \left[1 + \exp\left(-2\kappa h\right) \right] \right\} \tag{7-12}$$

where ψ_1 and σ_2 are the surface potential and surface charge of the interacting particles, which are separated by a medium with a dielectric constant ϵ, at a distance h apart. κ is the Debije-Hückel reciprocal length parameter.

It is evident that the free energy term is a strong function of the surface charge, the surface potential, and the distance of separation.

The van der Waals' or Hamaker free energy term is expressed as follows:

$$G_{disp.} = \frac{-A_{132}}{12\pi h^2} \tag{7-13}$$

The Hamaker constant (A) is not easily determinable for the clay—water—

oxide system, but the range of values is estimated to be (M.M. Sharma and T.F. Yen, unpublished results, 1982):

$$0.25 \times 10^{-13} \leqslant A_{132} \leqslant 3.0 \times 10^{-13} \tag{7-14}$$

Unfortunately, this range of values is about equal to the range of values in which the free energy barrier is most sensitive to this parameter. The overall free energy term as a function of the distance of separation is shown in Fig. 7-14. The maxima (ΔG_{max}) is the energy barrier that must be overcome to cause release of clays.

Fines migration in EOR processes

Laboratory experiments indicate that, in general, reservoirs are more compatible with low-pH fluids than with high-pH fluids (Mungan, 1965; Grey and Rex, 1966). Exposing a clayey consolidated sandstone to a high-pH fluid can reduce its permeability by 100-fold within a very short period of time (Monaghan et al., 1959). As discussed previously, at high pH values the

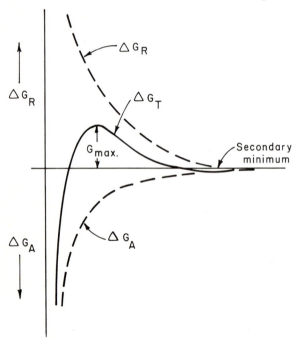

Fig. 7-14. The free energy of interaction as a function of distance of separation between two bodies. The repulsive term arises due to electrostatic interactions, whereas the attractive term arises due to dispersion or Hamaker interactions. The ΔG_{max} is the activation energy barrier for coagulation or adsorption. Subscripts: R = repulsive, A = attractive, T = total.

detrital grains acquire a net negative charge which gives rise to a large electrostatic repulsion term. This behavior, which can be studied semiquantitatively on the basis of DLVO-type theory, is very important in caustic flooding operations. Thus, caustic floods should be restricted only to sandstones and unconsolidated sands having a low clay content. Otherwise, severe plugging problems may result.

Another important parameter which determines the extent of fines migration is the salinity of the injected fluids. Low-salinity brines have a higher potential for formation damage and the effect of fresh water may be disastrous in many reservoirs. On the basis of the electrochemical behavior of the reservoir, a high ionic strength of the brine causes a lowering of the electrostatic potential of the surfaces, which results in a lower electrostatic repulsion term (Van Olphen, 1959). Many ionic species, besides sodium chloride, influence the surface potential. For example, the presence of divalent ions like Ca^{2+} or Mg^{2+} will in general cause a lowering of the surface potential of the clays even at high pH values, whereas ions such as phosphate dramatically increase the surface potential. Some data showing this behavior is shown in Fig. 7-15 for a kaolinite clay sample (Jang et al., 1982).

Caustic floods, surfactant floods, and even polymer floods may have fairly precise salinity requirements. For example, low interfacial tensions cannot be attained in caustic flooding at high sodium chloride concentrations (Shar-

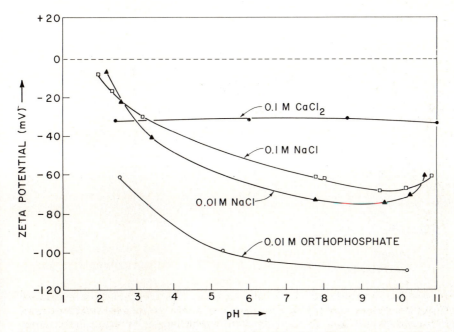

Fig. 7-15. Relationship between the zeta potential of kaolinite clay and pH in the presence of NaCl, $CaCl_2$, and orthophosphate. (From Jang et al., 1982, fig. 4.)

ma and Yen, 1983b). Surfactant precipitation and polymer rheological behavior is determined by the salinity of the brine. This limits the ranges of salinity under which such processes must be operated. In such cases, one must be careful that such salinity requirements do not imply formation damage due to the migration of fines. Low-salinity and high-pH conditions that are preferred in caustic floods and low-salinity conditions which are desirable in surfactant and polymer floods are potentially dangerous, because they can cause formation damage.

Other factors such as wettability and interfacial forces begin to play an important role in determining the movement of fines, when two or more immiscible phases are present in the pore spaces. Water-wet fines tend to remain in the thin film of water that surrounds the detrital grains and move only during the movement of the aqueous phase. They do not move if oil is the mobile phase. Upon becoming detached from the detrital grain surface, particles of mixed-wettability are located at the oil—water interface, and their movement seems to be confined to this interface (Muecke, 1979). Experiments showed that the presence of oil reduces the movement of fines in the porous medium, trapping them in thin wetting immobile films or at interfaces between fluids. When such thin films are disrupted or the equilibrium between the two phases is disturbed by wettability alterations, e.g., by any kind of miscible flooding that solubilizes both the oil and the water

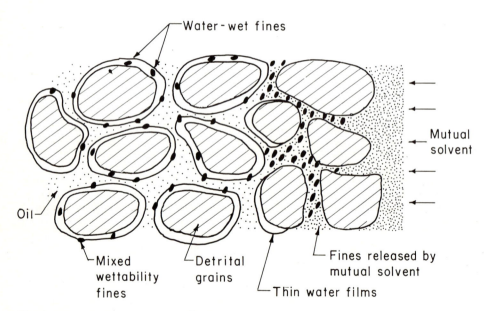

Fig. 7-16. A schematic illustration of the formation of a slug of suspended fines formed as the mutual solvent moves through the reservoir. The fines trapped by interfacial forces at the oil—water interface (and those in the wetting film) are released and suspended in the flowing stream and are potential plugging agents. (Modified after Muecke, 1979, fig. 8.)

causing single phase flow, the fines are released. Concentrated slugs of such fines, which have been observed in laboratory experiments, can cause severe plugging problems, as is shown schematically in Fig. 7-16. Such problems can also arise in other EOR methods involving wettability alteration as a mechanism of oil recovery, because fines which were trapped in the immobile wetting phase may become mobile in the new wetting phase. A considerable amount of research work remains to be done in order to understand this phenomenon better, especially in caustic and surfactant floods.

Dislodging of fines from detrital grains due to turbulent multiphase flow, which may occur in the case of CO_2 floods, may also induce migration of fines. Simultaneous flow of liquids and gases, if present under reservoir conditions, can result in turbulence and eddies, which may dislodge the fines from their stable locations. They may remain in suspension until the flow rate is reduced far away from the wellbore where the fines will precipitate and cause deep plugging.

In conclusion one can state that although fines migration potential may not be an important decision criterion, it should be evaluated for every EOR project.

REFERENCES

Allan, T.O. and Roberts, A.P., 1981. *Production Operations, Vol. 2*. Oil and Gas Consultants Inc., Tulsa, Okla., pp. 171—180.

Bilderback, C.A. and McDougal, L.A., 1969. Complete paraffin control in petroleum production. *J. Pet. Technol.*, 21 (9): 1151—1156.

Chilingarian, G.V. and Vorabutr, P., 1981. *Drilling and Drilling Fluids*. Elsevier, Amsterdam, 767 pp.

Eilers, J.J., 1949. The colloidal structure of asphalt. *J. Phys. Colloid. Chem.*, 53: 1195—1211.

Eldib, I.A., Dunning, H.N. and Bolen, R.J., 1960. Nature of colloidal materials in petroleum. *J. Chem. Eng. Data*, 5: 550—553.

Featherston, A.B., Mihram, R.G. and Water, A.B., 1959. Minimization of scale deposits in oil wells by placement of phosphates in producing zones. *J. Pet. Technol.*, 11 (3): 29—32.

Fertl, W.H., 1972. Status of shaly sand evaluation. *4th Formation Evaluation Symp., Calgary, Alta, May 9—10, 1972*.

Fertl, W.H., Chilingarian, G.V. and Yen, T.F., 1982. Use of natural gamma ray spectral logging in evaluation of clay minerals. *Energy Sources*, 6 (4): 335—360.

Gray, D.H. and Rex, R.W., 1966. Formation damage in sandstone caused by clay dispersion and migration. In: *Proceedings, 14th National Conference on Clay and Clay Minerals*. Pergamon, New York, N.Y., 335 pp.

Grim, R.E., 1968. *Clay Mineralogy*. McGraw-Hill, New York, N.Y., 596 pp.

Harvey, M.T., Jr., Shelton, J.L. and Kelm, C.H., 1977. Field injectivity experiences with miscible recovery projects using alternate rich-gas and water injection. *J. Pet. Technol.*, 29 (9): 1051—1055.

Hau, A.C., Collins, S.H. and Melrose, J.C., 1982. Stability of aqueous wetting films in Athabasca tar sands. *6th Int. Symp. on Oilfield and Geothermal Chemistry, Dallas, Texas, January 1982*. SPE 10626.

Hausler, R.H., 1978. Predicting and controlling scale from oil-field brines. *Oil Gas J.*, 76 (387): 146—154.

Hirschberg, A., de Jong, L.N.J., Schipper, B.A. and Meyers, J.G., 1982. Influence of temperature and pressure on asphaltenes flocculation. *57th Annu. Fall Tech. Conf. and Exhibition of SPE-AIME, New Orleans, La., September 1982*, SPE 11202.

Hogg, R., Healy, T.W. and Fuerstenau, D.W., 1966. Mutual coagulation of colloidal dispersions. *Trans. Faraday Soc.*, 62: 1638—1651.

Huang, E.T.S. and Tracht, J.M., 1974. The displacement of residual oil by carbon dioxide. *3rd SPE-AIME Symp. on Improved Oil Recovery, Tulsa, Okla., April 1974*, SPE 4735.

Jacoby, R.H. and Yarborough, L., 1967. PVT measurements on petroleum reservoir fluids and their uses. *Ind. Eng. Chem.*, 6 (10): 48—62.

James, C., 1970. Scale removal in the Virden Manitoba area. *J. Pet. Technol.*, 22 (6): 701—704.

Jang, L.K., Findley, J.E., Chang, P.W., Sharma, M.M. and Yen, T.F., 1982. Transport of bacteria through porous media. *Proc. Eng. Found. Symp. on Microbiological Enhanced Oil Recovery, Afton, Okla.*

Jorda, R.M., 1966. Paraffin depositions and prevention in oil wells. *J. Pet. Technol.*, 18 (2): 1605—1612.

Kar, G., Chander, S. and Mika, T.S., 1973. The potential energy of interaction between dissimilar double layers. *J. Colloid. Int. Sci.*, 44: 347—355.

Knowles, T.C., 1974. Wellbore gyp-scale control restores production in Means field. *Oil Gas J.*, 27 (48): 89—93.

Knox, J.A., Waters, A.B. and Arnold, B.B., 1962. Checking paraffin deposition by crustal growth inhibition. *37th Annu. Fall Meet. SPE-AIME, Los Angeles, Calif., October 1962*.

Langelier, W.F., 1946. Chemical equilibria in water treatment. *J. Am. Water Works Assoc.*, 38: 169—178.

Monaghan, P.H., Salathiel, R.A., Morgan, B.E. and Kaiser, A.D., 1959. Laboratory studies of formation damage in sands containing clays. *Trans. AIME*, 216: 209—213.

Muecke, T.W., 1979. Formation fines and factors controlling their movement in porous media. *J. Pet. Technol.*, 31 (2): 144—150.

Mungan, N., 1965. Permeability reduction through changes in pH and salinity. *J. Pet. Technol.*, 17 (12): 1449—1453.

Neasham, J.W., 1977. The morphology of dispersed clay in sandstone reservoirs and its effect on sandstone shaliness, pore space and fluid flow properties. *52nd Fall Meet. SPE-AIME, Denver, Colo., October 9—12, 1977*, SPE 6858.

Nellensteyn, F.J., 1938. In: A.E. Dunstam, A.W. Bash, B.T. Brooks and H.T. Tizard (Editors), *The Science of Petroleum*. Oxford University Press, Oxford, 2761 pp.

Ostroff, A.G., 1965. *Introduction to Oil-Field Water Technology*. Prentice Hall, Englewood Cliffs, N.J., 178 pp.

Pfeiffer, J.P.H. and Saal, R.N.J., 1940. Asphaltic bitumen as colloid system. *J. Phys. Chem.*, 44: 139—149.

Rieke, H.H., III, Chilingarian, G.V. and Vorabutr, P., 1981. Clays. In: G.V. Chilingarian and P. Vorabutr (Editors), *Drilling and Drilling Fluids*. Elsevier, Amsterdam, pp. 169—238.

Rohuvets, N. and Fertl, W.H., 1981. Digital shaly sand analysis based on Waxman-Smits model and log-derived clay typing. *7th European Formation Evaluation Symp., SPWLA, Paris, October 1981*.

Rutherford, W.M., 1962. Miscibility relationships in displacement of oil by light hydro-carbons. *Trans. AIME*, 225: 340—346.

Sarkisyan, S.G., 1972. Origin of authigenic clay minerals and their significance in petro-leum geology. *Sediment. Geol.*, 7: 1—22.

Schneider, F.N.M. and Owens, W.W., 1976. Relative permeability studies of gas-water flow following solvent injection in carbonate rocks. *Soc. Pet. Eng. J.*, 16 (2): 23—30.

Sharma, M.M. and Yen, T.F., 1982. Surface charge properties of sands and sandstones and shales. *Annu. Meet., Am. Chem. Soc., Div. Geochem., Las Vegas, Colo., 1982.*

Sharma, M.M. and Yen, T.F., 1983a. The interfacial electrochemistry of oxide surfaces in oil-bearing sands and sandstones. *J. Colloid. Interface Sci.*, 96 (2).

Sharma, M.M. and Yen, T.F., 1983b. A thermodynamic model for low interfacial tensions in caustic flooding. *Soc. Pet. Eng. J.*, 23 (1): 125—134.

Shelton, J.L. and Yarborough, L., 1977. Multiphase behavior in porous media during CO_2 or rich-gas flooding. *J. Pet. Technol.*, 29 (9): 1171—1179.

Shock, D.A., Sudbury, J.D. and Gockett, J.J., 1955. Studies of the mechanisms of paraffin deposition and its control. *J. Pet. Technol.*, 7 (9): 23—28.

Stiff, H.A., Jr. and Davis, L.E., 1952a. A method for predicting the tendency of oilfield waters to deposit calcium sulfate. *Trans. AIME*, 195: 25—28.

Stiff, H.A., Jr. and Davis, L.E., 1952b. A method for predicting the tendency of oilfield waters to deposit calcium carbonate. *Trans. AIME*, 195: 213—216.

Tillman, J., 1932. *Die chemische Untersuchung von Wassen und Abwassen.* Wilhelm Knap, Halle, 2nd ed. (scale).

Tynan, E.C. and Yen, T.F., 1969. Association of vanadium chelates in petroleum asphal-tenes as studied by ESR. *Fuel (London)*, 43: 191—208.

Van Olphen, H., 1959. *An Introduction to Colloid Chemistry.* Interscience, New York, N.Y., 149 pp.

Weaver, C.E. and Pollard, L.D., 1973. *The Chemistry of Clay Minerals.* Elsevier, Amster-dam, 213 pp.

Wiese, G.P. and Healy, T.W., 1970. Effect of particle size on colloid stability. *Trans. Fara-day Soc.*, 66: 490—499.

Wright, J.R. and Minesinger, R.R., 1963. The electrophoretic mobility of asphaltenes in nitromethane. *J. Colloid. Interface Sci.*, 18: 223—236.

Yen, T.F., 1974. Structure of petroleum asphaltene and its significance. *Energy Sources*, 1 (4): 447—436.

Yen, T.F. and Dickie, J.P., 1968. The compactness of the aromatic systems in petroleum asphaltenes. *J. Inst. Pet.*, 54: 50—53.

Yen, T.F. and Erdman, J.G., 1962. Investigation of the structure of petroleum asphaltenes by proton nuclear magnetic resonance. *Preprints, Div. Pet. Chem., Am. Chem. Soc.*, 7 (3): 99—111.

Yen, T.F., Erdman, J.G. and Hanson, W.F., 1961. Reinvestigation of densimetric methods of ring analysis. *J. Chem. Eng. Data*, 6: 443—448.

Yen, T.F., Sharma, M.M. and Jang, L.K., 1982. Bacteria transport through porous media. Quarterly report to the DOE.

Chapter 8

WATERFLOODING*

GERALD L. LANGNES, JOHN O. ROBERTSON, Jr., AMROLLAH MEHDIZADEH,
JALAL TORABZADEH, T.F. YEN, ERLE C. DONALDSON and
GEORGE V. CHILINGARIAN

INTRODUCTION

An excellent comparison of the various waterflood prediction techniques
(Dykstra, 1950; Guerrero and Earlougher, 1961) was presented by Schoeppel
(1968) in the *Oil and Gas Journal*. Table 8-1 summarizes the various prediction techniques presented, whereas Tables 8-2 and 8-4 provide details for the
primary techniques summarized in Table 8-7. Major assumptions, data requirements, and comparisons of methods are presented in these tables.
Tables 8-5 to 8-7 provide detailed comparisons of alternate predictive techniques within the major groupings. Reviewing these tables early in the life
of a field, provides direction to the future gathering of data.

Behavior of carbonate reservoirs is also discussed in this chapter.

WATERFLOODING "RULES OF THUMB"

When considering the use of waterflooding for a specific reservoir, it is
often valuable to get a quick look at the overall project economics. With this
in mind, the following "rules of thumb" were accumulated by the authors.
Two primary sources were used: N. van Wingen's lectures at the University
of Southern California and "Practical Waterflooding Shortcuts" published
in *World Oil* (Editorial, 1966).

Rules of thumb:

(1) Water requirements: 1½ to 2 pore volumes.

(2) Typical injection rates for wells average from 5 to 10 bbl/day ft of reservoir for pattern flooding; a rate of 3 bbl/day ft is considered minimum. A higher rate of 10—20 bbl/day ft can be anticipated for aquifer injection.

---

* This chapter is a revised version of Chapters 3 and 4 of G.L. Langnes, J.O. Robertson,
Jr. and G.V. Chilingar, *Secondary Recovery and Carbonate Reservoirs*, Elsevier, New
York, N.Y., 1972.

TABLE 8-1

Classification of 33 waterflood prediction methods (after Schoeppel, 1968; coutesy of *Oil and Gas Journal*)

Basic method	Modification
I. *Methods primarily concerned with permeability heterogeneity—injectivity*	
1. Dykstra-Parsons (1950)	(a) Johnson (1956)
	(b) Felsenthal-Cobb-Heuer (1962)[a]
2. Stiles (1949)	(a) Schmalz-Rahme (1950)[b]
	(b) Arps ("Modified Stiles") (1956)
	(c) Ache (1957)
	(d) Slider (1961)
3. Yuster-Suder-Calhoun (1949)	(a) Muskat (1950)
	(b) Prats et al. (1959)[c]
4. Prats-Matthews-Jewett-Baker (1959)	
II. *Methods primarily concerned with areal sweep efficiency*	
1. Muskat (1946)	
2. Hurst (1953)	
3. Atlantic-Richfield (1952—1959)	
4. Aronofsky (1952—1956)	
5. Deppe-Hauber (1961—1964)	
III. *Methods primarily concerned with the displacement process*	
1. Buckley-Leverett (1942)	(a) Terwilliger et al. (1951)
	(b) Felsenthal-Yuster (1951)
	(c) Welge (1952)
	(d) Craig-Geffen-Morse (1955)[c]
	(e) Roberts (1959)
	(f) Higgins-Leighton (1960—1964)[c]
2. Craig-Geffen-Morse (1955)	(a) Hendrickson (1961)
3. Higgins-Leighton (1960—1964)	
IV. *Miscellaneous theoretical methods*	
1. Douglas-Blair-Wagner (1958)	
2. Hiatt (1958)	
3. Douglas-Peaceman-Rachford (1959)	
4. Naar-Henderson (1961)	
5. Warren-Cosgrove (1964)	
6. Morel-Seytoux (1965)	
V. *Empirical methods*	
1. Guthrie-Greenberger (1955)	
2. Schauer (1957)	
3. Guerrero-Earlougher (1961)	

[a] Also applies to Stiles method.
[b] Also applies to Yuster-Suder-Calhoun and Schauer methods.
[c] Also concerned with areal sweep problem. Also recognized as basic method.

TABLE 8-2

Characteristics of the perfect waterflood prediction method (after Schoeppel, 1968; courtesy of *Oil and Gas Journal*)

Effect	Characteristic
Fluid flow	initial gas saturation is considered saturation gradient is considered varying injectivity is considered
Pattern	applies to linear systems applies to five-spot pattern applies to other patterns applicable to all mobility ratios considers areal sweep considers increased sweep after breakthrough does not require published laboratory data does not require additional laboratory data
Heterogeneity	considers stratified reservoirs considers crossflow considers spatial variations

(3) Injection rates for sandstone reservoirs vary from 0.35 to 1.5 bbl/acre-ft with an average of 0.5 bbl/acre-ft. Rates for carbonate reservoirs should run two to three times the average.

(4) Response to flood can be expected when two thirds of the fill-up has been achieved (i.e., when two thirds of the voidage created by primary production is filled by the water injected).

(5) Peak oil production rate is reached at the time of fill-up.

(6) Gross production rate equals 80% of water injection rate. The rest of the injected water is lost outside of the patterns or to the aquifer.

(7) Approximately one half of the secondary recovery oil (primary recovery oil left at time of flood plus increase in recovery) will be recovered prior to reaching peak production.

(8) Incremental recovery due to waterflooding is equal to the primary for crude oils having gravities above 30°API. For lower crude oil gravities (15—30°API), the incremental recovery ranges from 50 to 100% of primary recovery.

(9) The analysis of many pattern floods indicates that (Earlougher and Guerrero, 1965): (a) areal efficiency = 70—100%, (b) vertical efficiency = 40—80%, (c) combined efficiency = 28—80% (median of 60%), and (d) residual oil saturation = 15—30%.

TABLE 8-3

Basic assumptions in waterflood prediction methods (after Schoeppel, 1968; courtesy of *Oil and Gas Journal*)

Method and modification[a]	Date presented	Fluid-flow effects			Pattern effects									Heterogeneity effects		
		considers			applies to				considers		requires			considers		
		initial gas saturation	satura-tion gradient	varying injec-tivity	linear system	5-spot pattern	other patterns	mobility ratio used	areal sweep	increased sweep after break-through	pub-lished lab. data	addi-tional lab. data	stratified reservoirs	cross-flow	spatial variations
		(1)	(2)	(3)	(4)	(5)	(6)	(7)	(8)	(9)	(10)	(11)	(12)	(13)	(14)
The perfect method		yes	yes	yes	yes	yes	yes	any	yes	yes	no	no	yes	yes	yes
I. Dykstra-Parsons	1950	yes	no	no	yes	no	no	any	no	no	no	no	yes	no	no
Johnson	1956	yes	no	no	yes	no	no	any	no	no	no	no	yes	no	no
Felsenthal-Cobb-Heuer	1962	yes	no	no	yes	no	no	any	no	no	no	no	yes	no	no
Stiles	1949	no	no	no	yes	no	no	1.0	no	no	yes	no	yes	no	no
Schmalz-Rahme	1950	no	no	no	yes	no	no	1.0	no	no	no	no	yes	no	no
Arps	1956	no	no	no	yes	no	no	1.0	no	no	no	no	yes	no	no
Ache	1957	no	no	no	—	yes	no	1.0	no	no	no	no	yes	no	no
Slider	1961	yes	no	yes	yes	yes (?)	no	1.0	no	no	no	no	yes	no	no
Yuster-Suder-Calhoun	1949	yes	no	yes		yes	no	1.0	no	no	no	no	yes	no	no
Muskat	1950	no	no	yes	yes	no	no	any	no	no	no	no	yes	no	no
Prats et al.	1959	yes	no	yes	—	yes	yes	any	yes	yes	yes	yes	yes	no	no
II. Muskat	1946	no	no	no	—	yes	yes	1.0	yes	no	no	no	no	no	no
Hurst	1953	no	no	no	—	yes	no	1.0	yes	no	no	no	yes	no	no
Atlantic-Richfield	1952–1959	no	no	yes	—	yes	yes	any	yes	yes	yes	no	yes	no	no
Aronofsky	1952–1956	no	no	yes	—	yes	no	any	yes	no	yes	no	yes	no	no
Deppe-Hauber	1961–1964	yes	no	yes	—	yes	yes	any	yes	yes	yes	no	yes	no	no
III. Buckley-Leverett	1942	no	yes	no	yes	no	no	—	no	no	no	no	no	no	no
Welge	1952	no	yes	no	yes	no	no	—	no	no	no	no	no	no	no
Roberts	1959	no	yes	no	yes	no	no	—	no	no	no	no	yes	no	no
Craig-Geffen-Morse	1955	yes	yes	yes	—	yes	no	any	yes	yes	yes	no	yes	no	no
Higgins-Leighton	1960–1964	yes	yes	yes	—	yes	yes	any	yes	yes	yes	no	yes	no	no
Hendrickson	1961	no	yes	no	—	yes	no	any	yes	yes	yes	no	yes	no	no
IV. Douglas-Blair-Wagner	1958	no	yes	yes	—	yes	yes	any	yes	yes	yes	no	yes	no	no
Hiatt	1958	no	yes	no	yes	no	no	—	no	no	no	no	yes	yes	no
Douglas et al.	1959	no	yes	yes	—	yes	yes	any	yes	yes	yes	no	yes	no	no
Naar-Henderson	1961	no	yes	no	—	yes	no	any	yes	yes	yes	no	yes	no	no
Warren-Cosgrove	1964	no	yes	no	yes	no	no	—	no	no	no	no	yes	yes	no
Morel-Seytoux	1965	no	yes	yes	—	yes	yes	any	yes	yes	no	no	yes	no	no
V. Schauer	1957	yes	no	no	—	yes	no	any	yes	no	no	no	yes	no	no
Guerrero-Earlougher	1961	yes	no	no	—	yes	yes	—	no	no	no	no	yes	no	no

[a] Categorized according to classification presented in Table 8-1.

TABLE 8-4

Data required for waterflood prediction methods (after Schoeppel 1968; courtesy of Oil and Gas Journal)

Method and modification[a]	Absolute permeability	Stratified bed thickness	Effective oil and water permeability	Relative permeability saturation curve	Init. and resid. oil saturations	Initial gas saturation	Resid. gas saturation	Oil and water viscosity	Gas viscosity	Average injection rate	Injection history	Correlation for areal sweep
I. Dykstra-Parsons (1950)	X	X	X	—	X	—	—	X	—	X	—	—
Johnson (1956)	X	X	X	—	X	—	—	X	—	X	—	—
Felsenthal-Cobb-Heuer (1962)	X	X	X	—	X	X	X	X	—	X	—	—
Stiles (1949)	X	X	X	—	X	X	—	X	—	X	—	—
Schmalz-Rahme (1950)	X	X	X	X	X	—	—	X	—	X	—	—
Arps (1956)	X	X	X	X	X	—	—	X	—	X	—	—
Ache (1957)	X	X	X	—	X	—	—	X	—	X	—	—
Slider (1961)	X	X	b	X	X	—	—	X	—	X	—	—
Yuster-Suder-Calhoun (1949)	X	X	X	—	X	X	—	—	—	—	X	—
Muskat (1950)	X	X	—	X	—	X	—	—	—	—	—	—
Prats et al. (1959)	X	X	X	X	—	X	X	X	X	—	X	X
II. Muskat (1946)	—	—	—	—	—	X	—	—	—	X	—	—
Hurst (1953)	X	X	X	—	X	—	—	—	—	—	X	X
Atlantic-Richfield (1952)	X	X	—	—	X	—	—	X	—	X	X	X
Aronofsky (1952)	—	—	—	—	—	—	—	—	—	—	—	—
III. Buckley-Leverett (1942)	—	—	—	X	—	X	—	X	—	—	—	—
Welge (1952)	—	—	—	X	X	X	—	X	—	—	—	—
Craig-Geffen-Morse (1955)	X	X	X	X	X	—	X	X	—	—	X	X
Roberts (1959)	X	X	X	X	X	X	—	X	—	X	—	—
Higgins-Leighton (1960, 1962, 1963)	X	X	X	X	X	X	X	X	—	—	X	X
IV. Douglas-Blair-Wagner (1958)	X	X	X	X	X	X	X	X	X	—	—	—

[a] Categorized according to classification scheme of Table 8-1.
[b] k_{rw} only.

TABLE 8-5

Comparison of basic methods concerned with permeability–heterogeneity–injectivity problems (after Schoeppel, 1968; courtesy of *Oil and Gas Journal*)

	Dykstra-Parsons (1950)	Stiles (1949)	Yuster-Suder-Calhoun (1949)
Reservoir configuration	stratified-linear	stratified-linear	stratified five-spot
Permeability distribution	normal probability	rearranged actual data	average of permeability capacity
Injection controlled	mobility ratio	kh capacity	kh capacity
Total injection rate	constant	constant	variable (to fill-up), constant (after fill-up)
Layer injection rate	variable	constant	variable (to fill-up), constant (after fill-up)
Required mobility ratio	any	1.0	1.0
Gas fill-up	before oil production	initially	by individual beds
Displacement mechanism	pistonlike	pistonlike	pistonlike
Areal sweep	100% at water breakthrough	100% at water breakthrough	sweep efficiency factor[a]
Vertical sweep	proportional to permeability capacity and mobility ratio	proportional to permeability capacity	recovery factor[a]
Solution method	graphical	graphical and numerical	numerical

[a] Determined as a function of throughput.

(10) The costs of secondary recovery of oil in 1981 in the United States are broken down as equipment and operating costs for ten producing wells and eleven injection wells operating at a depth of 4000 ft as follows (Funk and Anderson, 1982, pp. 8, 15):

Additional lease equipment	1,012,200
Injection wells	2,366,100
	.$ 3,378,300

Normal daily operating costs	138,300
Surface repair	64,500
Subsurface repair	73,200
	$ 276,000

FLOOD DESIGN BY ANALOGY

The study of existing and completed waterflood projects and natural water-drive reservoirs provides the basis for the extension of the "rules of thumb" approach to flood design. A study of the reservoirs in the Denver Basin (Nebraska and Colorado) showed that natural water-drive reservoirs in the Nebraska portion of the basin had primary recoveries of 40—45% of the initial stock tank oil-in-place (Bleakley, 1965). In the Colorado portion of the basin, 18% was considered a good primary recovery (solution gas-drive mechanism) in similar types of reservoirs. The operators concluded that waterflooding would be feasible in the Colorado portion of the basin and that the ultimate recovery for a field could reasonably be expected to be 40%. The predicted performance of the West Lisbon waterflood in Loui-

TABLE 8-6

Comparison of basic methods concerned with areal sweep efficiency problems (after Schoeppel, 1968; courtesy of *Oil and Gas Journal*)

	Muskat (1946)	Hurst (1953)	Atlantic-Richfield (1952, 1955)
Reservoir configuration	stratified five-spot	stratified five-spot	stratified five-spot
Total injection rate	constant	constant	variable
Applicable mobility ratio	1.0	1.0	0.1—10
Initial gas saturation	none	none	none
Sweep efficiency at water breakthrough	72.3%	72.6%	variable
Sweep efficiency after water breakthrough	not available	not available	from correlation

TABLE 8-7

Waterflood prediction methods concerned with displacement mechanism (after Schoeppel, 1968; courtesy of *Oil and Gas Journal*)

Basic method	Highlights of method
Buckley-Leverett (1942)	material balance on element; variable-saturation flood front displacement mechanism; single layer—linear model; constant injection rate; no residual gas saturation; 100% sweep at water breakthrough
Modification	
1. Terwilliger et al. (1951)	saturations in flood front; stabilized zone concepts
2. Felsenthal-Yuster (1951)	radial case
3. Welge (1952)	average saturation at water breakthrough
4. Craig-Geffen-Morse (1955)	five-spot pattern; sweep efficiency correlated with mobility ratio at breakthrough
5. Roberts (1959)	allowance for stratification
6. Higgins-Leighton (1960)	channel and cell displacement; any well pattern

siana was based on the past performance of the waterflood of the Southwest Lisbon Pettit reservoir, Louisiana. It was estimated that the recovery would be raised from 14% under primary depletion to 32% under flood (Miller and Perkins, 1960).

Callaway (1959) suggested that by relating the results obtained from a waterflood to the reservoir parameters, which control the performance, a reliable set of experience factors can be obtained and the uncertainties can be greatly reduced. He divided the engineering factors involved in evaluating waterflood recovery into two sets of variables: (1) "primary variables," which are those factors bearing a direct mathematical relation to the amount of oil to be recovered; (2) "secondary variables," which operate indirectly through the primary variables to influence the oil recovery. The primary variables are (1) primary recovery efficiency, (2) connate water saturation, (3) sweep efficiency, (4) residual oil saturation, and (5) crude shrinkage. The secondary variables and the corresponding primary factors influenced (numbers in parentheses) are (a) oil viscosity (1, 3, 4); (b) permeability (1, 3, 4); (c) structural considerations (1, 3); (d) uniformity of reservoir rock (3); (e) type of flood (3); (f) time for start of flood (5); and (g) economic factors (1, 3, 4).

Callaway's (1959) equation for evaluating the recovery by waterflooding is as follows:

$$WR = 7758\phi \left(\frac{1-S_w}{B_{oi}}\right) \left\{1 - E_p - \frac{B_{oi}}{B_o}\left[1 - E_o\left(1 - \frac{S_{or}}{1-S_w}\right)\right]\right\} \quad (8\text{-}1)$$

where WR = waterflood recovery, bbl/acre-ft; B_{oi} = original formation volume factor for oil, bbl/STB; B_o = formation volume factor for oil during waterflood operations, bbl/STB; S_{or} = residual oil saturation, fraction: S_w = connate water saturation, fraction; ϕ = porosity, fraction; E_p = primary recovery efficiency, fraction of original oil-in-place; E_o = overall sweep efficiency, fraction of reservoir volume.

This equation is based on the assumption that the unswept portion of the reservoir at the time of flood abandonment is completely saturated with oil and connate water. It is also assumed that there is no gas cap and that a free gas saturation does not exist in the swept portion of the reservoir.

Goolsby (1967) suggested the following steps for flood design:

(1) Characterize the reservoir geologically as completely as possible.

(2) Determine the value for Callaway's (1959) five "primary variables". Most of these variables can be expressed with a range developed from field and laboratory data and the study of analogous fields.

(3) Calculate a range of waterflood recoveries using equation (8-1).

(4) Relate the maximum, average, and minimum recoveries to time by using fluid-in—fluid-out and water—oil production relationships taken from analogous fields. The effects of various injection rates can be incorporated in this step if felt necessary.

The "primary variables" can also be expressed as probability distributions. Equation (8-1) is then solved using these distributions yielding a probability distribution for the waterflood recovery instead of a simple range-average value.

SWEEP EFFICIENCY

The effectiveness of a secondary recovery process is dependent on the volume of the reservoir which will be contacted by the injected fluid. The latter, in turn, is dependent on the horizontal and vertical sweep efficiency of the process. The following factors control the sweep efficiency:

(1) Pattern of injectors.

(2) Off-pattern wells.

(3) Unconfined patterns.

(4) Fractures.

(5) Reservoir heterogeneity.

(6) Continued injection after breakthrough.

(7) Mobility ratio.

(8) Position of gas—oil and oil—water contacts.

Pattern selection

The selection of an injection pattern is one of the first steps in the design of secondary recovery projects. When making the choice, it is necessary to

260

Solid line indicates symmetry pattern of infinite well network.
Dashed line indicates symmetry element of infinite well network.

Code: ○ Injection wells, ● Producing wells

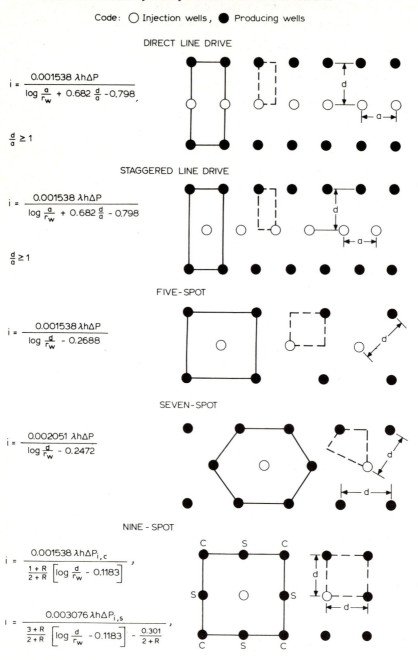

DIRECT LINE DRIVE

$$i = \frac{0.001538\,\lambda h \Delta P}{\log \frac{a}{r_w} + 0.682\,\frac{d}{a} - 0.798},$$

$$\frac{d}{a} \geq 1$$

STAGGERED LINE DRIVE

$$i = \frac{0.001538\,\lambda h \Delta P}{\log \frac{a}{r_w} + 0.682\,\frac{d}{a} - 0.798}$$

$$\frac{d}{a} \geq 1$$

FIVE-SPOT

$$i = \frac{0.001538\,\lambda h \Delta P}{\log \frac{d}{r_w} - 0.2688}$$

SEVEN-SPOT

$$i = \frac{0.002051\,\lambda h \Delta P}{\log \frac{d}{r_w} - 0.2472}$$

NINE - SPOT

$$i = \frac{0.001538\,\lambda h \Delta P_{i,c}}{\frac{1+R}{2+R}\left[\log \frac{d}{r_w} - 0.1183\right]},$$

$$i = \frac{0.003076\,\lambda h \Delta P_{i,s}}{\frac{3+R}{2+R}\left[\log \frac{d}{r_w} - 0.1183\right] - \frac{0.301}{2+R}},$$

Fig. 8-1. For description, see p. 262.

R = Ratio of producing rates of corner well (c) to side well (s),

$\Delta P_{i,c}$ = Difference in pressure between injection well and corner well (c),

$\Delta P_{i,s}$ = Difference in pressure between injection well and side well (s).

DIRECT LINE DRIVE BOUNDARY PATTERN

$$i = \frac{0.003076\,\lambda h \Delta P_{i1',q1'}}{\dfrac{q_1 + 4q_1'}{q_1 + 2q_1'}\left[\log\dfrac{d}{r_w} + 0.166 + 0.400\,R\right] + \dfrac{1.368q_1(1+R)}{q_1 + 2q_1'}},$$

$$R = \frac{q_2}{q_1} = \frac{q_2'}{q_1'},$$

$\Delta P_{i1',q1'}$ = Difference in pressure between injection well i_1' and
producing well q_1'.

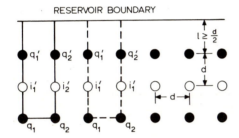

RESERVOIR BOUNDARY

FIVE-SPOT BOUNDARY PATTERN

$$i' = \frac{0.001538\,\lambda h \Delta P_{i',q'}}{\log\dfrac{d}{r_w} - 0.158 - 0.036\,\dfrac{q}{q'}},$$

$\Delta P_{i',q'}$ = Difference in pressure between injection well i' and
producing well q'

(Wells of outer row are alternating injection and producing wells.)

FIVE-SPOT BOUNDARY PATTERN

$$i' = \frac{0.003076\,h\lambda(q_1 + q_2 + 2q_1' + 2q_2')\Delta P_{i',q1'}}{2q_1'\left[(3+R)\left\{\log\dfrac{d}{r_w} + 0.815\right\} - 1.7216\right] + q_1(1+R)\left\{\log\dfrac{d}{r_w} - 0.098\right\}},$$

Fig. 8-1 (continued).

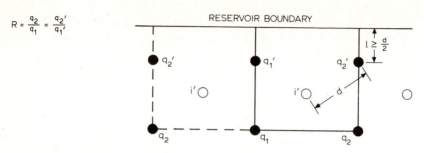

$$R = \frac{q_2}{q_1} = \frac{q_2'}{q_1'}$$

$\Delta P_{i',q'}$ = Difference in pressure between injection well i' and producing well q'.

(Wells of outer row are all producing wells.)

NINE - SPOT BOUNDARY PATTERN

$$i' = \cfrac{0.003076\, \lambda h \Delta P_{i',s'}\,(2q_s' + 2q_c' + 3q_s + q_c)}{2q_s'\left[(2+R)\left\{\log \frac{d}{r_w} + 0.548\right\} - 1.49\right] + q_s\left[(3+R)\left\{\log \frac{d}{r_w} + 0.885\right\} - 0.577\right]},$$

$$i' = \cfrac{0.003076\, \lambda h \Delta P_{i',c'}\,(2q_s' + 2q_c' + 3q_s + q_c)}{2q_s'\left[(1+2R)\left\{\log \frac{d}{r_w} + 0.204\right\} + 0.362\right] + q_s\left[(3+R)\left\{\log \frac{d}{r_w} - 0.460\right\} + 0.124\right]},$$

$$R = \frac{q_c}{q_s} = \frac{q_c'}{q_s'},$$

$\Delta P_{i',s'}$ = Difference in pressure between injection well i' and producing well s'.

$\Delta P_{i',c'}$ = Difference in pressure between injection well i' and producing well c'.

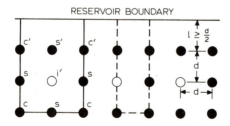

Fig. 8-1. Injectivities for regular patterns for mobility ratio equal to one. h = net pay thickness, ft; r_w = wellbore radius, ft; Δp = difference in pressure between the injection well and producer, psi; and λ = mobility of reservoir fluid, md/cp. (After Deppe, 1961; courtesy of the SPE of AIME.)

consider all the available information about the reservoir. The adverse effects of the other factors listed above can be partially offset if they are considered during the pattern selection. Other factors which should be considered in pattern selection are as follows:

(1) Flood life.
(2) Well spacing.
(3) Injectivity.
(4) Response time.
(5) Productivity.

The flood life depends on the availability of water, the rate at which water can be injected, well spacing, and proration policies. The performance and economics for various well spacings and pattern sizes should be analyzed in order to arrive at the economically optimum choice. These analyses, however, cannot be made without also considering injectivity, which is best determined from pilot operations. A well-controlled pilot operation is essential to understanding all the pattern selection factors.

Empirical methods for estimating water injectivity prior to an actual test for pattern floods have been worked out by Muskat (1946) and by Deppe (1961) for a mobility ratio of unity in the case of single fluid flow; they are presented in Fig. 8-1. Prats et al. (1959) have developed a plot relating dimensionless injectivity, I_D, to mobility ratio, $M_{w,o}$, for different stages of a flood in five-spot patterns, when the reservoir has an initial mobile gas saturation (Fig. 8-2). Dimensionless injectivity, I_D, is defined by the following equation:

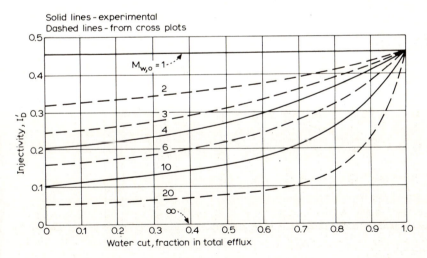

Fig. 8-2. Injectivity as a function of water cut and mobility ratio. (After Prats et al., 1959, p. 98; courtesy of the SPE of AIME.)

$$I_D = \frac{i_w \mu_w}{k_w h \Delta p} \qquad (8\text{-}2)$$

where i_w = water injection rate, bbl/day; μ_w = water viscosity, cp; k_w = permeability to water, md; h = thickness of injection zone, ft; Δp = differential injection pressure, psi.

Response time is dependent on injectivity and spacing. It is further influenced by reservoir heterogeneity and the oil, gas, and water saturations, which exist at the beginning of injection.

The pattern chosen must above all consider the physical characteristics of the reservoir. Formal patterns, such as the 5-, 7-, or 9-spot, and the direct or staggered line drives are useful only when the reservoir is generally uniform in character. Faulting and localized variations in porosity or permeability lead to irregular patterns or peripheral injection systems. For limestone reservoirs, the irregular pattern or peripheral system is more likely to be used.

Off-pattern wells

Prats et al. (1962) have calculated the effect of off-pattern wells (producers and injectors) on the performance of a regular five-spot waterflood. Data developed by their work are presented in Figs. 8-3 through 8-11. These results are strictly applicable only when the assumptions are (1) the reservoir is thin and horizontal; (2) porosity, permeability, and thickness are uniform; (3) only crude oil is mobile initially in the formation; (4) mobility ratio is one; (5) injection and production rates are the same for all wells throughout the field; (6) there is a sharp boundary between the oil and water banks;

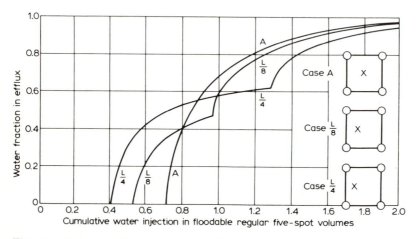

Fig. 8-3. Water-cut history of a laterally displaced production well. (After Prats et al., 1962, p. 173; courtesy of the SPE of AIME.)

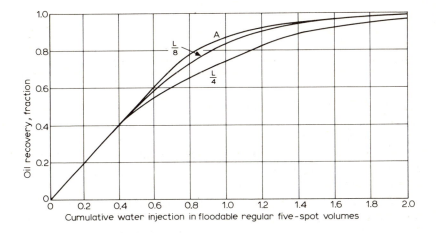

Fig. 8-4. Oil-production history of a laterally displaced production well. (After Prats et al., 1962, p. 173; courtesy of the SPE of AIME.)

(7) producing wells are kept flowing even after the individual wells have reached an economic limit cut of 98%. The data developed by Prats et al. (1962) can be used, however, to estimate the impact for other conditions. These estimates would generally represent the minimum impact to be expected for offset wells.

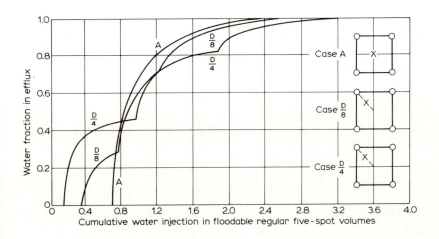

Fig. 8-5. Water-cut history of a diagonally displaced production well. (After Prats et al., 1962; courtesy of the SPE of AIME.)

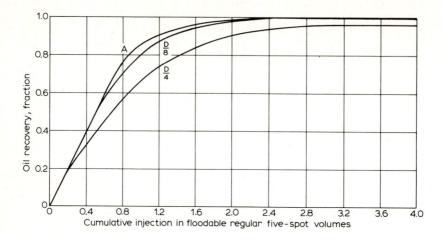

Fig. 8-6. Oil-production history of a diagonally displaced production well. (After Prats et al., 1962; courtesy of the SPE of AIME.)

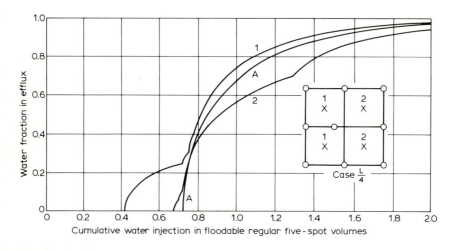

Fig. 8-7. Water-cut history of five-spot patterns surrounding laterally displaced injection well. (After Prats et al., 1962; courtesy of the SPE of AIME.)

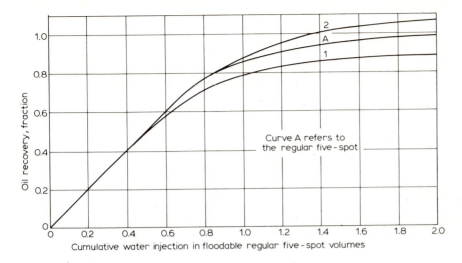

Fig. 8-8. Oil-production history of five-spot patterns surrounding laterally displaced injection well. (After Prats et al., 1962; courtesy of the SPE of AIME.)

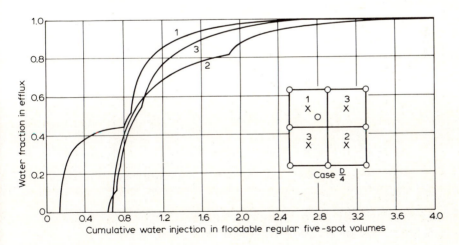

Fig. 8-9. Water-cut history of five-spot patterns surrounding diagonally displaced injection well. (After Prats et al., 1962; courtesy of the SPE of AIME.)

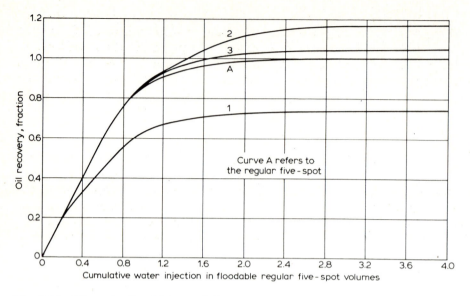

Fig. 8-10. Oil-production history of five-spot patterns surrounding diagonally displaced injection well. (After Prats et al., 1962; courtesy of the SPE of AIME.)

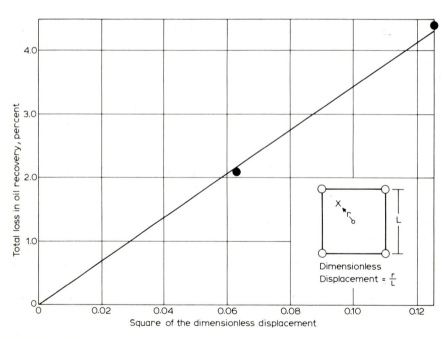

Fig. 8-11. Effect of displacement on total loss in oil recovery. (After Prats et al., 1962; courtesy of the SPE of AIME.)

Multiple irregular patterns in a single field can be evaluated with Figs. 8-3 through 8-11 as long as the irregular patterns are separated by at least one normal five-spot pattern. In Figs. 8-3 through 8-10, the A curves represent the performance of a normal five-spot flood. The other curves are keyed to the well diagrams included in the water-cut history figures. The D or L notations refer to the diagonal or lateral displacement of the wells.

Unconfined patterns

The principal problem associated with an unconfined pattern is the loss of injected energy to wells and/or aquifer outside of the injection pattern. The degree of pattern confinement is dependent upon the reservoir pressures around this pattern. Geologic barriers such as faults and pinchouts or the presence of high-pressure aquifers limit the loss of energy. Unconfined boundaries include boundaries adjacent to (1) leases where the reservoir is unpressured, and (2) low-pressure aquifers.

Exact percentages of energy loss or oil migration outside the confined pattern will vary with overall well configuration and estimates should be made as to the magnitude of the anticipated loss. A simple procedure often used to estimate the loss is based on the assumption of uniform radial flow of the injected water from the injector. The peripheral injectors are connected by lines on a map and the losses are calculated for each injector by measuring the exterior angle between the lines connecting the injector with the injectors on either side and dividing by 360° (an injector on the side of a project may lose around 50%, whereas a corner injector may lose about 75%).

Fractures

Dyes et al. (1958) studied the effect of fractures on the sweep efficiency of a five-spot pattern. Figs. 8-12 through 8-15 present a summary of their findings; in the figures, the fracture length, L, is expressed as the fraction of the distance between the fractured well and the boundary of the element in the flood pattern as shown in the small diagrams located in the figure.

Fig. 8-12 shows that a vertical fracture, when located in a favorable direction, has little effect on sweep efficiency. When the vertical fracture is located in an unfavorable direction, the breakthrough sweep efficiency drops with increasing fracture length, L. The ultimate sweep efficiency is unaffected if no restriction is placed on the amount of water injected or the water/oil ratio of the producer.

The work of Dyes et al. (1958) was directed toward investigating the impact of fracturing techniques on waterflooding operations. Their work, however, can also be used when considering natural fracture systems. By using the trends presented, patterns can be planned to minimize the impact of the fracture system.

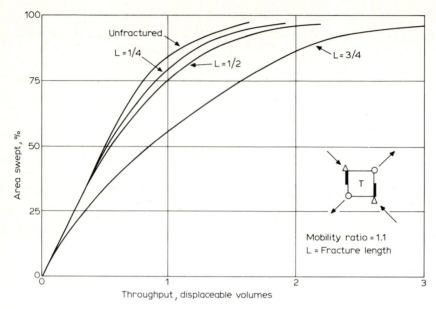

Fig. 8-12. Sweep-out with vertical fracture of favorable direction. (After Dyes et al., 1958; courtesy of the SPE of AIME.) Fracture length, L = fraction of distance between fractured well and boundary of element in flood pattern.

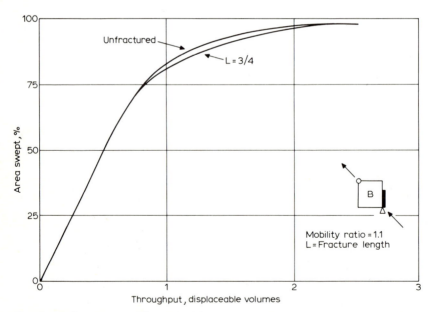

Fig. 8-13. Sweep-out with vertical fracture having unfavorable direction. (After Dyes et al., 1958; courtesy of the SPE of AIME.)

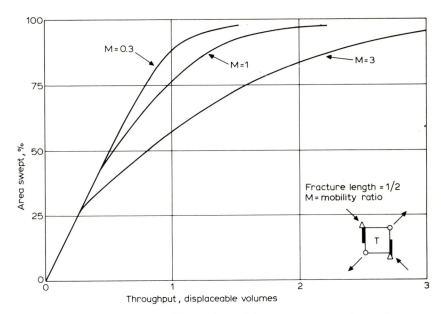

Fig. 8-14. Influence of mobility ratio and fracture on percentage of area swept. (After Dyes et al., 1958; courtesy of the SPE of AIME.)

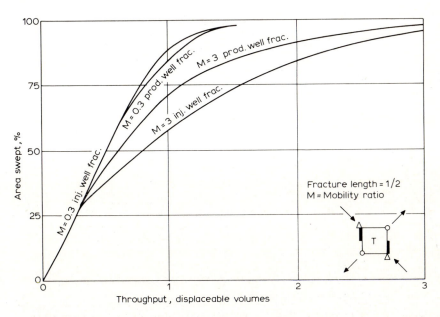

Fig. 8-15. Influence of injection or production well fracture on percentage of area swept. (After Dyes et al., 1958; courtesy of the SPE of AIME.)

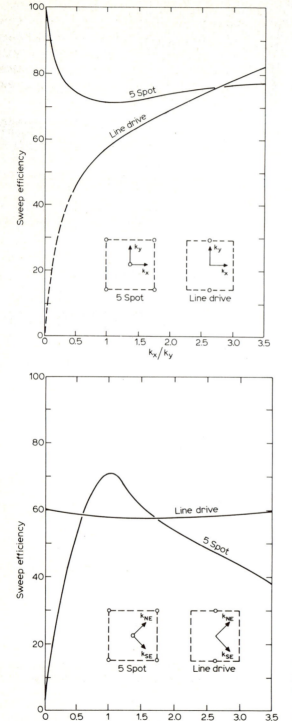

Fig. 8-16. The effect of directional permeability on sweep efficiency. (After Landrum and Crawford, 1960; courtesy of the SPE of AIME.)

Fig. 8-17. The effect of directional permeability on sweep efficiency. (After Landrum and Crawford, 1960; courtesy of the SPE of AIME.)

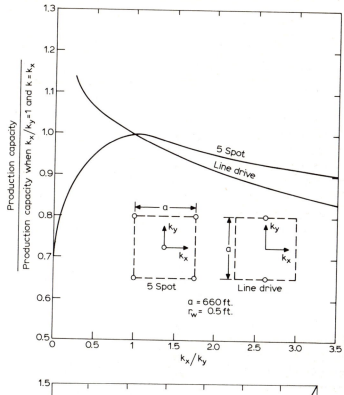

Fig. 8-18. The effect of directional permeability on production capacity. (After Landrum and Crawford, 1960; courtesy of the SPE of AIME.)

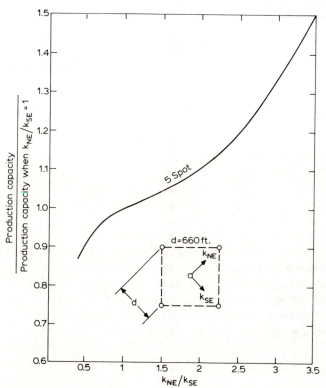

Fig. 8-19. The effect of directional permeability on production capacity. (After Landrum and Crawford, 1960; courtesy of the SPE of AIME.)

Reservoir heterogeneity

Landrum and Crawford (1960) have studied the effects of directional permeability on waterflood sweep efficiency and productive capacity. Figs. 8-16 and 8-17 illustrate the impact of directional permeability variations on sweep efficiency for a line drive and a five-spot pattern flood, whereas Figs. 8-18 and 8-19 present the variations in productive capacity for the same conditions. These data were generated through theoretical calculations and potentiometric studies assuming a mobility ratio of one and steady state flow. Gravity and capillary effects were neglected.

The effects of directional permeability can be minimized through adjustment of the geometry of the injector—producer system. Pattern distances should be lengthened in the direction of the greater permeability.

Continued injection after breakthrough

Continued injection after breakthrough can result in substantial increases in recovery, especially in the case of an adverse mobility ratio. Most of the published data on sweep efficiency beyond breakthrough have been obtained on porous-plate or sand-pack models without an initial gas saturation. The five-spot pattern is the one most extensively studied.

The work of Craig et al. (1955) has shown that the breakthrough of injected fluid at the producer is not the end of a successful flooding operation, because significant quantities of oil may be swept by water after breakthrough (Habermann, 1960). The higher the mobility ratio (crude oil gravity decreasing), the more important is the "after-breakthrough" production. Fig. 8-32 shows that the breakthrough efficiency changes from a low of 51% at a mobility ratio of 10 to a high of 100% at a mobility ratio of 0.17.

Mobility ratio

The mobility ratio is defined as the ratio of the mobility of the driving phase (such as water or LPG) to the mobility of the driven phase (such as oil). The mobility ratio, M, may be represented as:

$$M = \frac{k_w \mu_o}{\mu_w k_o} \tag{8-3}$$

where k_w = water permeability, md; k_o = oil permeability, md; μ_w = water viscosity, cp; and μ_o = oil viscocity, cp. The k_w refers to the effective water permeability behind the water—oil front and k_o refers to the effective oil permeability ahead of the front. This equation shows that high oil viscosities give rise to high, or unfavorable, mobility ratios (high mobility ratios generally result in poor recoveries). Fig. 8-20 is a schematic drawing of the regions of

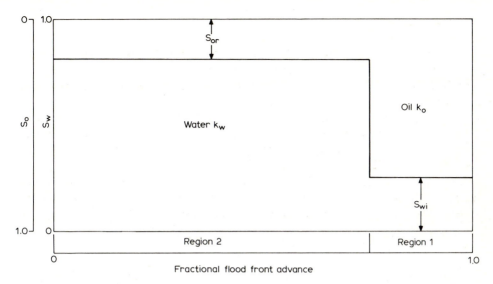

Fig. 8-20. Frontal displacement process with one movable phase behind the front.

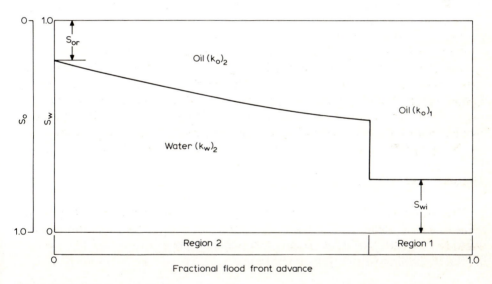

Fig. 8-21. Partial frontal displacement with two movable phases behind the front.

a reservoir being subjected to frontal displacement with one movable phase, water, behind the front and one, oil, in front. Fig. 8-21 is a schematic drawing of the regions of a reservoir being subjected to partial frontal displacement (two movable phases, oil and water, exist behind the front). The effective mobility, M_e, behind the front may be represented as:

$$M_e = \left(\frac{k_w}{\mu_w} + \frac{k_o}{\mu_o} \right)_2 \tag{8-4}$$

where the subscript 2 refers to the region behind the front. For partial frontal displacement, the mobility ratio, M, is given by:

$$M = \left(\frac{k_w}{\mu_w} + \frac{k_o}{\mu_o} \right)_2 \Big/ \left(\frac{k_o}{\mu_o} \right)_1 \tag{8-5}$$

The effective or relative permeability to oil and to water can be obtained from: (1) use of published data such as given by Leverett and Lewis (1941); (2) displacement tests from which k_w/k_o curves can be developed; (3) water-flood tests; oil permeability is measured before a core is flooded, and water permeability is measured at the end of the test.

For reservoirs that have a moderate or large interstitial gas saturation, displacement tests and/or waterflood tests should be made in a manner that accounts for this gas-occupied pore space. Favorable mobility ratios are equal to or are less than 1.0 and give rise to stable fronts and good recoveries. Unfavorable, or poor, mobility ratios are greater than 1.0 and have a tendency to cause unstable fronts. Recoveries become worse with increasing mobility ratios.

MAJOR PREDICTIVE TECHNIQUES

The major predictive techniques presented here in outline form include the Buckley-Leverett, Dykstra-Parsons (Johnson), and Stiles procedures.

Buckley-Leverett (1942, p. 107) predictive technique

Two significant variations on the basic Buckley-Leverett predictive techniques have evolved: (1) the layered Buckley-Leverett and (2) the double Buckley-Leverett. These variations are treated in detail immediately following the discussion of the basic technique.

Assumptions for basic Buckley-Leverett method
(1) A flood front exists, with only oil moving ahead of the front. Oil and water move behind the front.
(2) Reservoir is a single homogeneous layer. Cross-sectional area to flow is constant.
(3) Linear steady-state flow occurs and Darcy's law applies (q injected = q produced), where q is expressed in bbl/day.
(4) There is no residual gas saturation behind the front.

(5) Fractional flow of the displacing and displaced fluids after break-through is assumed to be a function of the mobility ratio of the two fluids (capillary and gravity effects are neglected) as expressed below:

$$f_w = \frac{1}{1 + (k_{ro}\mu_w/k_{rw}\mu_o)}$$
(8-6)

where k_{ro} = relative permeability to oil, fraction; k_{rw} = relative permeability to water, fraction; μ_o = oil viscosity, cp; and μ_w = water viscosity, cp.

(6) Fill-up occurs in all layers prior to flood response. The flood life should be increased to reflect the fill-up period.

Procedure for basic Buckley-Leverett method

(1) Organize relative permeability data into form suggested in Table 8-8. If several sets of relative permeability data exist for a reservoir, use the set which is representative of the portion of the reservoir to be flooded.

TABLE 8-8

Organization of relative permeability data for the Buckley-Leverett waterflood predictive technique

(1)	(2)	(3)	(4)	(5)	(6)
S_w	k_{ro}	k_{rw}	k_{ro}/k_{rw}	μ_w/μ_o	f_w

(2) Calculate the fractional flow, f_w, as a function of water saturation, S_w, using equation (8-6) and plot on cartesian coordinate paper as shown in Fig. 8-22.

(3) Draw tangent to fractional flow curve as indicated in Fig. 8-22. This gives the water saturation value at the flood front at breakthrough. The average saturation behind the front is read at $f_w = 1.0$.

(4) Determine graphically the rate of change in the fractional flow, f_w', as a function of the change in the floodfront water saturation:

$$f_w' = \frac{df_w}{dS_w} \approx \frac{\Delta f_w}{\Delta S_w}$$
(8-7)

(5) Draw 6 to 8 tangents to the fractional flow curve at S_w values greater than that at breakthrough. Determine the S_{wa} and f_w' values corresponding to these S_w points.

(6) Plot f_w' versus S_w at the flood front on cartesian coordinate paper and draw a smooth curve through the points. Read smoothed f_w' points for each of the S_w points.

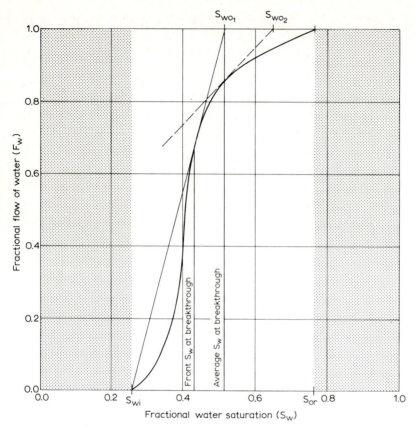

Fig. 8-22. Relationship between the fractional flow of water, f_w, and fractional water saturation, S_w.

(7) Calculate the recovery of oil, N_p, in barrels at breakthrough following the steps in Table 8-9:

$$N_p = 7758\, Ah\phi \left(\frac{S_{wa} - S_{wi}}{B_o} \right) \tag{8-8}$$

TABLE 8-9

Recommended steps in using the Buckley-Leverett (1942) waterflood predictive technique

(1)	(2)	(3)	(4)	(5)	(6)	(7)	(8)	(9)
$S_{w(front)}$	f_w	S_{wa}	$S_{wa}-S_{wi}$	N_p	f_w'	WOR	W_i	t

where A = areal extent of the reservoir, acres; h = average thickness of reservoir, ft; ϕ = average porosity, fraction; B_o = oil formation volume factor, bbl/STB; S_{wa} = average water saturation, fraction; and S_{wi} = initial water saturation, fraction.

(8) Calculate the recovery of oil to each of the S_w points using equation (8-8) and enter into Table 8-9.

(9) Calculate the water/oil production ratio, WOR, as follows for each of the S_w points and enter into Table 8-9:

$$WOR = \frac{B_o}{(1/f_w) - 1} \tag{8-9}$$

(10) Calculate the cumulative water injected, W_i, to each of the points as follows and enter into Table 8-9:

$$W_i = \frac{7758\,Ah\phi}{f_w'} \tag{8-10}$$

(11) Calculate the time, t, to reach each S_w point as follows and enter into Table 8-9:

$$t = \frac{W_i}{i_w} \tag{8-11}$$

If the injection rate, i_w, is not constant throughout the life of the flood, use a time-weighted average rate.

(12) Plot WOR versus N_p on cartesian coordinate paper. Select a WOR cutoff which is acceptable (90—98%, depending on lifting costs and other expenses).

(13) Plot WOR versus time on cartesian coordinate paper. Determine the life of the flood from the WOR cutoff point.

(14) Plot WOR versus W_i on cartesian coordinate paper. Determine total water injection from the WOR cutoff point.

Hovanessian (1960) discussed the impact on the Buckley-Leverett procedures in the case of multiple lines of producers, with a single line of injectors. For the field studied, he concluded there was no difference in the total oil recovered or the water injected. There was an increase, however, in the life of the flood as the first line of wells slows down the rate of advance of the front to the following lines.

Buckley-Leverett method applied to a layered system

Roberts (1959) published a paper describing how the Buckley-Leverett procedures could be applied to a layered system. Even with the modifications to the basic Buckley-Leverett procedure, the calculations could still be made with a desk calculator. As the value of the mobility ratio moves away from one, however, the results become more approximate. Snyder and Ramey (1967) presented a more complex method for applying the Buckley-Leverett method to a non-communicating layered system. Their method, however, requires the use of a digital computer. A short description of their approach is presented following the discussion of Robert's (1959) method. The following discussion is based on Roberts' method; however, certain modifications have been made along lines suggested in an Advanced Reservoir Engineering short course offered at Texas A&M University.

Assumptions for layered Buckley-Leverett method

(1) The assumptions for the basic Buckley-Leverett technique hold for each of the layers.

(2) The reservoir can be represented as a series of layers.

(3) Water enters each layer in direct proportion to its capacity, *kh*.

(4) There is no crossflow between layers.

Procedure for layered Buckley-Leverett method

The steps in the method described by Roberts (1959) consist of calculating the performance of each layer by using the Buckley-Leverett method and then summing the recoveries from the different layers as water breaks through each layer. The first six steps of the Roberts' procedure are identical to those previously presented for the basic Buckley-Leverett technique. The remainder of the steps can be summarized as follows:

(7) Segregate the core analysis data into permeability groups or layers and determine the average permeability, k, average porosity, ϕ, and average thickness, h, for each layer.

(8) For each layer, calculate the capacity, kh, and percent of total capacity.

(9) Calculate the injection rate, i_l, into each layer in bbl/day:

$$i_l = i_{total} \times (\% \text{ capacity}) \tag{8-12}$$

(10) Calculate, the cumulative water injection, W_{il}, in bbl, into each layer to each S_w point as follows:

$$W_{il} = \frac{7758\, A_l h_l \phi_l}{f_w'} \tag{8-13}$$

where A_l = areal extent of layer l, acres; h_l = average thickness of layer l, ft; and ϕ_l = average porosity of layer l, fraction.

(11) Calculate q_{ol} and q_{wl} in bbl/day for each layer to each S_w point as follows —

Before breakthrough:

$$q_{ol} = \frac{i_l}{B_o} \tag{8-14}$$

$$q_{wl} = 0 \tag{8-15}$$

After breakthrough:

$$q_{ol} = \frac{i_l}{B_o} (1 - f_w) \tag{8-16}$$

$$q_{wl} = i_l (f_w) \tag{8-17}$$

(12) Calculate the recovery, N_{pl}, and the time, t_l, at breakthrough for each layer using the following equations:

$$N_{pl} = 7758 \phi_l A_l h_l \frac{(S_{wa} - S_{wi})}{B_o} \tag{8-18}$$

where S_{wa} = average water saturation at t_l, fraction; and S_{wi} = water saturation at start of flood, fraction.

Time, t_l, at breakthrough is equal to:

$$t_l = \frac{W_{il}}{i_l} \tag{8-19}$$

(13) Calculate the recovery, N_p, and the time, t_l, to each S_w point for each layer using equations (8-18) and (8-19), respectively.

(14) Calculate the recovery in all tracts of lower permeability at breakthrough of each layer using the following equation:

$$N_{pl_2} = \frac{(N_p)_{BTl_2}}{(W_i)_{BTl_2}} (i_{l_2} t_{l_1}) \tag{8-20}$$

(15) Plot the oil production rate for each layer versus time on a cartesian coordinate graph. Determine the total oil production rate as a function of time, and plot.

(16) Plot the water production rate for each layer versus time on a cartesian coordinate graph. Determine the total water production rate as a function of time, and plot.

(17) Using the total oil and water production rates, calculate the water/oil ratio as a function of time. Plot total oil, total water, and WOR versus time on a cartesian coordinate graph.

(18) Determine the life of the flood by choosing an appropriate WOR cutoff point based on lifting costs and other expenses.

(19) Plot the recovery from each layer as a function of time on a carte-

sian coordinate graph. Determine the total recovery as a function of time, and plot.

(20) Determine the project's ultimate recovery using the *WOR*—time cuttoff from step 16.

Snyder-Ramey assumptions

The main differences between the Snyder-Ramey (1967) approach and the modified Roberts' method are given here.

(1) Different initial saturations, residual saturations, and relative permeability—saturation relationships can be used for each layer.

(2) Each layer is divided into cells, and calculations are carried from cell to cell (this permits close observation of the movement of the front in each layer).

(3) Injectivity into each layer is controlled by the resistances offered by the cells in series, which vary as a function of time.

The decision to use either the modified Roberts' or the Snyder-Ramey method must be made on the basis of the quality of the input data and the time available for the analysis.

Double Buckley-Leverett technique

The double Buckley-Leverett method of waterflood design provides for the analysis of the advance of a distinct interface behind the water—oil interface as shown in Fig. 8-23. The second interface moves at the same time as the water—oil interface, but at a different rate. The second front is as-

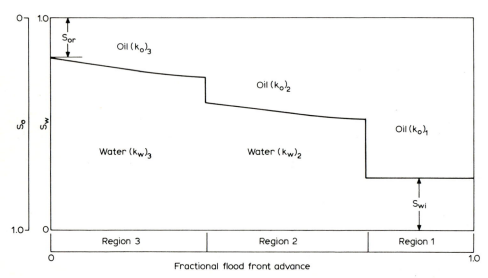

Fig. 8-23. Partial frontal displacement with two fronts and two movable phases behind each front.

sociated with improved waterflooding techniques involving the use of chemicals which tend to be adsorbed by the reservoir rock and alter the rock surface and/or fluid properties. The second front also occurs upon injection of hot water; in this case the heat interface travels slower than the water—oil interface. The following discussion is based on a lecture by C.E. Johnson presented at the University of Southern California in 1962.

Assumptions for double Buckley-Leverett method in the case of hot water injection

(1) The assumptions for the basic Buckley-Leverett technique also hold for the double Buckley-Leverett method.

(2) Temperature of the hot water is higher than that of the reservoir, and, as the thermal energy is transferred to the rock matrix, it is lowered to reservoir temperature.

(3) Behind the temperature front (region *3* in Fig. 8-23) the reservoir fluids and rocks are heated up to the temperature of the injected hot water (allowance is made for wellbore heat losses).

(4) There is no vertical heat loss from the reservoir. If included, the temperature front would advance slower.

(5) The rate of advance of the temperature front, with respect to the first oil—water front, is constant dependent upon the heat capacity of the reservoir rock.

(6) The increase in temperature results in an increase in the mobility of the oil and leads to more efficient displacement by the water.

Procedure for double Buckley-Leverett method in the case of hot water injection

(1) Determine the heat (in Btu), H_g, required to heat up the reservoir rock to the temperature of the injected water:

$$H_g = V_b (1 - \phi) \rho_r C_r (\Delta T) \tag{8-21}$$

where V_b = bulk volume, ft^3; $(1 - \phi)$ = fraction of bulk volume occupied by the reservoir rock; ρ_r = density or rock, lb/ft^3; C_r = heat capacity of rock, Btu/lb °F; and ΔT = temperature change, °F.

(2) Determine the pore volumes of hot water, PV_s, required to heat the reservoir rock:

$$PV_s = \frac{(1 - \phi)}{(\phi)} \frac{(\rho_r)}{(\rho_w)} \frac{(C_r)}{(C_w)} \tag{8-22}$$

where ϕ = porosity, fraction; ρ_w = density of water, lb/ft^3; and C_w = heat capacity of water, Btu/lb °F.

(3) Determine the pore volumes of hot water, PV_p, required to heat the reservoir rock and fill the pores:

$$PV_p = PV_s + 1 \tag{8-23}$$

(4) Calculate the ratio, r_v, of the velocity of the hot front to the velocity of the cold front:

$$r_v = \frac{1}{PV_p} \tag{8-24}$$

(5) Solve the general Buckley-Leverett equation for the reservoir saturation at water breakthrough for a waterflood at reservoir temperature (steps 1 to 4, p. 277). The slope at this point equals the relative velocity of the cold water front at breakthrough. Calculate the volume of injected water, W_i, using equation (8-10).

(6) Calculate the location of the heat front, in relationship to the location of the water front:

$$\frac{X}{L} = W_i r_v \tag{8-25}$$

where L = distance cold water has traveled, and X = distance hot water has traveled. Note that at breakthrough of cold water, L is equal to the distance from the injector to the producer and the distance X may be solved for directly. Or, by considering volumes of fluid injected, the distance traveled by the hot front may be calculated for any volume of total fluid injected.

(7) Calculate the waterflood recovery using the basic Buckley-Leverett procedures (steps 1 to 14, p. 277).

(8) Correct the oil recovery to account for additional oil recovered by the heat front as follows:

(a) Assume k_w/k_o is constant and that only μ_w/μ_o changes for the temperature changes to be considered.

(b) Calculate the waterflood response using the basic Buckley-Leverett procedures (steps 1 to 14, p. 277). In step 2, use the fluid viscosities corresponding to the temperature of the hot water injected. In step 7 use a formation volume factor for hot oil (take into consideration the expansion of the crude oil due to the higher injection temperature of the water).

(c) Plot oil recovery versus water injected in the following manner:

1. Use cold-front recoveries until breakthrough of cold front.
2. Plot hot-front recoveries after breakthrough of hot front.
3. Show a gradual transition from cold-front recoveries to hot-front recoveries (straight line between the two breakthrough recoveries).
4. Select the maximum producible WOR that is acceptable and calculate the water injected to reach this WOR. Obtain the value for oil recovered from the WOR versus N_p plot.

Assumptions for double Buckley-Leverett method in the case of improved waterfloods using chemicals

(1) The detergent or surfactant is soluble in water but not in oil.

(2) The detergent affects both the permeability and viscosity ratios.

(3) A certain percentage of detergent adheres to the rock surface, whereas the remainder stays in solution.

(4) The relative velocity, r_v, of the adsorbed additive to that of the water is given by:

$$r_v \approx \frac{c_w}{c_w + c_a} \approx \frac{1}{1 + (c_a/c_w)} \tag{8-26}$$

where c_w = concentration of additive in the water, lb/bbl; and c_a = concentration of additive adsorbed, lb/bbl.

Procedure of double Buckley-Leverett method for improved waterfloods using chemicals

(1) Determine the relative velocity of the additive to that of the water:

$$r_v = \frac{1}{1 + [(1 - \phi)/\phi](\rho_r/\rho_w)(A/c_d)} \tag{8-27}$$

where A = amount of additive adsorbed per unit weight of rock, lb/lb, or lb additive/bbl water; and c_d = concentration of additive dissolved in water, lb/lb, or lb/bbl.

(2) The remainder of the calculations to determine the oil recovery are similar to those discussed for hot-water floods (steps 5 through 8, p. 284).

Dykstra-Parsons method

In the Dykstra-Parsons (1950) technique, an oil reservoir is visualized as a layered system, and recovery is calculated taking into consideration the permeability variation of this layered system and the mobility ratio.

Assumptions for Dykstra-Parsons method

(1) The reservoir consists of isolated layers of equal thickness having uniform permeability with no cross flow between layers.

(2) Piston-like displacement occurs; only one phase is flowing in any given volume element.

(3) There is a linear and steady-state flow.

(4) The fluids are incompressible; there are no transient pressure effects.

(5) The pressure drop across every layer is the same.

(6) Fill-up occurs in all layers prior to flood response. The flood life should be increased to allow for the fill-up period.

(7) Excerpt for absolute permeability, the rock and fluid properties are the same for all layers.

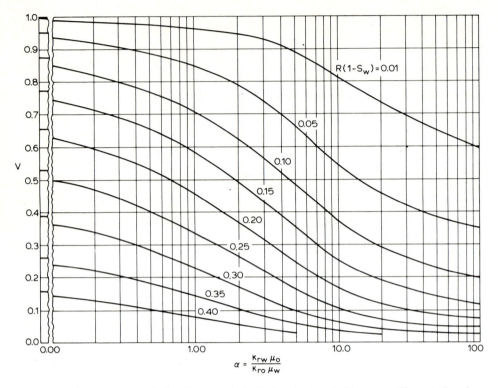

Fig. 8-24. Permeability variation (V or V_k) plotted against mobility ratio (M or α) showing lines of constant $[R(1 - S_w)]$ for a producing water/oil ratio of 1. (After Johnson, 1956; courtesy of the SPE of AIME.)

Procedure for Dykstra-Parsons method

The steps for calculating recovery with the aid of the coverage charts (Figs. 8-24 through 8-27) are as follows:

(1) Assemble permeability data in descending order. Develop cumulative frequency distribution for the permeability values. Convert this frequency distribution to a cumulative probability distribution.

(2) Plot the cumulative probability values versus log of permeability on probability paper, and draw a straight line (best fit) through the data points. Calculate permeability variation, V_k, using values from the straight line:

$$V_k = \frac{k_{50} - k_{84.1}}{k_{50}} \tag{8-28}$$

where k_{50} = the median permeability with 60% of the permeability values being greater than or equal to it, md; and $k_{84.1}$ = the permeability with 84.1% of the permeability values being greater than or equal to it, md.

(3) Calculate mobility ratio, M:

$$M = \frac{k_w}{\mu_w} \frac{\mu_o}{k_o} \qquad (8\text{-}29)$$

(4) From Johnson's (1956) charts (Figs. 8-24 through 8-27) determine the coverage, R, for a WOR equal to 1, 5, 25, and 100.

(5) Calculate the oil recovery, N_p, from the following equation:

$$N_p = \frac{7758\, Ah\phi R\, (S_{oi} - S_{or})}{B_o} \qquad (8\text{-}30)$$

where S_{oi} = initial oil saturation, fraction; and S_{or} = residual oil saturation, fraction.

(6) Plot N_p versus WOR. Decide on an acceptable WOR cutoff point and read the recovery from the graph.

(7) Integrate N_p-WOR curve graphically to get the volume of produced water, W_p.

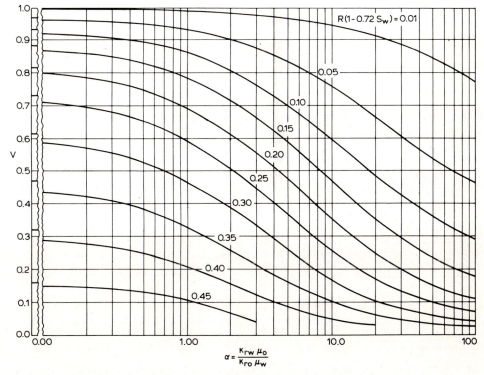

Fig. 8-25. Permeability variation (V or V_k) plotted against mobility ratio (M or α) showing lines of constant $[R(1 - 0.72S_w)]$ for a producing water/oil ratio of 5. (After Johnson, 1956; courtesy of the SPE of AIME.)

(8) Calculate the water injected, W_i:

$$W_i = N_p B_o + W_p \tag{8-31}$$

(9) Life, t, in years is given by:

$$t = \frac{W_i}{i_w \times 365} \tag{8-32}$$

where i_w = daily injection rate, bbl/day.

(10) Calculate oil production rate by dividing the differences in recovery by the corresponding differences in time.

Stiles method

In the Stiles (1949) method, an oil reservoir is visualized as a layered reservoir with each layer having a different permeability. Table 8-10 presents the steps in the Stiles method. The procedure steps presented here are based

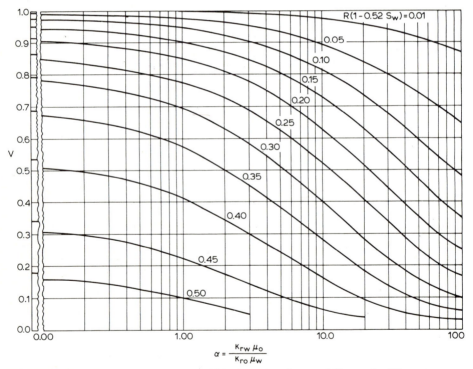

Fig. 8-26. Permeability variation (V or V_k) plotted against mobility ratio (M or α) showing lines of constant $[R(1 - 0.52S_w)]$ for a producing water/oil ratio of 25. (After Johnson, 1956; courtesy of the SPE of AIME.)

TABLE 8-10

Recommended steps in using the Stiles (1949) waterflood predictive technique[a]

(1)	Cumulative thickness	h
(2)	Fractional thickness	h_f
(3)	Permeability	k_i
(4)	Dimensionless permeability	K_i
(5)	Fractional permeability	F_p
(6)	Capacity	C_c
(7)	$(1-C_c)$	
(8)	Coverage	C_e
(9)	WOR	
(10)	Oil recovery	N_p
(11)	Water injected	W_i

[a] Must be presented in tabular form.

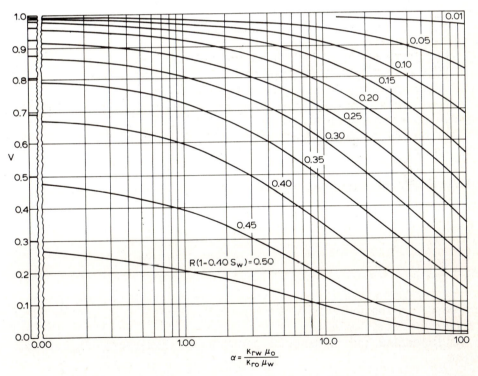

Fig. 8-27. Permeability variation (V or V_k) plotted against mobility ratio (M or α) showing lines of constant $[R(1 - 0.40S_w)]$ for a producing water/oil ratio of 100. (After Johnson, 1956; courtesy of the SPE of AIME.)

on a lecture by C.E. Johnson in a course on Secondary Recovery at the University of Southern California in 1962.

Assumptions for Stiles method
(1) Linear and steady-state flow occurs.
(2) The reservoir is composed of isolated homogeneous layers of equal thickness with varying permeabilities; there is no crossflow between layers.
(3) Except for absolute permeability, the rock and fluid properties are the same for all layers.
(4) The rates of oil or water production are proportional to the quantity of water injected. All fluid movement is pistonlike.
(5) Fluids are immiscible and incompressible; the pressure drop across each layer is the same.
(6) The distance of flood-front penetration for each layer is proportional to its permeability.
(7) Oil saturation in an unswept portion of a reservoir is unaffected by the flood front.
(8) Fill-up occurs in all layers prior to flood response. The flood life should be increased to allow for the fill-up period.

Procedure for Stiles method
(1) Prepare a profile of the absolute permeabilities versus depth.
(2) Divide the permeability profile into layers of equal thicknesses and select a representative permeability for each layer.
(3) Arrange the representative permeabilities in descending order with the highest value first and the lowest last.
(4) Calculate the cumulative thickness, h, and fractional thickness, h_f, for the layers.
(5) Determine the dimensionless permeability, K_i, for each layer and sum of all dimensionless permeabilities, K_s:

$$K_i = \frac{k_i}{k_{av}} \qquad (8\text{-}33)$$

$$K_s = \sum_{1}^{n} K_i \qquad (8\text{-}34)$$

where k_i = permeability to water for a layer, md; and k_{av} = average water permeability of all layers, md.
(6) Determine the incremental permeability or capacity, F_{pi}, of each layer:

$$F_{pi} = \frac{K_i}{K_s} \qquad (8\text{-}35)$$

(7) Determine the cumulative capacity, C_c:

$$C_c = \sum_1^n F_{pi} \qquad (8\text{-}36)$$

(8) Calculate the coverage, C_e:

$$C_e = \frac{K_i h_f + 1 - C_c}{K_i} \qquad (8\text{-}37)$$

(9) Calculate the water/oil ratio, WOR:

$$WOR = \frac{C_c}{(1 - C_c)} MB_o \qquad (8\text{-}38)$$

where M = mobility ratio, and B_o = oil formation volume factor, bbl/STB.
   (10) Calculate the recovery N_p:

$$N_p = \frac{7758\, Ah\phi C_e\, (S_{oi} - S_{or})}{B_o} \qquad (8\text{-}39)$$

where S_{oi} = initial oil saturation, fraction; and S_{or} = residual oil saturation (flood pot tests), fraction.
   (11) Pilot water/oil ratio versus net oil recovery, decide upon an acceptable WOR cutoff point, and read the anticipated oil recovery from the graph.
   (12) Calculate the amount of cumulative water injected, W_i, and life, t, of project following steps 7 to 9, p. 287.

CARBONATE RESERVOIR WATERFLOOD PREDICTIONS AND PERFORMANCE

Introduction

Carbonate reservoirs can be separated into three broad categories based on their porosity systems: (1) intercrystalline—intergranular, (2) fracture—matrix, and (3) vugular—solution. Each type of porosity system presents a different set of factors which must be considered in the design of a waterflood project. Prediction methods presented in previous sections may, therefore, require some modification. To increase the usability of the available reservoir performance data, they also are discussed separately for each porosity system.

Intercrystalline—intergranular porosity systems

Because of the similarity in distribution and movement of fluids within sandstone and carbonate rocks having intercrystalline—intergranular porosity, predictive techniques developed and used successfully for sandstone reservoirs can often be directly applied to this type of carbonate reservoir. Selection of the predictive technique depends upon the characteristics to be modeled and the available reservoir data. In a reservoir having distinct layers, the Dykstra-Parsons, Stiles, or layered Buckley-Leverett prediction technique can be used. If the mobility ratio is significantly greater than one, the Dykstra-Parsons or layered Buckley-Leverett technique is preferred. In a carbonate reservoir with a variety of permeabilities but no true layering, the Stiles method is suggested. The Buckley-Leverett technique gives excellent results in those cases where the interval under consideration has little permeability variation. If sufficient data are available and several techniques appear to be applicable, it is good practice to use at least two approaches for a comparison of predictions.

Prediction methods and comparison with actual performance

Goolsby and Anderson (1964) used both the Stiles and Dykstra-Parsons techniques in evaluating a five-spot pilot waterflood in the McElroy Field, Texas. The predicted water/oil ratios were much greater than the actual ones. The core permeability variation was calculated to be near 0.85, but the available injectivity profiles showed no indication of formation stratification in the reservoir. The Buckley-Leverett technique using hypothetical relative permeability curves gave a good approximation to the actual performance. Calhoun et al. (1950) also noted that efficient waterflooding is possible in a number of areas in the midcontinent despite large permeability variations within the pay zone, because the highly permeable streaks were not continuous between wells.

Henry and Moring (1967) used a modification of the Stiles technique to check the performance of a pilot waterflood in the Panhandle Field, Texas. They investigated the position of the flood front in relation to the producer for a five-spot pilot. Assuming a radial disposal of the injected water and knowing the volume of water injected, the distance of water movement in each layer was calculated using the following equation:

$$W_i = [\pi\phi \left(1 - S_{or} - S_{gr} - S_{wi}\right) C_f] \left(c^2 \sum_{i=1}^{i=n} k_i^2\right) \tag{8-40}$$

where W_i = cumulative water injected, bbl; ϕ = porosity, fraction; S_{or} = residual oil saturation, fraction; S_{gr} = residual gas saturation, fraction; S_{wi} = interstitial water saturation, fraction; C_f = conformance factor (correction factor for water lost outside flood interval, i.e., water bypassing part of reservoir); c = proportionality constant relating the radial distance for the advance of water in a layer to the permeability of the layer; k_i = permeability of a layer, md; and n = number of layers.

After substitution of values for S_{or}, S_{gr}, S_{wi}, C_f, and k_i, equation (8-40) can be reduced to:

$$W_i = \text{constant} \cdot c^2 \qquad (8-41)$$

Thus, knowing the amount of water injected at any point in time, c can be calculated.

The distance, r, the flood front moves in a radial system is related to the permeability of each layer by the following equation:

$$r_i = ck_i \qquad (8-42)$$

After determining c, r can be calculated for each layer. The calculated values of r can then be compared to the distance from the injector to the producer, and the number of layers which should have been watered out can be determined.

With use of the Stiles equation for the fractional flow of water, the theoretical well performance can be checked against the actual:

$$f_w = \frac{(kh)_w \, (B_o \mu_o k_{rw} / \mu_w k_{ro})}{(kh)_w (B_o \mu_o k_{rw} / \mu_w k_{ro}) + (kh)_o} \qquad (8-43)$$

where f_w = surface water cut, fraction; $(kh)_w$ = capacity of watered-out layers, md-ft; $(kh)_o$ = capacity of oil producing layers, md-ft; B_o = oil formation volume factor, bbl/STB; μ_o = oil viscosity, cp; μ_w = water viscosity, cp; k_{ro} = relative permeability to oil at connate water saturation; k_{rw} = relative permeability to water at residual oil saturation.

In a review of the performance of the pilot, calculations indicated that the producer should have been producing some water. Inasmuch as this was not the case, the well was fractured to overcome suspected skin damage. The well produced 10 bbl/day oil and no water before the treatment and 35 bbl/day oil and 12 bbl/day water after the treatment (Henry and Moring, 1967).

Abernathy (1964) compared the performance of pilot waterfloods in three carbonate reservoirs, with intercrystalline—intergranular porosity, using several prediction methods: (1) Stiles (1949), (2) Craig-Geffen-Morse (1955), (3) Craig-Stiles (Abernathy, 1964), and (4) Band (Hendrickson, 1961). The assumptions made in using these techniques are presented here.

A. Assumptions common to all multilayered methods

(1) The formation is considered to be composed of a number of strata, continuous from well to well and insulated from crossflow between wells.

(2) There is no segregation of fluids owing to gravity within any layer.

(3) A high water saturation zone does not exist which could permit bypassing of the injected water.

B. Stiles method assumptions

(1) The displacement occurs in a pistonlike manner.

(2) Fill-up occurs in all layers before oil production, resulting from the flood, begins in any layer.

(3) Sweep efficiency is constant after breakthrough.

(4) Other than specific permeability, all layers have the same properties.

C. Craig-Stiles method (multilayer) assumptions

(1) The information available from laboratory model studies on five-spot waterflooding applies (immiscible displacement) (Craig et al., 1955).

(2) Frontal advance theory applies.

(3) Each layer can have different absolute permeability, relative permeability, porosity, etc.

(4) Areas ahead of the flood front are resaturated with oil.

(5) After breakthrough, water injection is equal to total fluid production.

D. Band method assumptions

(1) The information available from laboratory model flow studies on five-spot waterflooding applies (miscible displacement) (Habermann, 1960).

(2) Frontal advance theory applies.

(3) Ten bands (layers) of equal pore volume, but having different capacities, are used.

E. Craig-Geffen-Morse method (single-layer) assumptions

(1) The formation can be considered to be composed of a single layer.

(2) The information available from laboratory model studies on five-spot waterflooding applies (immiscible displacements) (Craig et al., 1955).

(3) Frontal advance theory applies.

(4) Areas ahead of the flood front are resaturated with oil.

(5) After breakthrough, water injection is equal to total fluid production.

A summary of the reservoir data for the three fields studied by Abernathy (1964) and a comparison of the actual recoveries with the predictions are presented in Table 8-11. Figs. 8-28 to 8-30 provide a comparison of the actual results with the predictions (Abernathy, 1964). The Craig-Stiles prediction most closely matched the actual performance for all three reservoirs.

The steps for the Craig et al. (1955), Craig-Stiles (Abernathy, 1964), and Band (Hendrickson, 1961) methods are summarized below.

A. Craig-Geffen-Morse method procedures (Craig et al., 1955)

(1) Follow steps 1 to 6 from the Buckley-Leverett prediction procedure (p. 277).

(2) Determine k_{rw} at the average saturation behind the front at breakthrough, $(S_{wa})_{bt}$.

(3) Determine k_{ro} at the oil saturation ahead of the flood front.

(4) Calculate the mobility ratio, M:

$$M = \frac{k_{rw}\mu_o}{k_{ro}\mu_w} \qquad (8\text{-}44)$$

(5) Read the areal sweep efficiency at breakthrough, $(E_{as})_{bt}$, in percent from Fig. 8-31.

(6) Determine the manner in which E_{as} changes with continued water injection using Fig. 8-32. Construct a straight line parallel to the other lines

TABLE 8-11

Reservoir data for selected carbonate fields (after Abernathy, 1964; courtesy of the SPE of AIME)

Field and location	Panhandle, Texas	Foster, Texas	Welcha, Texas
Formation and age	Brown and White Dolomite, Permian	Grayburg-Brown Dolomite, Permian	San Andres Dolomite, Permian
Average depth, ft	3000	4200	4950
Net pay thickness, ft	65	129	75
Average porosity, %	11.7	8.6	10
Average permeability, md	7.2	2.6	6.3
Connate water saturation, %	39	23	23
FVF[a], bbl/STB	1.044	1.061 at 244 psi	1.088 at 325 psi
Oil viscosity, cp	2.33	2.91 at 244 psi	2.32 at 325 psi
Water viscosity, cp	0.83	0.80	0.82
Stock tank oil gravity, °API	40	35	34.4
BHP[a] at start of flood, psi	—	244	325
Primary recovery at start of flood, % oil-in-place	25.6 (includes effect of gas injection)	12	2.9
Current actual recovery, % oil-in-place	34.2 (98% water cut)	21 (55% water cut)	21.7 (45% water cut)
Craig-Stiles method (layered), % oil-in-place	35 (98% water cut)	28 (98% water cut)	38 (90% water cut)
Stiles method, % oil-in-place	—	31 (98% water cut)	29 (90% water cut)
	—	28 (55% water cut)	17.5 (45% water cut)
Band method, % oil-in-place	—	—	29 (65% water cut)
	—	—	21 (45% water cut)
Craig et al. method (single layer), % oil-in-place	37.5 (98% water cut)	32.5 (98% water cut)	—

[a] FVF = formation volume factor; BHP = bottomhole pressure.

through the value $(E_{as})_{bt}$ at $W_i/(W_i)_{bt}$ equal to 1.0, where W_i = volume of cumulative water injected and $(W_i)_{bt}$ = volume of water injected to break-through.

(7) Follow the steps outlined in Table 8-12. The values for the columns in the table can be determined as follows:

    1. Select $W_i/(W_i)_{bt}$ initially as 1.00, then add maximum increments of 0.10. Smaller incrementals yield more accurate results.

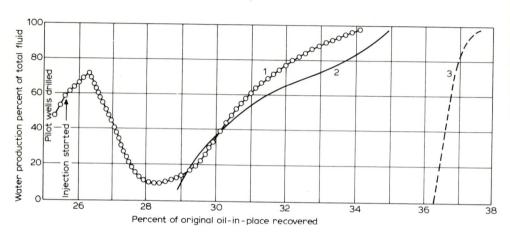

Fig. 8-28. Calculated versus actual performance of the Panhandle Field, Texas. Curve *1* = actual; curve *2* = calculated by the Craig-Stiles technique; and curve *3* = calculated using the Craig et al. method. (After Abernathy, 1964; courtesy of the SPE of AIME.)

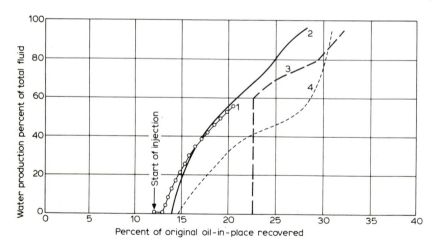

Fig. 8-29. Calculated versus actual performance of the Foster Field pilot in Texas. Curve *1* = actual; curve *2* = calculated by the Craig-Stiles technique; curve *3* = calculated by Craig et al. method; and curve *4* = calculated by using the Stiles method. (After Abernathy, 1964; courtesy of the SPE of AIME.)

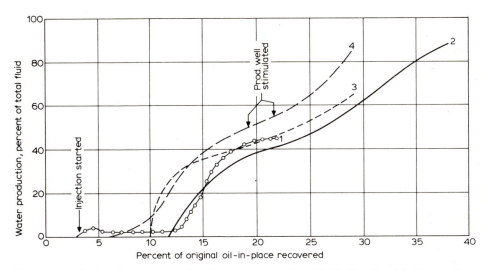

Fig. 8-30. Calculated versus actual performance of Welch Field pilot flood in Texas. Curve *1* = actual; curve *2* = calculated by Craig-Stiles technique; curve *3* = calculated by using the Band method; and curve *4* = calculated by Stiles method. (After Abernathy, 1964; courtesy of the SPE of AIME.)

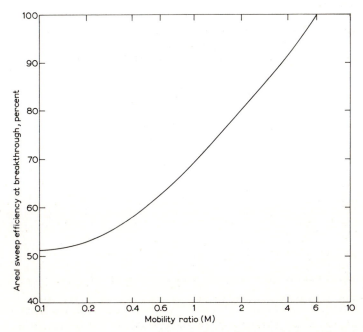

Fig. 8-31. Areal sweep efficiency at breakthrough for a five-spot well pattern. (After Craig et al., 1955, fig. 2, p. 9; courtesy of the SPE of AIME.) Mobility ratio is defined as $k_{ro}\mu_w/k_{rw}\mu_o$.

2. $(W_i)_{bt} = [(S_{wa})_{bt} - S_{wi}](E_{as})_{bt}(PV)$ (8-45)

where PV = pore volume of fice-spot pattern, bbl; $(S_{wa})_{bt}$ = average water saturation behind flood front at breakthrough, fraction; S_{wi} = initial water saturation, fraction.

3. (col. 1) × (col. 2).

4. Difference between present value of W_i and that found in previous step. First value is not determined.

5. Read values of E_{as} for values of col. 1 from Fig. 8-31.

6. (col. 4) ÷ [(col. 5) × (PV)]

7. First value = $(S_{wa})_{bt} - S_{wi}$. Subsequent values = (col. 6) + previous value.

8. 1.0 ÷ (col. 7).

9. Develop f_w versus S_w and f_w' versus S_w curves using steps 1 to 6,

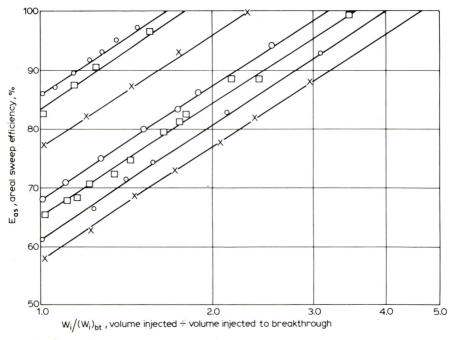

Fig. 8-32. Increase in areal sweep efficiency after breakthrough. (After Craig et al., 1955, fig. 4, p. 10; courtesy of the SPE of AIME.)

p. 277, for the basic Buckley-Leverett technique. Using the value of f_w' from col. 8, determine water saturation at the producing end of the invaded portion, S_{w2}.

10. Using the value of S_{w2} from col. 9, determine f_w from the f_w versus S_w curve.

11. $1.0 - (\text{col. } 10)$.

12. Difference between present and previous value of E_{as}. First value is not determined (blank).

13. $N_n = (\text{col. } 12) \times [(S_w)_{bt} - S_{wi}](PV)$, where $(S_w)_{bt}$ = the water saturation at the front at breakthrough, fraction; and N_n = incremental oil produced from newley invaded region, bbl.

14. $(\text{col. } 4) - (\text{col. } 13)$. First value is blank. N_p = incremental oil produced from previously invaded region, bbl; and W_p = incremental water produced from previously invaded region, bbl.

15. $(\text{col. } 11) \times (\text{col. } 14)$. First value is blank.

16. $(\text{col. } 14) - (\text{col. } 15)$. First value is blank.

17. $(\text{col. } 13) + (\text{col. } 15)$. First value is blank.

18. $(\text{col. } 17) \div B_o$. First value is blank.

19. $(\text{col. } 16) \div (\text{col. } 18)$. First value is blank.

20. First value = $\{[(S_{wa})_{bt} - S_{wi}] (E_{as})_{bt} - (S_g)_{ai}\}(PV)/B_o$, where $(S_g)_{ai}$ = the average initial gas saturation, fraction. Subsequent values = $(\text{col. } 18)$ + previous value of $(\text{col. } 20)$.

21. $(\text{col. } 3) \div$ average injection rate, i_w. Plot $(\text{col. } 20)$ versus $(\text{col. } 21)$.

22. Slope of curve of $(\text{col. } 20)$ versus $(\text{col. } 21)$.

(8) Plot $(\text{col. } 20)$ versus $(\text{col. } 19)$. Choose an acceptable *WOR* cutoff and read the oil recovery from the curve.

B. Craig-Stiles method procedures. Abernathy (1964) presented the layered Craig-Stiles prediction method. A Craig et al. (1955) prediction is made for each layer separately, and the results are superimposed to generate the overall reservoir response. The layers are chosen following the Stiles (1949) system. In order to determine the relative injection rate into each layer as a function of time, the five-spot conductance ratio data developed by Caudle and Witte (1959) were used (Fig. 8-33). The conductance ratio, γ, is the ratio of the water injection rate to the injection rate calculated by Muskat's (1950a) single fluid five-spot equation presented below (using the permeability to oil and oil viscosity):

$$ibase = \frac{3.541\, hkk_{ro}\Delta p}{\mu_o\, [\ln(d/r_w) - 0.619]} \tag{8-46}$$

where $ibase$ = oil injection rate, bbl/day; h = thickness of layer, ft; k = absolute permeability, md; k_{ro} = relative permeability to oil at initial conditions of saturation; Δp = pressure drop from the injection rock face to the producer rock face, psi; μ_o = viscosity of oil, cp; d = distance between injector and producer, ft; r_w = wellbore radius, ft.

The time necessary to inject a given volume of water can be determined for each layer independently. The results at specific time intervals are determined by summing the volumes of injected water for all the layers. This

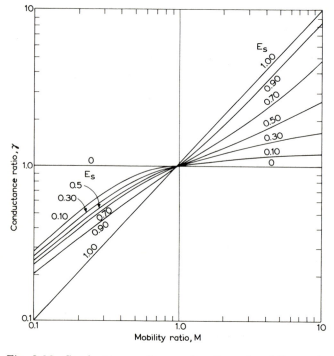

Fig. 8-33. Conductance ratio as a function of mobility ratio and the area swept, E_s (or E_{as}), for a five-spot well pattern. (After Caudle and Witte, 1959; courtesy of the SPE of AIME.)

procedure replaces steps 21 and 22 in Table 8-12. The following steps should be used for each layer to replace step 21.

(1) Using the calculated mobility ratio (equation 8-44), determine the conductance ratio, γ, as a function of E_{as} using Fig. 8-33.

(2) Convert the conductance ratio, γ, to a water injection rate, i_w, using the following equation:

$$i_w = \gamma(ibase) \tag{8-47}$$

This relationship holds when it is assumed that the pressure drop from the injector rock face to the producer rock face is constant during the life of the flood.

(3) Plot i_w versus W_i. Determine the average injection rate graphically.

(4) Determine time required using data from col. 4 of the Craig et al. (1955) procedure. First value is determined from $(W_i)_{bt}$ and initial i_w.

C. Band method. The Band method (Hendrickson, 1961) is a modification of the Craig et al. (1955) technique proposed by Hendrickson (1961). Ten bands or layers of equal volume are used in the analysis. The performance of

TABLE 8-12

Recommended steps in using the Craig et al. (1955) waterflood predictive technique for the five-spot water injection pattern, to be presented in tabular form

(1)	$W_i/(W_i)_{bt}$, ratio
(2)	$(W_i)_{bt}$, reservoir bbl
(3)	W_i, reservoir bbl
(4)	ΔW_i, reservoir bbl
(5)	E_{as}, fraction
(6)	ΔQ_i, pore volume
(7)	Q_i, pore volume
(8)	f'_w
(9)	S_{w2}, fraction
(10)	f_{w2}, fraction
(11)	f_{o2}, fraction
(12)	ΔE_{as}, fraction
(13)	N_n, reservoir bbl
(14)	$N_p + W_p$, reservoir bbl
(15)	N_p, reservoir bbl
(16)	W_p, reservoir bbl
(17)	$N_n + N_p$, reservoir bbl
(18)	Incremental oil production, STB
(19)	Average WOR over the increment, bbl/bbl
(20)	Cumulative oil recovered by waterflood, STB
(21)	Time after start of injection, days
(22)	Oil production rate after fill-up, bbl/day

each band is calculated, and the reservoir performance is determined by superimposition of the band results. The work of Habermann (1960) is used to estimate the performance after breakthrough within each band. Inasmuch as the use of the Craig-Stiles method (Abernathy, 1964) results in better agreement with the actual performance data than the use of the Band method (Hendrickson, 1961), the details of the latter method are not presented here.

Park (1965) used a combination of predictive techniques to evaluate the performance of the North Virden Scallion Field in Manitoba, Canada. Welge's (1952) method (simplification of the Buckley-Leverett technique) was used by Park to calculate displacement efficiencies versus water oil ratios. The Dykstra-Parsons technique was used to determine the permeability variation, median permeability, and the coverage or vertical sweep efficiency versus water/oil ratio. The graphs of Caudle et al. (1955) and Dyes et al. (1954) were used to determine the areal sweep efficiency as a function of the mobility and water/oil ratios. The mobility ratio used by Caudle et al. and Dyes et al. is the reciprocal of the Dykstra-Parsons mobility ratio. The three efficiencies (displacement, vertical, and areal) were combined to obtain the recoveries at different water/oil ratios. The recovery to breakthrough was calculated to be 14% of the oil-in-place at the start of the flood. The ultimate recovery was calculated to be 28.4%, as compared to the primary ultimate recovery of 12.8%. When reported in 1965 (Park, 1965), three years after the prediction, the actual performance was in close accord with that predicted.

Bleakley (1969a) reported on the evaluation system used in the Pennel and Lockout Butte Fields in Montana. The system featured computer generation of reservoir maps, calculation of reservoir volumes, and investigation of flooding patterns. The computer techniques were applicable because good core and log data were available to adequately describe the reservoirs. Data gathered during the operation of the South Pine waterflood, Montana, were also used to estimate injectivity and response time.

The model used considered the mobility ratio, vertical and horizontal displacement efficiencies, and piston-type displacement for each layer. Six layers were used. Five layers contained equal volumes of movable oil and had equal permeability capacity, kh, whereas the sixth layer contained only water to simulate the initial water cut and injected water lost to other zones.

In areas of the reservoir having good permeability, the recovery was calculated at 65—75% using a nine-spot pattern after the injection of 1.5—2.5 displaceable pore volumes. This was in close agreement with the 75% recovery observed in water-drive reservoirs and pilots on other portions of the Cedar Creek Anticline.

A production schedule for a 30-year period was determined for each reservoir. Injectivity decline owing to increased reservoir pressure was predicted for the South Pine flood. On the basis of the studies, the waterflood was started without the use of a pilot.

Waterflood performance

The McElroy Field, Texas, was pilot tested four times before full-scale flooding was started. The large number of pilots were necessary to evaluate the wide variations in reservoir characteristics and the applicability of flooding techniques. Pilot 1 was a single injection test in the area of the reservoir having good permeability. Good injectivity and banking of oil were demonstrated by this pilot test. Pilot 2 was also in the high-permeability area of the field. It was initiated to assess the susceptibility of the reservoir to pattern flooding; an irregular five-spot pattern was used. The maximum rates and recoveries were obtained at the center well (four-way push). The performance of wells subjected to a two-way push was less efficient; and the wells influenced by only one injector, one-way push, had least efficient performance.

Pilot 3 was conducted to test the susceptibility of the tight flank areas to waterflooding. Injection rates were acceptable at injection pressures of 900 to 950 psig, and the production rate was increased. Pilot 4 was started to investigate the applicability of wide-spaced patterns (160-acre). Pilot 2 used 20-acre spacing and Pilot 3 used 40-acre spacing.

Several additional operational techniques have been suggested for various carbonate reservoirs. To evaluate the performance of several injectors, tracers such as ammonium thiocyanate were added to injection water to identify which well was the source of injection water when breakthrough occurs (Park, 1965). The injector is identified by the presence or absence of the tracer. This technique is equally good for a pattern or line-drive flood.

Waterflooding in reservoirs with a primary gas cap can be undertaken as long as the loss of crude oil to the gas cap is minimized. In high-relief reservoirs, crestal gas injection in combination with peripheral water injection is the logical choice. In reservoirs having low relief, two alternatives are possible:

(1) Crestal gas injection plus peripheral water injection as used in the Haynesville Field, Louisiana-Arkansas (Akins, 1951).

(2) Water can be injected at the gas—oil contact (in low-dip reservoirs — 2° or less) along with peripheral injection. This technique has been used in the Rangely Field, Colorado (Editorial, 1964) and in the Sholem Alechem Fault Block A Unit, Oklahoma (Bleakley, 1969a). Wilson (1962) presented a method for analyzing the applicability of down-dip water displacement in the presence of a gas cap. The method requires a modification of the Dietz flow equations to handle the three mobile phases (oil, gas, and water). The criteria for stable flow in the presence of a gas cap are: (1) oil mobility, k_o/μ_o, must be greater than water mobility, k_w/μ_w, and (2) a flow rate must be greater than the critical velocities, v_c, for both the water—oil and water—gas displacements, but less than the critical velocity for the gas—oil displacement. The critical velocity may be calculated for each flowing phase from the following equation:

$$v_c = \frac{g\,(\rho_1 - \rho_2)\sin\alpha}{(\mu_1/k_1) - (\mu_2/k_2)} \tag{8-48}$$

where v_c = critical velocity, ft/day; g = gravitational constant, 32.174 ft/s^2; α = dip angle of formation, degrees; ρ_1 = density of displaced fluid, lb/ft^3; ρ_2 = density of displacing fluid, lb/ft^3; μ_1 = viscosity of displaced fluid, cp; μ_2 = viscosity of displacing fluid, cp; k_1 = permeability of displaced fluid, md; and k_2 = permeability of displacing fluid, md.

If water is injected into the up-structure region of a dipping reservoir containing a gas cap and the water—oil mobility ratio is less than one (oil more mobile than water), four distinct types of flow can occur (Wilson, 1962):

Type 1. At rates of flow ranging from zero to the critical velocity for the displacement of oil by water, fingering of water occurs along the bottom of the formation; the gas does not flow from the gas cap and oil recovery is low.

Type 2. At rates of flow lying between the critical velocities for the displacement of oil by water and gas by water, the gas does not flow from the gas cap and oil is displaced from under the gas in a uniform manner.

Type 3. At rates of flow lying between the critical velocities for the displacement of gas by water and oil by gas, both oil and gas are displaced in a uniform manner.

Type 4. At rates of flow greater than the critical rate for displacement of oil by gas, gas fingering occurs along the top of the formation faster than the stable displacement of oil by water.

A summary of waterflood performance in carbonate reservoirs with intercrystalline—intergranular porosity systems is presented in Table 8-13.

FRACTURE—MATRIX POROSITY SYSTEMS

Reservoirs with a fracture—matrix porosity system differ from those having intercrystalline—intergranular porosity in that the double porosity system strongly influences the movement of fluids. The double porosity can be the result of fractures, joints, and/or solution channels within the reservoir. The pores in the matrix are not highly interconnected and porosity cannot be correlated with permeability. Joints or fissures occurring in massive formations are commonly vertical and are attributed to relief of the tensile forces during faulting or folding. The dual porosity system gives rise to a complex reservoir performance that cannot be directly modeled by conventional predictive techniques. Prediction by analogy is also difficult. In reservoirs which are fractured or have directional permeability, recovery is dependent upon injector—producer location. Therefore, unless the two reservoirs are similarly developed and possess similar lithology variations, results ob-

TABLE 8-13

Summary of waterflood performance in carbonate reservoirs with intercrystalline—intergranular porosity systems

Field and state	Reference	Zone and age	Flood type	Recovery prim.	Recovery sec.	Status	Special problems handled
West Lisbon, Louisiana	Miller et al., 1960	Pettit A, Cretaceous	irregular pattern	14%	18% incr.	just starting when updated in 1960	extremely thin (8 ft max.) reservoir; irregular and streaky permeability
Haynesville, Louisiana—Arkansas	Akins, 1951	Pettit A and B, Cretaceous	peripheral water and crestal gas	18%	16% incr.	under flood for 4 years when reported in 1951	thin tight reservoir; multiple zone with varying degree of isolation
North Foster Unit, Texas	Gealy, 1966	Grayburg—San Andres, Permian	peripheral	—	—	under flood for 3 years when reported in 1966	limited basic data for flood evaluation
Goldsmith—Cummins, Texas	O'Briant, 1967	Grayburg—San Andres, Permian	peripheral	—	—	under flood for 3 years when reported in 1967	low permeability
Welch, Texas	(Hendrickson, 1961; Abernathy, 1964)	San Andres, Permian	5-spot	—	38% total	under flood for 5 years when reported in 1961; pilot successful	
Umm Farud, Libya	Allen et al., 1969	Dahra B and Bu Charma, Ordovician	peripheral	—	21–26% total	under flood for 2 years when reported in 1969	performance matched vs. 4-layer, 2D/2-phase simulator
Snyder, Texas	Wood et al., 1969	San Angelo, Permian	irregular pattern	—	—	under flood for 6 years when reported in 1969; secondary recovery running close to 95% of primary recovery	cooperative flood, not unitized, reservoir sensitivity to quality of injected water found very low
New Hope, Texas	(Trube, 1954; Trube and DeWitt, 1950)	Bacon, Cretaceous	peripheral	170 bbl/acre-ft	153 bbl/acre-ft	under flood for 8 years when reported in 1954	several zones of varying character
South Cowden, Texas	Fickert, 1965	Grayburg, Permian	5-spot	—	—	under flood for 10 years when reported in 1965; considered successful	directional permeability; tight reservoir
Elk Basin, Wyoming	Wayhan et al., 1969	Madison, Mississippian	peripheral	—	—	under flood for 7 years when reported in 1969	complex, heterogeneous, multi-zone reservoir; zonal separation of drive mechanisms
Waddell, Texas	Borgan et al., 1965	San Andres, Permian	peripheral bottom water	21%	21%	under flood for 6 years when reported in 1965; considered successful	aquifer underlying entire field

tained by using analogies should only be considered as general approximations.

A review of the literature shows no predictive techniques developed specifically for waterflooding reservoirs having a fracture—matrix porosity system. Computer modeling shows the most promise for performance prediction in these complex reservoirs, but these models require a considerable amount of reservoir data to be of any value. The most reliable method for the flood design and performance prediction at present is the pilot flood test.

In designing a pilot flood, however, extreme care must be taken that it is located in a part of the reservoir that is representative of the total field. If the reservoir characteristics vary in different parts of the field, it may be necessary to undertake more than one pilot test.

The anisotropic permeability and the orientation of principal axes of the fracture systems can be evaluated by multiwell interference tests or tracer tests. The analysis of pressure buildup or falloff data can help in determining the apparent permeability, completion damage, and, possibly, the nature of the primary and secondary porosity systems. These tests must, however, be used with caution, because it is possible that, with extremely low matrix porosity, the period of pressure buildup may not be long enough to obtain a true indication of the matrix pressure (Elkins, 1969). A good example of this type of performance is found in the West Edmond Field, Texas (Elkins, 1969).

The West Edmond Field is significantly different from most oil fields because its very tight matrix is substantially saturated with free gas. This was evidenced by (1) a lack of oil staining in the less permeable rocks, (2) gas evolution from cores which were not oil stained, (3) initial production of free gas with condensate from a mid-structure well, and (4) a lack of agreement between material balance and volumetric estimates of oil-in-place (Elkins, 1969).

A waterflood was initiated in the Bois d'Arc zone of the West Edmond Field in the middle of 1949 and continued to the end of 1953, when it was shut down because of a failure to increase the oil production. When the water injection was stopped, the oil production in the general area of the flood steadily increased from 700 to about 1500 bbl/day in 15 months and then leveled off at 900 to 1000 bbl/day for 4 years (Fig. 8-34). One former injector was put on production in 1956. By January of 1968, 131,500 bbl of oil were recovered from the reservoir by this well, which received more than one million bbl of water during injection. This reaction was thought to be the result of a combination of imbibition flooding, pressure-pulsing, and pressure control of the flow from the matrix to the fractures. An explanation of these methods of waterflooding follows (Elkins, 1969). Repetition of the cycle, however, was not considered economic.

When considering the application of waterflooding or any other type of

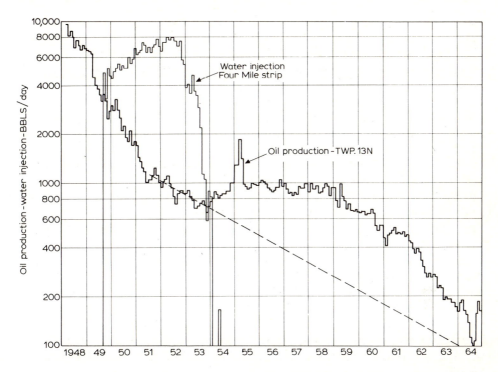

Fig. 8-34. Bois d'Arc waterflood performance, West Edmond Field, Township 13N, Texas. (After Elkins, 1969; courtesy of the SPE of AIME.)

secondary recovery process, the importance of having a complete understanding of the reservoir with a fracture—matrix porosity system cannot be overemphasized. Willingham and McCaleb (1967) investigated the failure of both a gas and a water injection project in the Cottonwood Creek Field in Wyoming. Both failures were attributed to the existence of an extensive fracture system, which caused the rapid channeling of the injected fluids. The fracture system was delineated through the analysis of cores, injected fluid movement, bottomhole pressure trends, and well performance. The properties of the rock matrix were determined from lithologic studies, log analysis, and production data. The detailed analysis provided guidance as to how the operating practices had to be revised. One major operating change involved starting a pressure-pulsing program of water injection. Initial results indicated that pressure-pulsing will increase oil production rates and improve ultimate recovery.

Imbibition flooding

The low primary oil recoveries experienced in reservoirs with a fracture—

matrix porosity system led operators to investigate secondary recovery techniques. As some of the world's largest reservoirs have fracture—matrix porosity systems, the development of a new method was imperative. Examples of fields with a double porosity system, created by the high degree of fracturing, are the Spraberry Field in Texas, the Kirkuk Field in Iraq, the Dukhan Field in Quatar, and the Masjid-i-Sulaiman and Haft-Gel Fields in Iran (Graham et al., 1959). Conventional methods of fluid injection, such as waterflooding, gas injection, or miscible flooding have limited applicability to this type of reservoir because of severe bypassing of reservoir fluids by the injected fluids. In a typical formation with a fracture—matrix porosity system, over 95% of the oil is stored in the matrix blocks.* On the other hand, the permeability of such a formation is mainly due to the presence of the fracture system. In a typical case, the permeability of the composite fracture—matrix system is more than 100 times that of the matrix alone (Graham et al., 1959). Stratified reservoirs or those having a fracture—matrix porosity system are difficult, if not impossible, to flood because of bypassing. The tendency of the injected fluid to bypass or channel through the more permeable zones is offset by the tendency of water to imbibe into the tighter zones. Imbibition may be defined as the spontaneous taking up of a liquid by a porous media (Graham et al., 1959). Common examples of this phenomenon are dry bricks soaking up water and expelling air, a blotter soaking up ink and expelling air, and reservoir rock (preferentially water-wet) soaking up water and expelling oil.

The two production mechanisms involved in recovery of oil from the matrix portion of the reservoir are displacement of oil by water flowing under (1) applied pressure gradient and (2) capillary pressure gradients (Graham et al., 1959). At high flooding rates, the applied pressure gradients tend to control the displacement process, whereas at very low injection rates, the capillary pressure gradients dominate. When the applied pressure gradient controls the displacement process, the injected water moves along the fracture system. Very little water imbibes into the matrix, and the bulk of the oil is bypassed. When the capillary pressure gradients dominate, water imbibes readily into the matrix and displaces oil to the fracture system by both displacement and countercurrent flow.

Graham and Richardson (1959) showed, through laboratory investigations, that the oil production varies directly with the oil—water interfacial tension and the square root of permeability. The rate of imbibition was found to be a complex function of the relative permeability and capillary pressure characteristics of the porous media. The presence of free gas saturation was found to decrease the rate of imbibition. Figs. 8-35 and 8-36 show

---

*Based on extensive experience, one of the writers uses the following rule of thumb: total secondary porosity (fractures, vugs, caverns, solution porosity) in carbonate rocks constitutes about 2.5%, of which 20—30% represent fracture porosity (G.V.C.).

the effects of injection rate and the fracture—matrix permeability ratio on the water/oil ratio during flood.

Important variables in the economically successful use of imbibition are (1) the fracture spacing, which must be sufficiently close so that the area provided for water imbibing at the natural rate is large enough to give an economical rate of production from the reservoir, and (2) the displacement efficiency by natural imbibition, which must be high enough to give a desirable ultimate oil recovery (Brownscombe and Dyes, 1952).

Three variations of imbibition flooding have been tested in the laboratory and in the field (primarily in the Spraberry Field, Texas). In the first variation, *low-rate injection*, part of the wells are converted to injectors. The injection rate is kept low enough to permit the capillary forces to predominate; the water is slowly and cautiously injected. In the second variation, *cyclic injection*, part of the wells are converted to injectors, and water is injected at high rates for several months and then stopped. The water production rates of the producers drop and the oil rates increase owing to the capillary forces which are acting during the period when injection is suspended. The cycle is

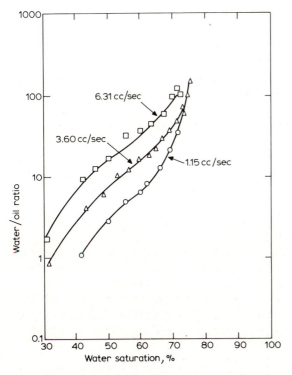

Fig. 8-35. Effect of injection rate on produced water/oil ratio for a constant fracture size. (After Graham and Richardson, 1959; courtesy of the SPE of AIME.)

repeated as long as economical oil rates are obtained. In the third variation, *pressure-pulsing*, all the wells are used as injectors and producers. All wells are first used as high-rate injectors for several months. They are then shut in for 2 to 6 months before being placed on production. The cycle is repeated as long as economical oil rates are obtained.

Low-rate injection

A pilot test of the imbibition process was conducted in the Spraberry Field by Atlantic Refining Company using low injection rates during the 1952-55 period. It showed the process to be technically feasible but economically unsuccessful owing to the low production rates. The second pilot water-flood in the Spraberry area was undertaken by Humble Oil and Refining Company in 1955 (Barfield et al., 1959). A single 80-acre, five-spot pattern and an injection rate of around 500 bbl/day per well were used. The center well produced 151,000 bbl of oil by May 1962. Two former injection wells were returned to production after the injection of about 1 million bbl of water each. They subsequently each produced about 30,000 bbl of oil. The

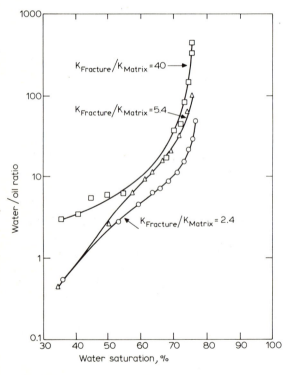

Fig. 8-36. Effect of fracture width on produced water/oil ratio for a constant injection rate. (After Graham and Richardson, 1959; courtesy of the SPE of AIME.)

pilot test also indicated a preferential movement of water parallel to the major fault trend (Barfield et al., 1959). This preferential water movement was also observed in the pilot test conducted by the Atlantic Refining Company.

A major conclusion reached on the basis of the pilot and computer simulation studies was that a relatively high sweep efficiency can be achieved with pattern flooding if operations are conducted in a manner which takes advantage of the anisotropic permeability conditions (Barfield et al., 1959). It is necessary however, that the producing wells be located on an axis midway between and parallel to the rows of injection wells which are located on the fracture trends. The injected water moves preferentially toward the adjacent injectors, and mutual interference occurs in a short time. Pressure buildup occurs after the fill-up of the fractures, and, consequently, the water must move in the direction perpendicular to the major fracture trends along a broad front.

The next flood started in the Spraberry area was initiated by Mobil Oil Company in 1959. As in the case of Humble's test, only the upper zone was flooded. Mobil's choice of injector locations was based on the fracture orientation. The injection rates were, for most of the test, around 400 bbl/day. Until October 1966, a "dump" flood operation (gravity injection) was used. In October, the injectors were equipped for pressure injection. By January 1967, the incremental oil production owing to fluid injection was equal to 83% of the primary ultimate recovery. The ultimate secondary recovery was expected to exceed the primary by the time of reaching the flood economic limit (Guidroz, 1967).

Cyclic injection

The next pilot flood was started by Sohio in April 1961 (Elkins and Skov, 1963). The injectors were placed on the fault trend in the same manner as in the preceding pilots. Early breakthrough with no oil bank ahead of the water caused a reassessment of the proposed flood. A study of the pilot response and the properties of the Spraberry area lead to the conclusion that too high an injection rate had been used and that the capillary forces were not effective. Sohio engineers postulated that stopping injection might permit the capillary forces to become dominant, and that the expansive force of the fluids and rock under pressure would act to squeeze the oil out of the rock into fractures from which it can be produced (Elkins and Skov, 1963). This proved to be correct, and "cyclic waterflooding" was born. With the cyclic process, the performance is still dependent on capillary forces, but water is actually squeezed into the rock during the injection or pressuring-up part of the cycle. Capillary forces hold the water in the rock while the expansion forces are pushing the oil out (assuming the rock is water-wet).

An evaluation of the cyclic injection process leads to the conclusion that injection equipment investment could be reduced substantially by using the same equipment in different parts of the field (Elkins and Skov, 1963). As the injection cycle was generally 2 months long, whereas the duration of the production cycle was close to 6 months, one set of injection equipment could handle three flood patterns sequentially.

In 1968, a comparison study of the Mobil and the Sohio floods was made by Sohio engineers (Elkins et al., 1968). They concluded that the long-term performances of the two floods were comparable and that lower investment and fewer operating problems were involved with the slow and steady-rate waterflooding. The oil was recovered faster with cyclic injection, but at a higher cost. Another conclusion reached was that the rate of imbibition of water into the matrix is completely inflexible and cannot be increased.

Pressure-pulsing

In 1966, a new process, "pressure-pulsing" was initiated by Pan American Petroleum Corporation (Owens and Archer, 1966). The pulsing approach appears to have several advantages over the cyclic type flooding. First, all wells may be used for water injection which should hasten fill-up and develop a more rapid and uniform increase in reservoir pressure. Second, because of increased injection pressures and rates, flooding gradients will force water into the reservoir matrix more rapidly than by imbibition alone. Third, during the pressure depletion cycle of the flood, the increased compression of the reservoir fluids and gas caused by repressuring provides energy for the displacement of the oil at a greater rate than by cyclic flooding. Fourth, during the pressure depletion cycle, all wells are used as producers resulting in a higher oil withdrawal rate than is possible with the cyclic process (Felsenthal and Ferrel, 1967).

Felsenthal, in laboratory tests on limestone and sandstone fractured blocks, noted that either conservation of reservoir gas or supplementing reservoir energy by gas injection appears to be necessary for the success of pressure-pulsing (Felsenthal and Ferrel, 1967). Energy is needed to expel the oil out while capillary action holds the water in (this is similar to the use of a gas with acid stimulation). In the laboratory tests, primary recovery was 20% of the tank oil initially in place. The first pressure-pulse cycle produced an additional 15% of the oil initially in place, but the gain in oil from pressure-pulsing dropped sharply on the following cycles.

There are also similarities between pressure-pulsing and cyclic steam injection: both need energy for oil expulsion, a soak period appears to be beneficial, and the oil recovery declines with each succeeding cycle (less oil and energy available with each succeeding cycle). This decline was also observed in the case of cyclic waterflooding.

Pressure-pulsing pilot tests have been conducted in the Grayburg Limestone reservoir in the Permian Basin, Texas (Owens and Archer, 1966), and Austin Chalk and Buda Limestone Formations of the Darst Creek and Salt Flat Fields, Texas (Hester et al., 1965). In the case of the Austin and Buda

Formations, only the first and second cycles were successful. The causes of the failure of subsequent cycles were not determined at the time of the report; however, it was concluded that the failure was most likely owing to (a) insufficient energy to expel the released oil to the wellbore, (b) uncontrolled water entry into zones of high water saturation, and (c) wellbore damage. It was also concluded that the process would have been applicable to the Buda and Austin Formations if the above problems did not exist. The cost of obtaining the necessary oil saturation data and maintaining adequate control over injection, however, may be high enough to prohibit the use of pressure-pulsing process in complex marginal reservoirs.

VUGULAR—SOLUTION POROSITY SYSTEMS

The carbonate reservoirs with a vugular—solution porosity system exhibit a wide range of permeabilities. The permeability distribution may be relatively uniform or quite irregular. The reservoirs with a uniform permeability distribution will probably respond to waterflooding in a manner similar to that of reservoirs with an intercrystalline—intergranular porosity system. The reservoirs with the irregular permeability distributions may respond in a manner similar to that of the reservoirs having a fracture—matrix porosity system.

The Pegasus Ellenburger Field, Texas, is a good example of a vugular reservoir which responds to waterflooding in a similar manner as the reservoirs having a fracture—matrix porosity system (Cargile, 1969). The vugular porosity system of the Ellenburger, however, is due to the random occurrence of breccia fragments rather than to solution processes. The peripheral injection system which was used initially could not supply sufficient energy, and, consequently, the reservoir pressure in the center of the field continued to decline. To overcome this problem, two injectors were placed in the center. As a result, the pressure decline was arrested, but severe channeling was experienced and poor injectivity profiles were obtained. Thus, the water injection program was severely curtailed (Cargile, 1969).

The Canyon Reef reservoir, Texas, exhibits a more uniform distribution of permeabilities (Black et al., 1964). In 1955, a peripheral water injection system was installed in the Sharon Ridge Canyon Unit, which is one of several fields producing from the Canyon Reef. Injectors were deepened prior to the start of the flood to permit injection into the aquifer as well as laterally into the oil zone. On the average, 48% of the injected water entered the formation above and 52% below the oil—water contact. About 100% of the reef's gross thickness was open to the injection. In 1964, 9 years after the start of injection, the flood was considered successful. Recoveries were running slightly better than indicated by the Stiles prediction calculations. Black and Lacik (1964) estimated that the ultimate waterflood recovery would be near 50% or twice the primary ultimate recovery of 25.2%.

Arps (1956) presented the "permeability-block" method of waterflood prediction, which is a simplified version of the Stiles technique. Table 8-14 shows the details of the application of this method to the Tensleep Sandstone reservoir in Wyoming. It seems that this procedure could be utilized for the carbonate reservoirs having vugular—solution porosity system only when the permeability distribution is relatively uniform. A larger number of "blocks" may be required, however; this will provide greater detail in the WOR versus recovery curve. The Stiles assumptions (p. 288) hold true for the permeability-block method. An explanation of some of the calculations shown in Table 8-14 follows:

(10) Unit recovery factor is calculated with the following equation:

$$UR = 7758\,\phi\,\left(\frac{1 - S_{wi}}{B_o} - S_{or}\right) \qquad (8\text{-}49)$$

where UR = unit recovery factor, bbl/acre-ft; ϕ = porosity, fraction; S_{wi} = interstitial water, fraction; S_{or} = residual oil saturation, under reservoir conditions, fraction; and B_o = oil formation volume factor, bbl/STB.

(13) The water/oil ratio of the produced stream is calculated using the following equation:

$$WOR = \frac{\mu_o}{\mu_w}\frac{k_{rw}}{k_{ro}}\frac{\Sigma C_w}{\Sigma C_o} \qquad (8\text{-}50)$$

where μ_o = oil viscosity, cp; μ_w = water viscosity, cp; k_{rw} = relative permeability to water at residual oil saturation, md; k_{ro} = relative permeability to oil at interstitial water saturation, md; ΣC_w = cumulative wet capacity (line 11, Table 8-14) d-ft; ΣC_o = cumulative oil capacity (line 12), d-ft.

(14) Cumulative oil recovery in bbl/acre-ft when group 1 is watered out. For group 1, the recovery is the product of line 2 and line 10. For the other groups, the recovery is the product of line 2 and line 10, multiplied by the ratio of its average permeability (line 3) to the minimum wet permeability.

(15 to 18) Cumulative oil recovery in bbl/acre-ft as groups 2 to 5 water out. As each group waters out, its recovery becomes simply (line 2) × (line 10). Recoveries for all other groups are multiplied by the ratio of the average permeability to the prevailing minimum wet permeability for each group.

The recovery is then plotted versus WOR. The economic limit WOR is calculated and the corresponding recovery is read from the curve. Unless the continuity of the permeability blocks across the reservoir is definitely established, the calculated recovery should be considered conservative (Arps, 1956).

In carbonate pools, producing under a bottom water drive (natural or artificial), such as some of the vugular D-3 Reef reservoirs in Alberta, extreme ranges of permeabilities are found. Here the permeability-block method

TABLE 8-14

Computation of the water-drive recovery factor for Tensleep Sandstone reservoir in Wyoming by the permeability-block or modified Stiles method (after Arps, 1956, p. 187; courtesy of the SPE of AIME)

	Group					Total
	1	2	3	4	5	
(1) Permeability range, md	>100	50–100	25–50	10–25	0–10	
(2) Fraction of samples	0.085	0.109	0.145	0.212	0.449	
(3) Average permeability, md	181.3	69.0	34.4	16.1	2.4	
(4) Capacity, d-ft	1.543	0.752	0.499	0.341	0.108	
(5) Average porosity (ϕ), fraction	0.159	0.150	0.152	0.130	0.099	
(6) Average resid. oil saturation (S_o), fraction	0.173	0.195	0.200	0.217	0.222	
(7) Relative water perm. (k_{rw})	0.65	0.63	0.60	0.56	0.54	
(8) Average interstitial water saturation (S_w), fraction	0.185	0.154	0.131	0.107	0.185	
(9) Relative oil perm. (k_{ro})	0.745	0.53	0.61	0.66	0.47	
(10) Est. unit recovery factor, bbl/acre-ft	725	693	721	623	414	
(11) Cumulative wet cap. = Σ (4)	1.543	2.295	2.794	3.135	3.243	
(12) Cumulative clean oil cap. = 3.243 — (11)	1.700	0.948	0.449	0.108	0	
(13) Water/oil ratio	15.5	36.0	76.5	307.9	∞	
(14) Cum. rec.[a], $WOR = 15.5$, min. $k_{wet} = 100$ md	61.6	52.1	35.9	21.3	4.5	175.4
(15) Cum. rec.[a], $WOR = 36.0$, min. $k_{wet} = 50$ md	61.6	75.5	71.9	42.5	8.9	260.4
(16) Cum. rec.[a], $WOR = 76.5$, min. $k_{wet} = 25$ md	61.6	75.5	104.5	85.1	17.8	344.5
(17) Cum. rec.[a], $WOR = 307.9$, min. $k_{wet} = 10$ md	61.6	75.5	104.5	132.1	44.6	418.3
(18) Cum. rec.[a], $WOR = \infty$, min. $k_{wet} = 0$	61.6	75.5	104.5	132.1	185.9	559.6

a Cumulative recovery in bbl/acre-ft.

yields results that are much too low. This is caused by two factors: buoyancy and imbibition (Arps, 1956).

The water advances through the highly permeable sections of the reef and bypasses oil in the tighter sections. When the bypassing occurs, a buoyancy gradient is set up across the tight sections due to the density difference between the oil and water, and this gradient tends to drive the oil into the more permeable sections. The imbibition of water from the high-permeability areas into the tighter sections adds to the buoyancy effect.

The SACROC Unit in Kelly Snyder Field, Texas, is the site of an unusual approach to waterflooding of a reef (vugular—solution porosity system) (Allen and Thomas, 1959). The injectors are run across the crest of the reef. This is not as extreme a procedure as seems at first. The Canyon Reef reservoir has a low mound shape and gently-dipping flanks. Its thickness varies from 0 to 795 ft, and it is 7 miles wide. The use of a peripheral injection system was not feasible because the zone is thin at the edges. Allen and Thomas (1959) listed several advantages for the center-to-edge line drive flood: (a) the water moves in the same direction as that of the natural pressure gradients, (b) the water is injected into the thickest portion of the reef, permitting injection of large volumes of water and reducing the number of injectors required, and (c) the injection system costs less than a peripheral system due to fewer injectors, shorter injection line, and fewer pump stations. After 5 years of operation (reported in 1959), the flood was considered a success. The ultimate recovery was estimated at over twice the primary recovery of 23.6%.

IMPROVED WATERFLOOD PROCESSES

In the discussion of the double Buckley-Leverett predictive technique, the subject of the improved waterflood processes was mentioned briefly. This section provides additional information on two major groups of chemicals used to increase waterflood recovery: polymers and surfactants. The chemicals either improve the mobility ratio of the process by changing the properties of the water (polymers) or reducing the interfacial tension (surfactants).

Polymers

Polymers are used in waterflooding to reduce the mobility of the water (k_w/μ_w) and thus improve its displacement efficiency. The reduction in mobility is caused by both an increase in the water viscosity and a decrease in the permeability to water. A reduced driving phase mobility results in improvements in the areal and vertical sweep efficiencies.

During the early attempts to alter the characteristics of the water, such materials as glycerin, sugar, glycols, and some naturally occurring polymers were employed. The high concentrations of these agents required to achieve the desired changes in water properties, however, made them economically unattractive.

With the advent of the less expensive synthetic organic polymers, polymer flooding became a reality instead of a theory. Polyethylene oxides and poly-acrylamides are the most common of polymers being used. The polyacryl-amides are sometimes partially hydrolyzed, which further increases their molecular weight. These chemicals cause a large reduction in water mobility at low concentrations and are adsorbed only negligibly. Mungan et al. (1966) found that little reduction in residual oil saturation should be expected from polymer flooding. The increase in recovery is mainly the result of increasing the volume of the reservoir swept. Polymer flooding is best suited for reservoirs in which water sweep efficiency is very low owing to an unfavorable mobility ratio (e.g., because of low crude oil gravity) or wide permeability variations. Both conditions are common in the carbonate reservoirs.

Adsorption studies on the reservoir rock for any proposed polymer flood are absolutely essential in order to determine the polymer requirements and, therefore, the economics of the operation. Mungan et al. (1966) suggested that in pilot testing a polymer flood, it would be useful to utilize (1) a tracer ahead of the polymer slug, (2) a tracer with the polymer, and (3) a third tracer following the polymer. Comparison of the results obtained on using these three tracers may provide answers on the development and the transport rate of the polymer bank.

Excellent discussion of the details of polymer flooding has been presented by Pye (1964), Burcik (1968), Sandiford (1969) and Gogarty (1967).

Armstrong (1967) in reporting on polymer flooding, made the following comments:

(1) A field with a large gas cap, or which is highly fractured, or has its permeability controlled by vugs is not a good candidate for polymer flooding.

(2) The polymer flood is not a tertiary tool; therefore, if a waterflood is approaching the economic limit, a polymer flood should not be started.

(3) Large volumes of bottom water can strip the polymer flood of chemicals.

(4) Thickness of the section does not limit polymer application.

(5) Depth presents no problem if the reservoir temperature is below 300°F.

(6) The quality of the injected water does not affect the polymer injection.

The double Buckley-Leverett predictive technique can be used for polymer flood predictions. Laboratory work and pilot testing, however, should definitely be undertaken before a field waterflood is initiated.

A unique form of pusher flooding was used in the Crane zone of the Northeast Hallsville Field, Harrison County, Texas (Snell and Schurz, 1966). The reservoir is a slightly dipping monocline with a large associated gas cap.

The reservoir rock is an oolitic limestone (intergranular porosity) of varying permeability. The crude oil gravity is very high (57.1°API), and withdrawals from the gas cap were causing crude oil migration into the gas cap and consequent loss. To correct this, two strategic wells were chosen for polymer injection to form a viscous barrier at the gas—oil contact. After the barrier was placed, a water injection project was started in the oil zone. Water breakthrough occurred earlier than expected owing to the permeability variation. Polymer injection was tested next, and it reduced the produced water/oil ratio within 2 months from the beginning of injection.

Surfactants

The use of surfactants (surface-active chemicals) in order to improve oil recovery started in the late 1920's and early 1930's. Water-soluble surfactants were described by De Groot (1929, 1930) as an aid to improve oil recovery. Holbrook (1958) proposed other water-soluble compounds, such as organic perfluoro compounds, fatty acid soaps, polyglycol ether, salts of fatty or sulfonic acid, and polyoxyalkylene compounds for surfactant flooding. Holm and Bernard (1959) filed for a patent in which they proposed injecting 0.1—3% surfactant dissolved in low-viscosity hydrocarbon solvent which reduced surfactant adsorption in water-wet formations. Gogarty and Olsen (1962) filed for a patent describing the use of microemulsions in a new miscible-type recovery process known as Maraflood^{TM}. Gogarty and Tosch (1968) suggested addition of cosurfactants or electrolyte to improve recovery.

There has been much laboratory work and some field testing of surfactants. These tests show a reduction of the interfacial tension between the oil and water which results in a substantial reduction of the residual oil content of the reservoir rock after waterflooding (Andersen et al., 1950). Earlougher and Guerrero (1965) estimated that use of surfactants might result in additional recoveries of about 10% of the original oil-in-place above that obtained by conventional waterflooding.

Two different concepts have been developed for using surfactants for improving oil recovery (Gogarty, 1976). In the first concept, a solution containing a low concentration of a surfactant is injected. The surfactant is dissolved in either water or oil and is in equilibrium with aggregates of the surfactant known as micelles. Large pore volumes of the solution are injected into the reservoir to reduce interfacial tension between oil and water and, thereby, increase oil recovery. Oil may be banked using the surfactant solution process, but residual oil at a given position in the reservoir will approach zero only after passage of large volumes of surfactant solution.

---

^{TM} Trademark of Marathon Oil Company.

In the second process, a relatively small pore volume of a higher-concentration surfactant solution is injected into the reservoir. With the higher surfactant concentration, the micelles become a surfactant-stabilized dispersion of either water in hydrocarbon or hydrocarbon in water. The high surfactant concentration allows the amount of dispersed phase in the microemulsion to be high as compared to the low value in the dispersed phase of the micelles in the low-concentration surfactant solutions (Gogarty, 1977). The injected slug is usually formulated with three or more components. The basic components (i.e., hydrocarbon, surfactant, and water) are sufficient to form the micellar solutions. The fourth component of cosurfactant can be added to enhance reduction of interfacial tension. Electrolytes, normally inorganic salts, form a fifth component that may be used in preparing the micellar solutions or microemulsions (Gogarty, 1977). The high-concentration surfactant solutions displace both oil and water and rapidly displace all the oil contacted in the reservoir. As the high-concentration slug moves through the reservoir, it is diluted by formation fluids and adsorption to the rock, and the process reverts to a low-concentration flood.

Adsorption of surfactants on reservoir rocks is a deterrent to oil recovery by chemical means. As the injected phase traverses the reservoir, surfactants are depleted from solution. As a result, the surfactant loses its ability to lower the interfacial tension between oil and water. In addition to adsorption, surfactant loss in reservoir can occur also by retention and/or precipitation, and by partitioning of the surfactant into the oil phase. Many factors affect the amount of surfactant loss by adsorption and other means. These include the relative molecular mass distribution of the surfactant, concentration of the surfactant, the electrolytic state of the dissolving medium and the reservoir fluids, constitution of the adsorbing material, and temperature of the adsorbing environment. The main cause of the failures on using surfactants is that they tend to adsorb on the surfaces of the solids; this depletes the quantity of surfactant available to work on the oil—water interface. Johnson (1960) derived an equation to estimate the weight of surfactant which must be injected so that its concentration would not drop below some predetermined minimum value before reaching the end of the reservoir:

$$[(1 + kc_s)/kc_s]^2 = (X/L)\,(a/w) \tag{8-51}$$

where X = distance of movement of the surfactant having concentration c_s, ft; L = total length of the reservoir, ft; c_s = concentration of surfactant in solution in equilibrium with the adsorbed material, lb/bbl; a = a constant, the maximum adsorption capacity per unit volume of pore space, lb/ft^3; k = the reciprocal of that solution concentration, which is in equilibrium at an adsorption capacity of $\frac{1}{2}a$, bbl/lb; and w = specific weight of surfactant, lb/ft^3.

A graphic representation of the equation is shown in Fig. 8-37. This graph

320

can be used for a radial system by changing the label on the horizontal axis to $(X/L)^2$ (a/w) (Johnson, 1960). To use the curve, k and a must first be measured, c_m (minimum concentration effective in removing oil) must be determined experimentally, and X/L must be obtained from the physical system. The graph is entered from the left with the calculated kc_m value and the corresponding (X/L) (a/w) value is read; w is then calculated.

Inks and Lahring (1968) found that nonionic surfactants tend to adsorb to a lesser extent than the anionic and cationic surfactants. A field test of a non-ionic surfactant yielded an apparent increase in oil recovery of about 9% (Inks and Lahring, 1968, p. 1320). G.A. Babalyan and E.K. Kovalenko presented an analysis of the utilization of the non-ionic surfactants to increase the recovery from fissured reservoir rocks, when the oil is contained mainly in the fissures or possibly large vugs. In this application, surfactants are used to lower the surface tension at the oil—water interface and the force of adhesion of the oil to the solid surface. This increases both the amount of oil recovered and the rate at which it is recovered. The actual consumption of surfactant is small, as only the fissure (fracture) walls are available for surfactant adsorption.

When the reservoir has a fracture—matrix porosity system with the majority of the oil in the matrix, or a vugular pore system with large and small vugs, with the majority of the oil residing in the small vugs, it was recommended that a surfactant (anionic or cationic) be used to make the walls of the fractures or large vugs hydrophobic (G.A. Babalyan and E.K. Kovalenko,

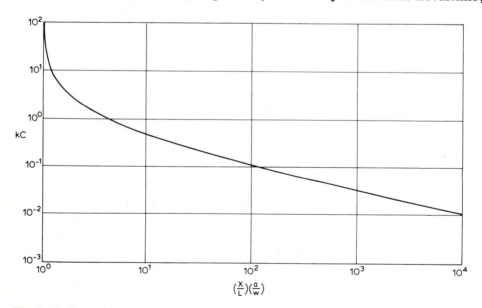

Fig. 8-37. A graphic representation of equation (8-5). (After Johnson, 1960; courtesy of the *Oil and Gas Journal*.) (The writers prefer term c_s to C.)

personal communication, 1968)*. This will increase the rate of flow of the oil into fractures or large vugs.

Mobility control is important in both low- and high-concentration surfactant flooding. Conditions have been described by Gogarty et al. (1970) for obtaining mobility control with miscible-type waterfloods using micellar solutions. High-molecular-weight, water-soluble polymers have been used for mobility control in both types of surfactant processes. Gogarty and Davis (1972) reported that oil-in-water emulsions also have been used for mobility control with high-concentration surfactant flooding. With low-concentration surfactant injection, mobility control of the surfactant slug is accomplished by dissolving polymer in the surfactant solution. Results reported by Gogarty and Tosh (1968) indicate that, generally, the mobility of the high-concentration surfactant slug is fixed by adjusting the composition of such micellar components as the cosurfactant and electrolyte. Continued stability in both high- and low-concentration displacements requires the use of a mobility buffer.

Work is underway both in the laboratory and the field to select the optimum method of injecting surfactant to enhance oil recovery. Pope (1980) has applied the concept of fractional flow analysis (discussed earlier) to EOR methods using polymers, surfactants, etc. Saturation and recovery profiles are given in Figs. 8-38 and 8-39.

Status of micellar—polymer field tests was presented by Lake and Pope (1979). A micellar—polymer flood is a combination of surfactant and polymer flood in which trapped oil is mobilized by injecting a surfactant slug driven successively by a mobility control agent (i.e., polymer-thickened water) and then by chase water.

Table 8-15 is an extensive summary of several important variables pertaining to planned, current, and completed micellar—polymer field tests. The variables are those which have the greatest influences on oil recovery of the flood. Table 8-15 also includes general information, such as operator identity and formation age, and estimates derived from secondary sources, i.e., laboratory studies and reports of other projects in the same reservoir.

Lake and Pope (1979) used the field data of Table 8-15 and correlated the oil recovery data with various independent variables. Their studies showed strong correlations between oil recovery and (1) mobility, (2) buffer size, and (3) the ratio of viscous to capillary forces. Poor correlations were noted between recovery and surfactant slug size. Results of Lake and Pope (1979) indicate that micellar—polymer flooding is less restricted than previously thought. In addition, better screening of reservoirs for application of the process is possible. They recommended that for micellar—polymer flood

---

* G.A. Babalyan and E.K. Kovalenko. The possibility of using surfactants to increase the petroleum yield from fissured traps (unpublished report).

322

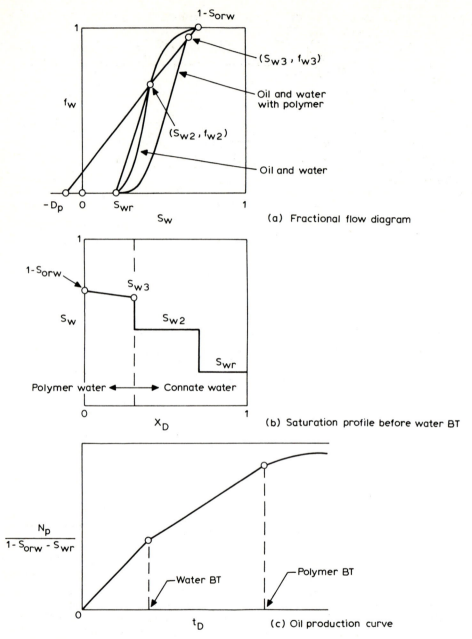

(a) Fractional flow diagram

(b) Saturation profile before water BT

(c) Oil production curve

Fig. 8-38. Fractional flow diagram, saturation profile before water breakthrough, and oil production curve for polymer flooding. (After Pope, 1980, fig. 2, p. 193; courtesy of the SPE of AIME.) f_w = fractional flow, the flux of water phase divided by the total flux; S_w = water saturation; S_{wr} = water residual saturation; S_{orw} = residual oil saturation after waterflood; X_D = non-dimensional distance; X/L; t_D = non-dimensional time or injected pore volumes of fluid, $qt/AL\phi$; N_p = oil production, pore volumes; D_p = adsorption of polymer in pore volumes per pore volume injected; q = flow rate; L = length of porous medium.

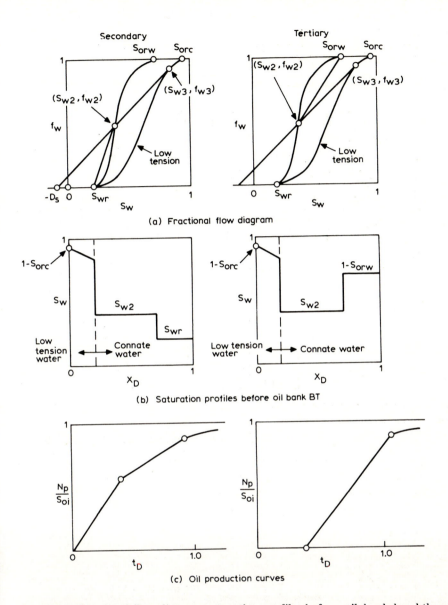

(a) Fractional flow diagram

(b) Saturation profiles before oil bank BT

(c) Oil production curves

Fig. 8-39. Fractional flow diagram, saturation profiles before oil bank breakthrough, and oil production curves for low-tension flooding. (After Pope, 1980, fig. 3, p. 194; courtesy of the SPE of AIME.) f_w = fractional flow, the flux of water phase divided by the total flux; S_{oi} = initial oil saturation; S_w = water saturation; S_{orw} = residual oil saturation after waterflood; S_{orc} = residual oil saturation after chemical or surfactant flood; S_{wr} = water residual saturation; X_D = non-dimensional distance, X/L; t_D = non-dimensional time or injected pore volumes; N_p = oil production, pore volumes; D_s = adsorption of surfactant in pore volumes per pore volume injected.

TABLE 8-15

Summary of field test results for micellar—polymer floods (after Lake and Pope, 1979, table 1; courtesy of *Petroleum Engineer International*)

Field	State	Operator	Project	Begin Date	Project area, acres	Pattern type	No. of patterns	Depth, ft	Temp. °F	Formation	Clay fr.
Aux Vases[5,6,9]	Ill.	Marathon		5/70	4.3	N5	1	3,000		Aux Vases	
Batesville Pool[16]	Kans.	Bird-Hanley &Sheedy	—	10/56	331	IRR	—	1,420	80	Quincy-B'ville	
Bell Creek[1,3,7,11-15,17-(24)]	Mont.	Gary	Unit A	2/79	40	CN5	1	4,550	110	Muddy	0.07
1. Benton[1,25,(26)-(29)]	Ill.	Shell	Pilot	5/68	1	IRR	1	2,120	95	Tar Sprgs.	(0.05)
Benton[1-3,(27)-(29),30]	Ill.	Shell	Stage I	1972	160	IRR	—	2,100	95	Tar Sprgs.	(0.05)
2. Big Muddy[1,4,7,11,31-(34)]	Wyo.	Conoco	Pilot	8/73	1.25	CN5	1	3,050	120	Frontier	(0.11)
Big Muddy[1,15,35]	Wyo.	Conoco	Demo.	1979	90	15	9	3,050	115	Frontier	<0.11
3. Borregos[1,36,(37)]	Tex.	Exxon	—	1965	1.25	N5	1	4,998	165	Frio	0.085
4. Bradford[5,6,9,38]	Penna.	Pennzoil	Bingham 533	12/68	0.75	I5	1	1,860	68	Bradford	
5. Bradford[1-7,9,38]	Penna.	Pennzoil	Bingham Exp.	3/71	46.5	CN5	16	1,866	68	Bradford	
Bradford[1,3,7,12,13,39-41]	Penna.	Pennzoil	Lawry	5/77	24	N5	16	1,280	64	Bradford	
6. Bridgeport[2,5,6,9]	Ill.	Marathon	118K	9/69	2.4	R5	1	1,500		Kirkwood	
Chateaurenard[11,42]	(France)	Elf Aquit.	—	2/78	2.5	I5	1	1,970	86	Neocomian	0.15
7. Dela.-Childers[1,3,7,11-15,43]	Okla.	B & N Oil Co.	Costen	11/75	2.5	CI5	1	620	86	B'ville	0.10
El Dorado[1,4,7,11-15,24,44-47,(48)-(52)]	Kans.	Cities Serv.	Chesney	11/76	25.6	N5	4	650	69	Waubaunsee	0.21
El Dorado[1,4,7,11-15,24,44-47,(48)-(52)]	Kans.	Cities Serv.	Hegberg	6/76	25.6	N5	4	650	69	Waubaunsee	0.21
Glenn[3]	Okla.	Gulf		1980	45	N5	6	5,000	100	Glen	
Goodwill Hill[2,5,9]	Penna.	Quaker St.		5/71	10	I5	9	600		Venango	
Griffin Cons.[3,4,7]	Ind.	Conoco		11/73	0.8	N5	1	2,400	85	U. Cypress	
Guerra[53]	Tex.	Sun		(1970)	2	N5	1	2,200	122	Jackson	
8. Jones City Reg.[1,2,54]	Tex.	Union	Higgs Unit	7/69	8.2	IRR	1	1,870	95	Bluff Creek	
Loma Novia[1,55]	Tex.	Mobil			5	N5	1			Frio	0.095
9. Louden[1,(36),56]	Ill.	Exxon		1969	0.625	N5	1	1,460	95	Ches. Cyp.	0.09
Main Cons.[3]	Ill.	Getty		1980	10	N5	9	1,000		Robinson	
Manvel[1,57]	Tex.	Texaco		6/77	10	SLD	1	5,400	165	Frio	0.12
Montague Co. Reg.[2,31-33]	Tex.	Conoco									
10. N. Burbank[1,4,7,11-15,58-64]	Okla.	Phillips	Tract 97	8/76	90	I5	9	2,900	120	Cherokee	
Rincon[31-33]	Tex.	Conoco									
Robinson[5,6,8-10,65,66]	Ill.	Marathon	Dedrick	11/62	2.5	N5	1	1,000	(72)	Robinson	
11. Robinson[5-10,65-67]	Ill.	Marathon	Henry-W	11/65	0.75	I5	1	1,000	72	Robinson	
12. Robinson[5,6,8-10,66,67]	Ill.	Marathon	Henry-E(I)‡‡	11/65	0.75	I5	1	1,000	72	Robinson	
13. Robinson[5,6,8-10,66,67]	Ill.	Marathon	Henry-E(II)‡‡	6/66	1.5	I5	1	1,000	72	Robinson	
14. Robinson[5,6,8-10,66,67]	Ill.	Marathon	Henry-E(III)‡‡	6/67	3	I5	1	1,000	72	Robinson	
15. Robinson[5,6,8-10,66]	Ill.	Marathon	Wilkin	1/64	2.5	N5	1	1,000		Robinson	
16. Robinson[1,6,8-10,65,66]	Ill.	Marathon	119R	9/68	40	LD	16	1,000	72	Robinson	
17. Robinson[1,4,8,66-68]	Ill.	Marathon	219R	10/75	113	C5	37	1,000	72	Robinson	
Robinson[1,7,13-15,69-71]	Ill.	Marathon	M-1	2/77	407	N5	88	1,000	72	Robinson	0.085
18. Salem[1,3,72-75]	Ill.	Texaco		12/74	5	CN5	1	1,750	80	U. Benoist	
Sales[3]	Tex.	Conoco	Pilot	1963	2.5			1,900		Flippen	
Salt Creek[1,3,(76),77]	Wyo.	Amoco		Prop.	3	N5	1	2,300	110	Frontier	0.24
Slaughter[3]	Tex.	Texaco		10/73		4		5,000	109	San Andres§§	
19. Sloss[1,3,7,11,34,78,79]	Nebr.	Amoco		2/77	9	N5	1	6,256	165	Muddy	(0.11)
20. Wichita Co. Reg.[1,7,11]	Tex.	Mobil	West Burk Co-op	10/75	209	N5	4	1,700	89	Gunsight	
Wilmington[1,7,11,13-15,80-82]	Calif.	Long Beach		7/77	10.7	SLD	2	2,900	125	Puente	

Footnotes

°Includes co-surfactant ‡Based on assumed residual oil saturation of 0.40 ¶Multiple preflushes ††Polymer preflush

†Slug sizes and recoveries based on zone B only §Secondary flood °°Sacrificial agent in preflush ‡‡Henry E test was expanded twice; treated as three tests here

The following notations are used in this table (for references cited in the table refer to the original paper):

Pattern type: N5 = normal five-spot (four injectors—one producer); older publications refer to this as "inverted"; CN5 = confined normal five-spot; I5 = inverted five-spot (four producers—one injector); CI5 = confined inverted five-spot; LD = line drive; SLD = staggered line drive; IRR = irregular.

Clay fr: weight fraction of clay in rock matrix.

Perm var: permeability variation as defined by Dykstra-Parsons.

Net pay, ft	Perm., Md	Perm. var. (D-P)	Porosity fraction	Oil vis., cp	Resid. oil sat., fr.	Resident salinity, mg/l	Resident hardness, mg/l	PF size, fr.V_p	Slug size, fr.V_p	Surf. conc., fr.	MB conc., ppm	MB size, fr.V_p	Oil rec., fr. S_{or}
									0.025			0.24	
20	55		0.19	1	0.39	31,600	2,750		(1.14)	0.00025–	0		0.09
								—		0.000025		—	
8.1	1,050	0.7	0.29	4.6	0.28	2,680	16-27	0.1**	0.035	0.081*	950	0.6	—
4†	90	0.42	0.17	3.5	0.32	(41,000)	(4,800)	(1.30)	(1.11)	0.013	275	(3.30)	0.29
44	(55)	(0.42)	0.19	3.5	0.32	(41,000)	(4,800)	(0.08)	0.10	(0.50)	(650)		—
65	52	0.6	0.19	4	0.32	3,050	23	0.8	0.25	0.025	200	0.3	0.35
65	52	0.61	0.19	5.5		3,050	23						
3.8	434	(0.7)	0.21	0.36	0.307	19,600	1,671		0.47	0.023	0	0.51	0.20
23.7	82	(0.65)	0.18	5	0.308	2,800		0.064	0.064	0.13	1,700	1.4	(0.57)‡
23	82		0.18	5	0.362	2,800		—	0.05	0.12	1,500	1.2	0.43
29	8	(0.65)	0.13	5	0.32(L)	3,000(TDS)		0.10	0.094	0.09			—
	90		0.18	5.5	0.3-0.45			—	0.085			0.87	(0.48)¶¶
8	1,000		0.3	40	0.55	400(TDS)	70	—	0.10		1,700	1.25	
52	100	(0.48)	0.21	6.9	0.34	5,200	695	0.113	0.085	0.054	1,350	0.4	0
18.4	265	0.78	0.243	5.2	0.333	53,000	4,200	(0.41)¶	0.113	0.026	(2,000)	0.66	—
17.5	208	0.85	0.245	4.8	0.307	53,000	4,200	(0.16)**	0.04	0.12	1,600	0.6	
	130		0.20	5									
				(4.5)				—	0.05			0.95	
	75		0.2	3.6	0.30								
9	2,500		0.33	1.6	(0.39)	20,000	300	—	0.13	0.04	0	—	
14.25	500		0.229	4.3	0.23	54,000		0.017††	0.02			0.34	(0.19)
10-12			0.2			6,060	38	0.1**	0.12	0.019	0	0.1	
15.5	103	(0.42)	0.206	14	0.291	13,000		0.098**	0.4	0.023	558	0.32	0.153
	200		0.20	10	0.465								
17	500		0.3	4	0.3	107,000(TDS)	2,400	—	0.25	0.025	1,400	0.5	
						150,000(TDS)		—			0	—	
47	52	0.61	0.155	3	0.35	53,400	7,260	(0.46)¶	0.05	0.06	2,500	0.465	0.11
						25,000(TDS)					0	—	
			(0.2)	7	(0.7)§			—	0.035	0.072		0.066	0.35
10	200	(0.26)	0.20	7	0.4	10,000		—	0.09	(0.10)	1,200	1.91	0.63
10	200		0.2	7	0.4	10,000		—	0.40			0.44	0.27
10	200		0.2	7	0.4	10,000		—	0.20			0.83	0.33
10	200		0.2	7	0.4	10,000		—	0.10			0.55	0.26
	(200)		(0.2)	9	0.4	3,700		—	0.035	0.072		0.068	0.14
25	211		0.193	7	0.40	10,000		—	0.07		1,200	1.0	0.38
16.6	165	0.62	0.208	7	0.40	9,400		—	0.102	0.10	1,156	1.05	0.18
27.8	102	(0.59)	0.19	6	0.40			—	0.10		1,200	1.05	
26	87	(0.34)	0.148	3.6	0.3	40,000	3,300	0.519¶	0.285	0.0195	700	0.3	0.43¶¶
	457		0.217		0.30								
(76)	(58)	(0.64)	0.18	3.1		13,367(TDS)	356						
	4		0.11	1.8									
12.2	93	(0.45)	0.17	0.8	0.3	2,500(TDS)	50	(3)	0.155	0.065	800	(1)	0.29
13	53		0.22	2.3	0.30	160,000(TDS)		0.4	0.15	0.015	500	0.3	<0.01
56	180	(0.64)	0.27	31.7	(0.30)	30,000(TDS)		—	0.07			0.93	

§§Limestone formation   ***Reservoir temperature 200°F

¶¶Limits averaged for data analysis   Note: References enclosed by parentheses indicate a secondary data source.

Resident salinity: chloride concentration in mg chloride/liter liquid of produced samples. When salinities are reported as total dissolved solids, the entry is designated with "TDS".

Resident hardness: sum of the divalent cation concentrations.

PF size: preflush size reported as fraction pore volume (true for all slug sizes).

MB conc: mobility buffer polymer concentration in initial portion of the usually tapered mobility buffer.

MB size: total amount (including taper) of polymer water injected.

Oil rec: oil recovery reported as fraction of original oil in place before micellar—polymer flood.

designs more attention should be focused on those factors that affect mobility control and capillary number. Less weight should be given to the effects such as the oil content of the slug, large versus small slug size, low versus high surfactant concentration, low versus high salinity, temperature, etc. This is not meant to suggest that the proper amount of the most cost-effective chemicals should not be used, nor that formulation chemistry is not important. The factors which affect the very important capillary number should be examined more carefully.

PERFORMANCE OF SOME RECENT IMPORTANT WATERFLOODS

Jay/Little Escambia Creek Field

Jay/Little Escambia Creek (Jay/LEC) Field Smackover reservoir was discovered in June of 1970, and 102 development wells were cored conventionally from 1970 to 1974. During 1973, as planning for waterflood operations progressed, it became apparent that comprehensive surveillance was needed to complement the reservoir description data and, thereby, optimize recovery through effective management. During 1977, a comprehensive study was made of waterflood performance. It was concluded that areal sweep efficiency is better than vertical sweep efficiency. In order to improve sweep efficiency, maps of unswept oil areas were prepared for each zone and the following work programs were initiated: (1) infill drilling, (2) workovers, (3) injection balancing, and (4) surface facility modifications. Although the program was costly, it proved to be highly profitable for this field. According to Langston et al. (1981), the increased oil production attributable to this program has cost less than \$2/bbl (\$11.2/m^3). Jay/LEC unit performance is shown in Fig. 8-40.

Bell Creek Field

Bell Creek Field is located on the northeastern flank of the Powder River Basin. It consists of six Muddy Sand reservoirs unitized as six units. The field is 15 miles long and $3\frac{1}{2}$ miles wide, with 17,000 productive acres containing 305 active producing and injection wells. A line-drive waterflood from the west edge of each reservoir has been in operation since 1970. Anticipated high injection rates, excellent reservoir conformance, and reduced capital expenditures led to selection of a line-drive flood pattern. This pattern proved successful except for one of the six units, where reservoir heterogeneity and sand quality resulted in limited waterflood response. Sand production has been controlled by use of "Variperm" sand screens. Oil production in November of 1974 was 26,000 bbl and cumulative production to December of 1974 was 66.9 million bbl. The ultimate recovery for all units is 98.2 mil-

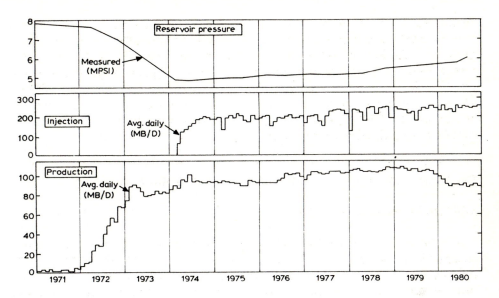

Fig. 8-40. Jay/LEC unit performance. (After Langston et al., 1981, fig. 9, p. 791; courtesy of the SPE of AIME.)

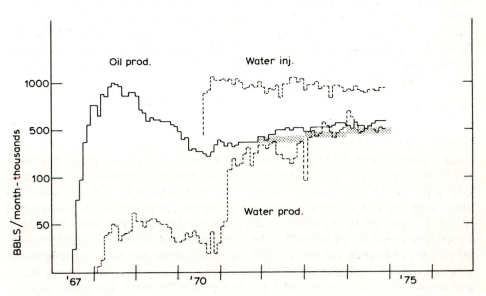

Fig. 8-41. Production performance curve for Unit A of Bell Creek Field. (After Burt et al., 1975, fig. 6, p. 1446; courtesy of the SPE of AIME.)

lion bbl, or 40.4% of the original oil-in-place. According to Burt et al. (1975), tertiary recovery is under active consideration, and a micellar process is believed to have the greatest potential. The performance of Unit A is presented in Fig. 8-41.

West Delta Block 73 Field

The West Delta Block 73 Field is located 27 miles southeast of Grand Isle, La., and 17 miles west of the mouth of the Mississippi River's Southwest Pass. The field was discovered in 1962 and developed during 1963—1966. Waterflood operations were initiated in 1967 at three small West Delta 73 waterflood plants. Filtered and deaerated Gulf of Mexico water was used, because there were no conductors available to drill source-water wells and economic incentive existed for using seawater if feasible. By 1967, injection-well performance and falloff tests revealed damage to the injection formation. Casing leaks indicated the existence of corrosion, whereas water quality studies showed that large amounts of fine solid particles and organic matter were not removed from the injected water. Also core studies showed the probability of clay swelling and redistribution of fines. Most West Delta 73 reservoirs have a high (10—15%) clay content. In original clay-sensitivity study, permeability losses ranged from 5.3% for the higher-permeability specimens to 94.2% for the more shaly, low-permeability cores. The 1972 study resulted in changing from seawater to a higher-salinity subsurface water. Operating expenses for subsurface water at West Delta Block 73 Field are about 65% below those for seawater, because of lower cost for manpower, materials and supplies, and fewer injection pump repairs. According to Ogletree and Overly (1977), conversion to subsurface source water at West Delta Block 73 Field has arrested permeability declines and reduced injection-well workover frequency. At the end of 1975, cumulative oil production was 128 million bbl and oil production was 19,600 bbl/day.

Wichita County Regular Field

Mobil's West Burkburnett waterflood is about 4 miles southwest of Burkburnett, Texas, and is part of the Wichita County Regular Field. The pool was discovered in June of 1912. Waterflooding began in 1944 with a pilot flood developed on 20-acre five-spot patterns. Full development occurred during the period 1948—1950. In 1971, all the leases in the waterflood project either had become uneconomical to operate or were being projected to reach the economic limit by 1972 or 1973. The low-tension waterflood (LTWF) project began in November of 1973 with a fresh-water preflush. Injection of the surfactant slug started in October of 1975 and ended in June of 1976. This was followed by a polymer drive until April of 1978. Since then, fresh water has been continuously injected. Tertiary oil production response first was noted near the completion of surfactant injection and, to date, 17 out of the original 20 producers are yielding incremental oil. The total tertiary oil production to the end of 1980 and to the anticipated project termination in 1984 (projected) is estimated to be 238,000 and 320,000 bbl, respectively, which are equivalent to recovery factors of 17.6 and 24% of its oil-in-place (Talash and Strange, 1982). The cumulative oil production and projections in LTWF area are presented in Fig. 8-42.

Salem Unit, Marion County, Illinois

The Lake Centralia—Salem Pool is located between cities of Salem and Centralia in Marion County, Ill., about 70 miles east of St. Louis. The Benoist Sand in the pilot area is separated into upper and lower segments by a thin shale stringer (Widmyer et al., 1979). In 1974, a joint Texaco Inc.—Mobil Research and Development Corp. test to evaluate a LTWF process was initiated. In this pilot test, the process used a sequence of fresh-water preflush, chemical pretreatment, surfactant solution, a mobility-control (polymer) solution, and field brine. Oil recovery was substantially less than expected, i.e., only about 25—30% of the recovery predicted using the concept of symmetrical displacement in all pattern quadrants, with recovery factors based on earlier laboratory investigation (Widmyer et al., 1979). The production shortfall can be explained by many reasons, including lower-than-expected oil displacement efficiency for the chemical system or a smaller pat-

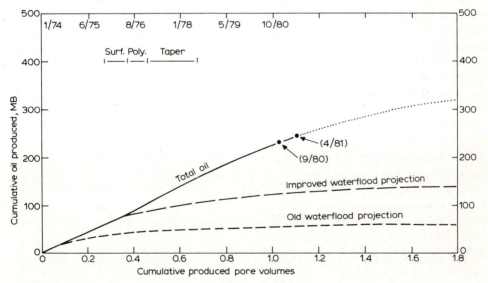

Fig. 8-42. Relationship between cumulative produced pore volumes and cumulative oil produced in LTWF area. (After Talash and Stange, 1982, p. 2501, fig. 12; courtesy of SPE of AIME.)

tern reservoir volume that actually was flooded and contributed to oil production (Widmyer et al., 1979).

REFERENCES

Abernathy, B.F., 1964. Analysis of various waterflood prediction methods using actual performance of pilot waterfloods in carbonate reservoirs. *J. Pet. Technol.*, 16 (3): 276—282.

Ache, P.S., 1957. Inclusion of radial flow in use of permeability distribution in waterflood calculations. *AIME Tech. Pap.*, 935—G.

Akins, D.W., Jr., 1951. Primary high pressure waterflooding in the Pettit Lime Haynesville Field. *Trans. AIME*, 192: 239—248.

Allen, H.H. and Thomas, J.B., 1959. Pressure maintenance in SACROC Unit operations, January 1, 1959. *J. Pet. Technol.*, 11 (11): 42—48.

Allen, W.W., Herriot, H.P. and Stiehler, R.D., 1969. History and performance prediction of Umm Farud Field, Libya. *J. Pet. Technol.*, 21 (5): 570—578.

Andersen, K.H., Torrey, P.D. and Dickey, P.A., 1950. Capillary and surface phenomena in secondary recovery. In: *Secondary Recovery of Oil in the United States.* API, 2nd ed., pp. 233—239.

Armstrong, T.A., 1967. Polymer floods attacking recovery gap. *Oil Gas J.*, 65 (1): 46—48.

Aronofsky, J.S., 1952. Mobility ratio, its influence on flood patterns during water encroachment. *Trans. AIME*, 195: 15—24.

Aronofsky, J.S. and Ramey, H.J., 1956. Mobility ratio — its influence on injection patterns and production histories in five-spot floods. *Trans. AIME*, 207: 205—210.

Arps, J.J., 1956. Estimation of primary oil reserves. *Trans. AIME*, 207: 182—191.

Barfield, E.C., Jordan, J.K. and Moore, W.D., 1959. An analysis of large scale flooding in fractured Spraberry trend area reservoir. *J. Pet. Technol.*, 11 (4): 15—19.

Black, J.L. and Lacik, H.A., 1969. History of a Scurry County, Texas, reef unit. In: *Proceedings of the Southwest Petroleum Short Course*, pp. 35—39.

Bleakley, W.B., 1965. A case for total engineering. *Oil Gas J.*, 63 (3): 74—76.

Bleakley, W.B., 1969a. Sound planning = successful flood. *Oil Gas J.*, 67 (3): 154—159.

Bleakley, W.B., 1969b. Computers calculate flood potential. *Oil Gas J.*, 67 (10): 147—149.

Borgan, R.L., Frank, J.R. and Talkington, G.E., 1965. Pressure maintenance by bottom water injection in a massive San Andres Dolomite reservoir. *J. Pet. Technol.*, 17 (8): 883—888.

Brownscombe, E.R. and Dyes, A.B., 1952. Water-imbibition displacement — A possibility for the Spraberry. *Drill. Prod. Pract., API*, pp. 383—390.

Buckley, S.E. and Leverett, M.C., 1951. Mechanism of fluid displacement in sands. *Trans. AIME*, 146: 107—116.

Burcik, E.J., 1968. What, why and how of polymers for waterflooding. *Pet. Eng.*, 40 (8): 60—64.

Burt, R.A., Haddenhorst, F.A. and Hartford, J.C., 1975. Review of Bell Creek waterflood performance — Powder River, Montana. *J. Pet. Technol.*, 27 (12): 1443—1449.

Calhoun, J.C., Jr., McCarthy, J.C. and Morse, R.A., 1950. Effects of permeability on secondary recovery of oil. In: *Secondary Recovery of Oil in the United States.* API, 2nd ed., pp. 214—221.

Callaway, F.H., 1959. Evaluation of waterflood prospects. *J. Pet. Technol.*, 11 (10): 11—16.

Cargile, L.L., 1969. A case history of the Pegasus Ellenburger reservoir. *J. Pet. Technol.*, 21 (10): 1330—1336.

Caudle, B.H. and Witte, M.D., 1959. Production potential changes during sweep-out in a five-spot system. *Trans. AIME*, 216: 446—448.

Caudle, B.H., Erickson, R.A. and Slobod, R.L., 1955. The encroachment of injected fluids beyond the normal well pattern. *J. Pet. Technol.*, 7 (5): 79—85.

Craig, F.F., Jr., Geffen, T.M. and Morse, R.A., 1955. Oil recovery performance of pattern gas or water injection operation from model tests. *Trans. AIME*, 204: 7—15.

De Groot, M., 1929. Flooding process for recovering oil from subterranean oil-bearing strata. *U.S. Patent*, No. 1,823,439.

De Groot, M., 1930. Flooding process for recovering fixed oil from subterranean oil-bearing strata. *U.S. Patent*, No. 1,823,440.

Deppe, J.C., 1961. Injection rates — the effect of mobility ratio, area swept and pattern. *Soc. Pet. Eng. J.*, 1 (6): 81—91.

Douglas, J., Blair, P.M. and Wagner, R.J., 1958. Calculation of linear waterflood behavior including the effects of capillary pressure. *Trans. AIME*, 213: 96—102.

Douglas, J., Peaceman, D.W. and Rachford, H.H., 1959. A method for calculating multidimensional immiscible displacement. *Trans. AIME*, 216: 209—308.

Dyes, A.B., Caudle, B.H. and Erickson, R.A., 1954. Oil production after breakthrough as influenced by mobility ratio. *J. Pet. Technol.*, 4 (6): 27—32.

Dyes, A.B., Kemp, C.E. and Caudle, B.H., 1958. Effects of fractures on sweep-out pattern. *Trans. AIME*, 213: 245—249.

Dykstra, H. and Parsons, R.L., 1950. The prediction of oil recovery by waterflood. In: *Secondary Recovery of Oil in the United States*. API, 2nd ed., pp. 160—174.

Earlougher, R.C. and Guerrero, E.T., 1965. New development in waterflooding, Part II. Newer waterflooding process and equipment. *Prod. Mon.*, 29 (3): 9—16.

Editorial, 1964. Rangely waterflood. *Oil Gas J.*, 3 (March 2).

Editorial, 1966. Practical waterflooding shortcuts. *World Oil*, 163 (12): 89—92.

Elkins, L.F., 1969. Internal anatomy of a tight fractured Hunton Lime reservoir revealed by performance — West Edmond Field. *J. Pet. Technol.*, 21 (2): 221—232.

Elkins, L.F. and Skov, A.M., 1963. Cyclic waterflooding the Spraberry utilizes "end effects" to increase oil production rate. *J. Pet. Technol.*, 15 (4): 877—884.

Elkins, L.F., Skov, A.M. and Gould, R.C., 1968. Progress report on Spraberry waterflood reservoir performance, well stimulation and water treating and handling. *J. Pet. Technol.*, 20 (9): 1039—1049.

Felsenthal, M. and Ferrel, H.H., 1967. Pressure pulsing — an improved method of waterflooding fractured reservoirs. *Soc. Pet. Eng. Permian Basin Oil Recovery Conf., Midland, Texas, September 1967, Soc. Pet. Eng of AIME, Pap.*, 1788.

Felsenthal, M. and Yuster, S.T., 1951. A study of the effect of viscosity on oil recovery by waterflooding. *AIME Meet., Los Angeles, Calif., October 1951*.

Felsenthal, M., Cobb, T.R. and Heuer, G.J., 1962. A comparison of waterflood evaluation methods. *Soc. Pet. Eng. of AIME, Pap.*, 1332.

Fickert, W.E., 1965. Economics of waterflooding the Grayburg Dolomite in South Cowden Field. In: *Proceedings of the Southwest Petroleum Short Course*, pp. 21—31.

Funk, V.T. and Anderson, T.C., 1982. Costs and indexes for domestic oil and gas field equipment and production operations, 1981. *U.S. Dep. Energy, Energy Inf. Admin.*, DOE/EIA-0185 (81).

Gealy, F.D., Jr., 1966. North Foster Unit — Evaluation and control of a Grayburg—San Andres waterflood based on a primary oil production and waterflood response. *Soc. Pet. Eng., 41st Annu. Fall Meet., Dallas, Texas, October 2—5, 1966*, SPE Preprint 1474.

Gogarty, W.B., 1967. Mobility control with polymer solutions. *Soc. Pet. Eng. J.*, 7 (6): 161—173.

Gogarty, W.B., 1976. Status of surfactant or micellar methods. *J. Pet. Technol.*, 28 (1): 93—102.

Gogarty, W.B., 1977. Oil recovery with surfactants: history and a current appraisal. In: D.O. Shah and R.S. Schechter (Editors), *Improved Oil Recovery by Surfactant and Polymer Flooding*. Academic Press, New York, N.Y., pp. 27—54.

Gogarty, W.B. and Olson, R.W., 1962. Use of microemulsions in miscible-type oil recovery procedure. *U.S. Patent*, No. 3,254,714.

Gogarty, W.B. and Tosch, W.C., 1968. Miscible-type water flooding: oil recovery with micellar solutions. *J. Pet. Technol.*, 20 (12): 1407—1414.

Gogarty, W.B. and Davis, J.A., Jr., 1972. Field experience with the Maraflood process. *SPE-AIME Improved Oil Recovery Symp., Tulsa, Okla., April 16—19, 1972, Soc. Pet. Eng. of AIME, Pap.*, 3806.

Gogarty, W.B., Meabon, H.P. and Milton, H.W., Jr., 1970. Mobility control design for miscible-type waterfloods using micellar solutions. *J. Pet. Technol.*, 22 (2): 141—147.

Goolsby, J.L., 1967. Here's the relation of geology to fluid injection in Permian carbonate reservoirs, West Texas. *Oil Gas J.*, 65 (July 31): 188—190.

Goolsby, J.L. and Anderson, R.C., 1964. Pilot waterflooding in a dolomite reservoir, the McElroy Field. *J. Pet. Technol.*, 14 (12): 1345—1350.

Graham, J.W. and Richardson, J.G., 1959. Theory and application of imbibition phenomena in recovery of oil. *J. Pet. Technol.*, 2 (11): 65—69.

Guerrero, E.T. and Earlougher, R.C., 1961. Analysis and comparison of five methods used to predict waterflood reserves and performance. *Drill. Prod. Pract., API*, p. 78.

Guidroz, G.M., 1967. E.T. O'Daniel Project — a successful Spraberry flood. *J. Pet. Technol.*, 19 (9): 1137—1140.

Guthrie, R.K. and Greenberger, M.H., 1955. The use of multiple-correlation analyses for interpreting petroleum-engineering data. *Drill. Prod. Pract., API*, pp. 130—137.

Habermann, B., 1960. The efficiency of miscible displacement as a function of mobility ratio. *Trans. AIME*, 219: 264—272.

Hauber, W.C., 1964. Prediction of waterflood performance for arbitrary well patterns and mobility ratio. *Trans. AIME*, 231: 95—103.

Hendrickson, G.E., 1961. History of the Welch Field San Andres pilot waterflood. *J. Pet. Technol.*, 13 (8): 745—748.

Henry, J.C. and Moring, J.D., 1967. Pilot waterflood evaluation — Pandhandle Field. *Soc. Pet. Eng. Reg. Secondary Recovery Symp., Pampa, Texas, October 26—27, 1967.* SPE Preprint 1801.

Hester, C.T., Walker, J.W. and Sawyer, G.H., 1965. Oil recovery by imbibition water flooding in the Austin and Buda Formations. *J. Pet. Technol.*, 17 (8): 919—925.

Hiatt, W.N., 1958. Injected-fluid coverage of multiwell reservoirs with permeability stratification. *Drill. Prod. Pract., API*, pp. 165—194.

Higgins, R.V. and Leighton, A.J., 1960. Waterflood performance in stratified reservoirs. *USBM*, RI-5618.

Higgins, R.V. and Leighton, A.J., 1962. Computer prediction of waterdrive of oil and gas mixtures through irregularly bounded porous media—three phase flow. *Trans. AIME*, 225: 1048—1054.

Higgins, R.V. and Leighton, A.J., 1963. Waterflood prediction of partially depleted reservoirs. *Soc. Pet. Eng., Tech. Pap.*, 757.

Higgins, R.V., Boley, D.W. and Leighton, A.J., 1964. Aids to forecasting performance of waterfloods. *J. Pet. Technol.*, 16 (9): 1076—1082.

Holbrook, O.C., 1958. Surfactant-water secondary recovery process. *U.S. Patent*, No. 3,006,411.

Hovanessian, S.A., 1960. Waterflood calculations for multiple sets of producing wells. *J. Pet. Technol.*, 12 (8): 65—68.

Hurst, W., 1953. Determination of performance curves in five-spot waterflood. *Pet. Eng.*, 25: B-40.

Inks, C.G. and Lahring, R.I., 1968. Controlled evaluation of a surfactant in secondary recovery. *J. Pet. Technol.*, 11: 1320—1324.

Johnson, C.E., Jr., 1956. Prediction of oil recovery by waterflood — a simplified graphical treatment of the Dykstra-Parsons method. *Trans. AIME*, 207: 345—346.

Johnson, C.E., Jr., 1960. Surfactant slugs in waterflooding — how much? what concentration? *Oil Gas J.*, 58 (9): 220—226.

Lake, L.W. and Pope, G.A., 1979. Status of micellar-polymer field tests. *Pet. Eng.*, 13: 38—60.

Landrum, B.L. and Crawford, P.B., 1960. Effect of directional permeability on sweep efficiency and production capacity. *J. Pet. Technol.*, 12 (11): 67—71.

Langston, E.P., Shirer, J.A. and Nelson, D.E., 1981. Innovative reservoir management — key to highly successful Jay/LEC waterflood. *J. Pet. Technol.*, 33 (5): 783—791.

Leverett, M.C. and Lewis, W.B., 1941. Steady flow of oil—gas—water mixtures through unconsolidated sands. *Trans. AIME*, 142: 107.

Miller, F.H. and Perkins, A., 1960. Feasibility of flooding thin, tight limestones. *Pet. Eng.*, 32: B-55 to B-75.

Morel-Seytoux, H.J., 1965. Analytical-numerical method in waterflooding predictions. *Soc. Pet. Eng. J.*, 5 (9): 147—157.

Mungan, N., Smith, F.W. and Thompson, J.L., 1966. Some aspects of polymer floods. *J. Pet. Technol.*, 18 (9): 1143—1150.

Muskat, M., 1946. *Flow of Homogeneous Fluids*. J.W. Edwards, Ann Arbor, Mich., 763 pp.

Muskat, M., 1950a. *Physical Principles of Oil Production*. McGraw-Hill, New York, N.Y., 922 pp.

Muskat, M., 1950b. The effect of permeability stratifications in complete water drive systems. *Trans. AIME*, 189: 349—358.

Naar, J. and Henderson, J.H., 1961. An imbibition model — its application to flow behavior and the prediction of oil recovery. *Soc. Pet. Eng. J.*, 1 (6): 61—70.

O'Briant, J.F., 1967. Operation and performance review of the Goldsmith-Cummins (San Andres) Unit water flood. In: *Proceedings of the Southwest Petroleum Short Course*, pp. 43—51.

Ogletree, J.O. and Overly, R.J., 1977. Sea-water and subsurface-water injection in West Delta Block 73 waterflood operations. *J. Pet. Technol.*, 29 (7): 623—628.

Owens, W.W. and Archer, D.L., 1966. Water flood pressure-pulsing for fractured reservoirs. *J. Pet. Technol.*, 18 (6): 745—752.

Park, R.A., 1965. Pressure maintenance by waterflooding North Virden Scallion Field, Manitoba. *Soc. Pet. Eng. Reg. Meet., Bakersfield, Calif., November 4, Soc. Pet. Eng. of AIME, Pap.*, 1321.

Pope, G.A., 1980. The application of fractional flow theory to enhanced oil recovery. *Soc. Pet. Eng. J.*, 20 (6): 191—205.

Prats, M., Matthews, C.S., Jewett, R.L. and Baker, J.D., 1959. Prediction of injection rate and production history for multifluid five-spot floods. *Trans. AIME*, 216: 98—101.

Prats, M., Hazebrook, P. and Allen, E.E., 1962. Effect of off pattern wells on the performance of a five-spot flood. *J. Pet. Technol.*, 14 (2): 173—178.

Pye, D.J., 1964. Improved secondary recovery by control of water mobility. *J. Pet. Technol.*, 16 (8): 911—916.

Roberts, T.G., 1959. A permeability block method of calculating a water drive recovery factor. *Pet. Eng.*, 31: B-45.

Sandiford, B.B., 1969. Laboratory and field studies of waterfloods using polymer solutions to increase oil recovery. *J. Pet. Technol.*, 16 (8): 917—922.

Schauer, P.E., 1957. Application of empirical data in forecasting waterflood behavior. *AIME Tech. Pap.*, 934.

Schmalz, J.P. and Rahme, H.S., 1950. The variation in waterflood performance with variation in permeability profile. *Prod. Mon.*, 14: 9.

334

Schoeppel, R.J., 1968. Waterflood prediction methods. *Oil Gas J.*, Jan. 22: 72—75; Feb. 19: 98—106; March 18: 91—93; April 8: 80—86; May 6: 111—114; June 17: 100—105; July 8: 71—79.

Slider, H.C., 1961. New method simplifies predicting waterflood performance. *Pet. Eng.*, 33: B-68.

Slobod, R.L. and Caudle, B.H., 1952. X-ray shadowgraph studies of areal sweepout efficiencies. *Trans. AIME*, 195: 265—270.

Snell, G.W. and Schurz, G.F., 1966. Polymer chemicals aid in unique recovery process. *Pet. Eng.*, 38: 53—59.

Snyder, R.W. and Ramey, H.J., Jr., 1967. Application of Buckley-Leverett displacement theory to noncommunicating layered system. *J. Pet. Technol.*, 19 (11): 1500—1506.

Stiles, W.E., 1949. Use of permeability distribution in waterflood calculations. *Trans. AIME*, 186: 9—13.

Suder, F.E. and Calhoun, J.C., 1949. Waterflood calculations. *Drill. Prod. Pract., API*, pp. 260—270.

Talash, A.W. and Strange, L.K., 1982. Summary of performance and evaluation in the West Burkburnett chemical waterflood project. *J. Pet. Technol.*, 34 (11): 2495—2502.

Terwilliger, P.L., Wilsey, L.E., Hall, H.N., Bridges, P.M. and Morse, R.A., 1951. An experimental and theoretical investigation of gravity drainage performance. *Trans. AIME*, 192: 285.

Trube, A.S., Jr., 1954. Oil production by primary artificial frontal water drives in the New Hope Field, Franklin County, Texas. In: *Proceedings of Seventh Oil Recovery Conference*. Texas Petroleum Research Committee, pp. 57—75.

Trube, A.S., Jr. and DeWitt, S.N., 1950. High-pressure water injection for maintenance reservoir pressures, New Hope Field, Franklin County, Texas. *Trans. AIME*, 189: 325—334.

Warren, J.E. and Cosgrove, J.J., 1964. Prediction of waterflood behavior in a stratified system. *Soc. Pet. Eng. J.*, 4 (6): 149—157.

Wayhan, D.A. and McCaleb, J.A., 1969. Elk Basin Madison heterogeneity — its influence on performance. *J. Pet. Technol.*, 21 (2): 153—159.

Welge, H.J., 1952. A simplified method for computing oil recovery by gas or water drive. *Trans. AIME*, 195: 91—98.

Whiteley, R.C. and Ware, J.W., 1977. Low-tension waterflow pilot at the Salem Unit, Marion County, Illinois, 1. Field implementation and results. *J. Pet. Technol.*, 29 (8): 925—932.

Widmyer, R.H., Frazier, G.D., Strange, L.K. and Talash, A.W., 1979. Low-tension waterflood at Salem Unit, post pilot evaluation. *J. Pet. Technol.*, 31 (9): 1185—1190.

Willingham, R.W. and McCaleb, J.A., 1967. The influence of geologic heterogeneities on secondary recovery from the Permian Phosphoria reservoir, Cotton Creek, Wyoming. *Rocky Mountain Reg. Soc. Pet. Eng. Meet., Casper, Wyo., May 1967*, SPE Preprint 1770.

Wilson, J.F., 1962. Waterflooding — down structure displacement in presence of a gas cap. *J. Pet. Technol.*, 16 (12): 1383—1388.

Wood, B.O. and McShane, J.B., Jr., 1969. A successful Glorieta — San Angelo waterflood, Snyder Field, Howard County, Texas. In: *Proceedings of the Southwest Petroleum Short Course*, pp. 49—57.

REFERENCES INDEX*

---

* The help extended by Mehmet Parlar and Saeed Mogharabi in preparing the indexes is indeed greatly appreciated by the editors.

336

Braitsch, O., 159, *218*
Brandner, C.F., 55, 56, 61, *72*, 106, 108, *115*
Braun, E.M., 54, 67, *71*
Braun, P.H., 109, *116*
Bray, E.E., 12, 31, *42*
Bredehoeft, J.D., 179, *218*
Breger, I.A., 29, *45*
Brenner, H., 78, *113*
Bridges, P.M., 88, 98, *117*, 252, 258, *334*
Brigham, W.E., 146, 147, *148*
Brissaud, F., 104, *112*
Brons, F., 129, 313, 142, *148*, *149*
Brooks, R.H., 85, 87, *112*
Brown, R.J.S., 80, *112*
Brownscombe, E.R., 89, 90, 91, *112*, 309, *330*
Bruce, W.A., 81, *116*
Buckley, S.E., 88, 93, 94, 95, 105, 107, *112*, 180, *218*, 252, 254, 258, 276, 277, 279, 280, 282-285, 292, 294, 299, 302, 314, 315, *330*
Burcik, E.J., 317, *330*
Burdine, N.T., 79, 81, 82, 84, 102, *112*
Burt, R.A., 327, 328, *330*

Calhoun, J.C., Jr., 109, *117*, 252, 254, 255, 256, 292, *330*, *334*
Califet-Debyser, Y., 16, *45*
Callaway, F.H., 258, 259, *330*
Cargile, L.L., 313, *331*
Carlberg, B.L., 198, *218*
Carman, P.C., 77, 78, *112*
Carothers, W.W., 175, 176, *218*
Carpenter, A.B., 163, 164, 167, *218*
Carpenter, C.W., Jr., 191, *219*
Cassan, J.P., 209, *218*
Catchpole, J.P., 107, *112*
Caudle, B.H., 88, 89, 90, 91, 100, 102, 105, *112*, 269, 270, 300, 302, *331*, *334*
Cerini, W.F., 199, *218*
Chander, S., 243, *248*
Chang, P.W., 245, *248*
Chatiudompunth, S., 84, *115*
Chatzis, I., 49, 53, 54, 57, 58, *71*, *73*
Chavent, G., 96, *112*
Cheshire, S., 111
Chilingarian, G.V., 99, *114*, 240, *247*, *248*, 251
Chuoke, R.L., 70, *71*, 88, 95, 105, *112*
Cinco, H., 142, *148*

Claypool, G.E., 18, *42*, *44*
Clayton, R.N., 164, *218*
Clayton, J.L., 22, *42*
Coats, K.H., 65, *71*
Cobb, T.R., 252, 254, 255, *331*
Coggeshall, N.E., 180, *221*
Cohen, G., 96, *112*
Coleman, H.J., 35, *44*, 99, *114*
Collins, A.G., 163—165, 174, 177, 183, 187, 206, 207, *219*, *220*
Collins, R.E., 68, *71*
Collins, S.H., 228, *248*
Collonna, J., 104, *112*
Combaz, A., 14, 15, 17, 19, *42*, *45*
Comer, A.C., 104, *113*
Connan, J., 17, 22, *42*
Cook, F.D., 26, *43*
Cooper, J.E., 176, *219*
Cordell, R.J., 19, 22, *42*, *44*
Cornell, D., 121, *148*
Corey, A.T., 81—83, 85—87, 89, 94, 97, 99, 100, 102, *112*
Cosgrove, J.J., 252, 254, *334*
Costantinides, G., 36, *42*
Cotton, F.O., 99, *114*
Cowan, J.C., 206, *221*
Craig, F.F., Jr., 47, *71*, 98, 108, *112*, 252, 254, 255, 258, 274, 293—299, *331*
Craig, H., 165, *219*
Cram, P.J., 60, *73*, 110, *115*
Crane, F.E., 104, 105, *113*
Crank, J., 63, *71*
Crawford, P.B., 272, 273, 274, *333*
Crocker, M.E., 108, *115*
Croes, G.A., 95, *112*
Crowell, D.C., 88, 89, *114*
Cuiec, L., 98, *112*

Darcy, H., 75, 104, *112*
Datta, P., 88, *113*
Davidson, L.B., 110, *112*
Davis, G.T., 108, *113*
Davis, H.T., 49, *72*, *73*, 103, *115*
Davis, J., 198, *219*
Davis, J.A., Jr., 321, *332*
Davis, J.B., 187, 213, *219*
Davis, J.W., 206, *219*
Davis, L.E., 233—237, *249*
Dean, G.W., 97, 100, 102, *113*
Degens, E.T., 176, *219*
De Groot, M., 318, *331*
De Jong, L.N.J., 230, *248*

338

SUBJECT INDEX

354